The Advanced Calculus
of One Variable

The Appleton-Century Mathematics Series

Raymond W. Brink and John M. H. Olmsted, Editors

The Advanced Calculus of One Variable

Don R. Lick

Western Michigan University

APPLETON-CENTURY-CROFTS
EDUCATIONAL DIVISION
New York MEREDITH CORPORATION

In memory of my mother
Florence May Lick

Preface

The intention of this book is to present a thorough treatment of the material usually covered in a course of one-dimensional advanced calculus. The two principal goals of this text are a rigorous foundation in the basic concepts and tools of analysis and an intuitive feeling for analysis. These goals are accomplished through the presentation of well-motivated definitions, many illustrative examples, ample explanation of the methods and techniques, simplified proofs of the theorems, and many appropriate exercises. For this reason, most of the problems are not of a computational nature but are chosen to help extend the students' understanding of the concepts presented, as well as to extend the material of the text.

This book is intended for students who have taken the ordinary freshman and sophomore courses in calculus. Any student completing this book should be prepared for the beginning courses in real variables.

It would be difficult to present all the material in this text in one semester because of the abundance of material available. Thus, if the entire book is to be covered, at least two semesters should be allowed. If the text is to be used for only one semester, some thought should be given to the emphasis of the course and to the selection of material suitable for that emphasis.

I would like to express my deep appreciation for the help and suggestions given me by Professors John M. H. Olmsted, Southern Illinois University, and Dale W. Lick, Drexel University. I am also indebted to Darlene Lard for the excellent typing and clerical assistance in the preparation of this manuscript. Finally, I would like to express my gratitude to my wife and family for their patience and understanding while I completed this book.

D.R.L.

Contents

Chapter One: BASIC SET THEORY

Chapter Two: THE REAL NUMBER SYSTEM

Chapter Three: TOPOLOGY OF THE REAL NUMBER SYSTEM

Chapter Four: LIMITS AND CONTINUITY

Chapter Five: DIFFERENTIATION

Chapter Six: INTEGRATION

Chapter Seven: INFINITE SERIES

Chapter Eight: SEQUENCES AND SERIES OF FUNCTIONS

Chapter Nine: POWER SERIES

Chapter Ten: IMPROPER INTEGRALS

Chapter Eleven: STIELTJES INTEGRATION

The Advanced Calculus
of One Variable

Chapter One

BASIC SET THEORY

1.1. Sets

The main purpose of this text is to study the properties of the real number system and of the real functions of a real variable. Since the fundamental concepts of mathematics may be expressed in the terminology of set theory, in this chapter we state and prove some of the basic facts about sets. Even though the properties of the real number system will be developed rigorously in Chapter 2, it is useful to assume, for the purpose of illustrative examples in this chapter, that the reader is reasonably familiar with the elementary properties of the real number system.

In mathematics, the word *set* refers to an undefined concept, but we intuitively think of a set as a collection of objects and use the words set and collection interchangeably. Other synonyms for set are "aggregate," "bundle," "class," "ensemble," and "family." The objects that make up a set are called *elements* or *members* of the set. An element is said to *belong* to the set or to be a *member* or an *element* of the set. Some examples of sets are the following: (1) the set consisting of the numbers 3, 5, 7, and 9; (2) the set consisting of the U.S. Senators; (3) the set of rational numbers; and (4) the set consisting of Joe, Walt, Tony, Dalton, and Don. In examples (1), (2), and (4), the elements can be listed or itemized easily; for instance, the set in example (1) may be represented by

$$\{3, 5, 7, 9\}$$

where the braces notation is understood to mean "the set consisting of the elements 3, 5, 7, and 9." Likewise, the set in example (4) may be written

$$\{\text{Joe, Walt, Tony, Dalton, Don}\}.$$

1

For some sets it is difficult to itemize the elements, and so the set is described by a statement or a rule telling exactly what elements are in the set. For instance, the set in example (3) may be expressed as

$$\{x: x \text{ is a rational number}\}$$

or

$$\{x \mid x \text{ is a rational number}\},$$

both of which are read: "The set of all x such that x is a rational number." In general, if $P(x)$ is a statement involving the object x, then we denote by

$$\{x: P(x)\}$$

the set.of all objects x such that the statement $P(x)$ is true.

Sets will usually be denoted by capital letters A, B, C, etc., while elements of a set will usually be denoted by small letters a, b, c, etc. We use the notation

$$x \in A$$

to mean "x is an element of the set A" or "x belongs to A." The notation

$$x \notin A$$

means "x is not an element of A" or "x is not a member of A."

Two sets A and B are said to be **equal** if and only if every element of A is an element of B and every element of B is an element of A; that is, A and B contain the same elements. Equality of sets A and B is denoted by

$$A = B.$$

The notation

$$A \neq B$$

indicates that set A is not equal to set B, that is, either there is an element $x \in A$ such that $x \notin B$, or there is an element $y \in B$ such that $y \notin A$.

A set A is a **subset** of a set B if an only if every element of A is an element of B. This is denoted by

$$A \subseteq B$$

and is read: "A is contained in B." If $A \subseteq B$ and there is a $y \in B$ such that $y \notin A$, then A is a **proper subset** of B. This is denoted by

$$A \subset B$$

and is read: "A is a proper subset of B." Note that the set of integers is a proper subset of the set of rational numbers.

If two sets A and B are equal, then $A \subseteq B$ and $B \subseteq A$. Conversely, if $A \subseteq B$ and $B \subseteq A$, then every element of A is an element of B and every element of B is an element of A, and so $A = B$. It is important in mathematics to determine when two sets are equal. We see that one method of determining when the sets A and B are equal is to show that $A \subseteq B$ and $B \subseteq A$. We state this equivalence as our first theorem.

Theorem 1.1.1. *Let A and B be sets. Then $A = B$ if and only if $A \subseteq B$ and $B \subseteq A$.*

The **empty set**, or **null set**, is the set that contains no elements. The empty set is denoted by "$\varnothing$." For every set A, $A \subseteq A$ and $\varnothing \subseteq A$.

We now discuss operations on sets.

Definition 1.1.1. *Let A and B be sets. The set of all objects that belong to either A or B or possibly to both A and B is called the **union** of A and B. The union of A and B is denoted by $A \cup B$. The set of all objects that belong to both A and B is called the **intersection** of A and B. The intersection of A and B is denoted by $A \cap B$.*

In the previous notation, $A \cup B = \{x : x \in A \text{ or } x \in B\}$, where "or" is used in the mathematical or nonexclusive sense; that is, the set $A \cup B$ includes elements that belong to both A and B, as well as elements in either set alone. It is easy to see that y is a member of $A \cup B$ if and only if y is a member of at least one of the sets A and B. The intersection of A and B may be written $A \cap B = \{x : x \in A \text{ and } x \in B\}$. The right-hand term may also be written $\{x : x \in A, x \in B\}$, where the comma is read as "and." The intersection of A and B is the set of elements that are common to both A and B.

Example 1.1.1. Let $A = \{-3, -1, 1, 3, 5, 7\}$ and $B = \{0, 1, 2, 3, 4\}$.

Then $A \cup B = \{-3, -1, 0, 1, 2, 3, 4, 5, 7\}$ and $A \cap B = \{1, 3\}$.

For any two sets A and B,

$$A \cap B \subseteq A \subseteq A \cup B,$$

$$A \cap B \subseteq B \subseteq A \cup B.$$

The following three statements are equivalent: $A \subseteq B$, $A \cup B = B$, and $A \cap B = A$. That is, $A \subseteq B$ if and only if $A \cup B = B$, $A \subseteq B$ if and only if $A \cap B = A$, and $A \cup B = B$ if and only if $A \cap B = A$.

Theorem 1.1.2. *For any sets A, B, and C, the following properties hold:*

(a) $A \cup B = B \cup A,$

    $A \cap B = B \cap A,$ *(Commutative Laws)*

(b) $A \cup (B \cup C) = (A \cup B) \cup C$,

$A \cap (B \cap C) = (A \cap B) \cap C$,   (*Associative Laws*)

(c) $A \cup (B \cap C) = (A \cup B) \cap (A \cup C)$,

$A \cap (B \cup C) = (A \cap B) \cup (A \cap C)$.   (*Distributive Laws*)

We will prove the first parts of (a) and (c) and leave the remaining statements as Exercise 6 of Problem Set 1.1.

Proof.   (a) Let x be any element of $A \cup B$. Then x is an element of A or x is an element of B. Hence x is an element of $B \cup A$, and $A \cup B \subseteq B \cup A$. In a similar manner we can show that $B \cup A \subseteq A \cup B$. Applying Theorem 1.1.1, we have $A \cup B = B \cup A$.

(c) Let x be an element of $A \cup (B \cap C)$. Then x is an element of A or x is an element of $(B \cap C)$. If x is an element of A, then x is an element of $A \cup B$ and $A \cup C$, and so x is an element of $(A \cup B) \cap (A \cup C)$. If x is an element of $B \cap C$, then x is an element of both B and C. But then $x \in A \cup B$ and $x \in A \cup C$, and so $x \in (A \cup B) \cap (A \cup C)$. Therefore, any element of $A \cup (B \cap C)$ is an element of $(A \cup B) \cap (A \cup C)$, or

$$A \cup (B \cap C) \subseteq (A \cup B) \cap (A \cup C).$$

Now let x be an element of $(A \cup B) \cap (A \cup C)$. Then x is an element of both $A \cup B$ and $A \cup C$. If x is an element of A, then x is an element of $A \cup (B \cap C)$. If x is not an element of A, the remaining possibility then, since x belongs to both $A \cup B$ and $A \cup C$, is that x is a member of both B and C. Hence x belongs to $B \cap C$, and so x belongs to $A \cup (B \cap C)$. In any case, if x belongs to $(A \cup B) \cap (A \cup C)$, then x belongs to $A \cup (B \cap C)$, and so

$$(A \cup B) \cap (A \cup C) \subseteq A \cup (B \cap C).$$

Applying Theorem 1.1.1, we have

$$A \cup (B \cap C) = (A \cup B) \cap (A \cup C).$$

In most mathematical discussions, all the sets that we encounter are subsets of a given reference set, usually called a **universal set.** For such cases, we make the following definition.

Definition 1.1.2. *For any universal set S, if $A \subseteq S$, then the* *complement of A relative to S is the set*

$$\mathbf{C}(A) = \{x: x \in S \text{ and } x \notin A\}.$$

Observe that $\mathbf{C}(S) = \emptyset$ and $\mathbf{C}(\emptyset) = S$.

The operations of union, intersection, and complement are now related by the following laws, called **DeMorgan's Laws**. These rules are named for the English mathematician, Augustus DeMorgan (1806–1871),who discovered them.

Theorem 1.1.3. *If A and B are any subsets of S, then*

$$\mathbf{C}(A \cup B) = \mathbf{C}(A) \cap \mathbf{C}(B),$$

$$\mathbf{C}(A \cap B) = \mathbf{C}(A) \cup \mathbf{C}(B);$$

that is, the complement of the union is equal to the intersection of the complements, and the complement of the intersection is equal to the union of the complements.

We prove the first property; the second follows in a similar manner and is left as Exercise 7 of Problem Set 1.1.

Proof. If x is an element of $\mathbf{C}(A \cup B)$, then, by the definition of complement, $x \notin A \cup B$. Thus $x \notin A$ and $x \notin B$. Since $x \notin A$ and $x \notin B$, we have that $x \in \mathbf{C}(A)$ and $x \in \mathbf{C}(B)$. Hence $x \in \mathbf{C}(A) \cap \mathbf{C}(B)$ and $\mathbf{C}(A \cup B) \subseteq \mathbf{C}(A) \cap \mathbf{C}(B)$. If x is an element of $\mathbf{C}(A) \cap \mathbf{C}(B)$, then $x \in \mathbf{C}(A)$ and $x \in \mathbf{C}(B)$. Thus $x \notin A$ and $x \notin B$, which implies that $x \notin A \cup B$. Therefore, $x \in \mathbf{C}(A \cup B)$ and $\mathbf{C}(A) \cap \mathbf{C}(B) \subseteq \mathbf{C}(A \cup B)$. Applying Theorem 1.1.1, we have that $\mathbf{C}(A \cup B) = \mathbf{C}(A) \cap \mathbf{C}(B)$.

The concepts of union and intersection are now extended to include more than two sets. For example, if A, B, and C are sets, then the **union** of A, B, and C, denoted by $A \cup B \cup C$, is the set of all objects y such that y belongs to at least one of the sets A, B, or C. The **intersection** of A, B, and C, denoted by $A \cap B \cap C$, is the set of all objects y such that y is common to all three sets A, B, and C. The ideas of union and intersection can be extended further.

Definition 1.1.3. *Let $A_1, A_2, ..., A_n$ be sets. The set*

$$\{x : x \in A_i \text{ for at least one } i, 1 \leqslant i \leqslant n\}$$

*is called the **union** of the sets $A_1, A_2, ..., A_n$ and is denoted by*

$$\bigcup_{i=1}^{n} A_i \quad \text{or} \quad A_1 \cup A_2 \cup ... \cup A_n.$$

The set
$$\{x : x \in A_i \text{ for each } i, 1 \leqslant i \leqslant n\}$$

*is called the **intersection** of the sets $A_1, A_2, ..., A_n$ and is denoted by*

$$\bigcap_{i=1}^{n} A_i \quad \text{or} \quad A_1 \cap A_2 \cap ... \cap A_n.$$

Example 1.1.2. Let $n = 4$, and let $A_1 = \{a, b, c\}$, $A_2 = \{b, d, f\}$, $A_3 = \{b, x, y\}$, and $A_4 = \{b, p, q\}$. Then

$$\bigcup_{i=1}^{4} A_i = A_1 \cup A_2 \cup A_3 \cup A_4$$

$$= \{a, b, c, d, f, x, y, p, q\},$$

and

$$\bigcap_{i=1}^{4} A_i = A_1 \cap A_2 \cap A_3 \cap A_4 = \{b\}.$$

Example 1.1.3. Let n be a positive integer. Let $A_i = \{x : x$ is a real number and $5(i-1) \leqslant x < 5i\}$, where $i = 1, 2, \ldots, n$. Then

$$\bigcup_{i=1}^{n} A_i = \{x : x \text{ is a real number and } 0 \leqslant x < 5n\},$$

and

$$\bigcap_{i=1}^{n} A_i = \varnothing.$$

Using arguments analogous to those in the proof of Theorem 1.1.3, it can easily be verified that DeMorgan's Laws hold for more general unions and intersections, that is.

$$\mathbf{C}\left(\bigcup_{i=1}^{n} A_i\right) = \bigcap_{i=1}^{n} \mathbf{C}(A_i),$$

$$\mathbf{C}\left(\bigcap_{i=1}^{n} A_i\right) = \bigcup_{i=1}^{n} \mathbf{C}(A_i).$$

The proofs are left as Exercise 16 of Problem Set 1.1.

Problem Set 1.1

1. Let $A = \{a, b, c, d\}$. List all the subsets of A.
2. Give an example of two sets A and B such that A is not a subset of B and B is not a subset of A.
3. Does $\varnothing = \{\varnothing\}$? Why or why not?
4. Show that if $A \subseteq B$ and $B \subseteq C$, then $A \subseteq C$ [Transitive Law].
5. Let $A = \{a, b, 1, 2\}$, $B = \{b, d, 1, 3, 5\}$, and $C = \{b, 1\}$.
    (a) Find $A \cup B$ and $B \cup A$.
    (b) Find $A \cap B$ and $B \cap A$.
    (c) Find $A \cup (B \cap C)$ and $(A \cup B) \cap (A \cup C)$.
    (d) Find $A \cap (B \cup C)$ and $(A \cap B) \cup (A \cap C)$.
6. Prove the remaining parts of Theorem 1.1.2.

7. Prove the second part of Theorem 1.1.3.

8. Prove that $A \cup (B \cup C) = A \cup B \cup C$.

9. Prove that $A \cup (A \cap B) = A$.

10. Prove that $A \cap (A \cup B) = A$.

11. Prove that if $A \subseteq S$, then $A \cup \mathbf{C}(A) = S$.

12. Prove that $A = B$ if and only if $[A \cap \mathbf{C}(B)] \cup [\mathbf{C}(A) \cap B] = \emptyset$.

13. Prove that $B \subseteq \mathbf{C}(A)$ if and only if $A \cap B = \emptyset$.

14. Let A be a proper subset of B. Is it necessarily true that $C \cup A$ is a proper subset of $C \cup B$? Explain.

15. Let A and B be subsets of S. Prove that $A \subseteq B$ if and only if $\mathbf{C}(B) \subseteq \mathbf{C}(A)$.

16. Prove that $\mathbf{C}\left(\bigcup_{i=1}^{n} A_i \right) = \bigcap_{i=1}^{n} \mathbf{C}(A_i)$ and $\mathbf{C}\left(\bigcap_{i=1}^{n} A_i \right) = \bigcup_{i=1}^{n} \mathbf{C}(A_i)$.

17. Prove that the following statements are equivalent:

    (I) $A \subseteq B$;

    (II) $A \cap B = A$;

    (III) $A \cup B = B$.

[*Hint*: One method of proving that these statements are equivalent is to show that (I) implies (II), (II) implies (III), and (III) implies (I).]

18. Let A and B be subsets of S. The **difference** of A and B (or the **complement of B relative to A**) is the set
$$A \setminus B = \{x : x \in A \text{ and } x \notin B\}.$$
For the sets A, B, and C defined in Exercise 5, find $A \setminus B$, $A \setminus C$, $B \setminus A$, $B \setminus C$, $C \setminus A$, and $C \setminus B$. Is the operation of taking differences of two sets commutative? That is, is $A \setminus B = B \setminus A$ for all sets A and B?

19. Show that $A \setminus B = A \cap \mathbf{C}(B)$.

20. Show that $A \setminus B = A \setminus (A \cap B)$.

21. Show that $A \setminus B = \emptyset$ if and only if $A \subseteq B$.

22. Let A be a set with n elements. Show that there are 2^n subsets of A.

1.2. Relations and Functions

The purpose of this section is to define relations and functions and to prove some results about them. In order to do this we need to introduce a few new terms and to prove some preliminary results.

Consider the set with objects a and b, where $a \neq b$. We can write either $\{a, b\}$ or $\{b, a\}$ to identify this set; the order in which the elements are written is not important. If we are interested in another set involving the elements a and b where the order of the elements is important, we want an *ordered pair*. The distinction here is that each element has a special place. The idea of ordered pairs is encountered, for example, in the arithmetic of fractions, where numerators and denominators need to be distinguished from one another. With this intuitive notation in mind, we now define an ordered pair.

Definition 1.2.1. *An **ordered pair** (a, b) is the set $\{\{a\}, \{a, b\}\}$. In the ordered pair (a, b), a is called the **first coordinate**, and b is called the **second coordinate**.*

Theorem 1.2.1. *Let (a, b) and (c, d) be two ordered pairs. Then $(a, b) = (c, d)$ if and only if $a = c$ and $b = d$.*

Theorem 1.2.1 states that two ordered pairs are equal if and only if the first coordinates are equal and the second coordinates are equal. For example, $(a, b) = (2, 3)$ if and only if $a = 2$ and $b = 3$. The proof of Theorem 1.2.1 is left as Exercise 14 of Problem Set 1.2.

The concept of ordered pairs permits a way of forming another set, called the *Cartesian product*, from two given sets.

Definition 1.2.2. *Let A and B be two nonempty sets. Then the set of ordered pairs*

$$\{(a, b): a \in A \text{ and } b \in B\}$$

*is called the **Cartesian product** of A and B and is denoted by $A \times B$.*

The Cartesian product $A \times B$ is the set of all ordered pairs (a, b) where $a \in A$ and $b \in B$. For example, if

$$A = \{x, y\} \quad \text{and} \quad B = \{1, 2, 3\},$$

then

$$A \times B = \{(x, 1), (x, 2), (x, 3), (y, 1), (y, 2), (y, 3)\}.$$

If R denotes the real number system, then $R \times R$ corresponds to the Euclidean or xy plane.

Definition 1.2.3. *Let A and B be any two nonempty sets. Then any subset of $A \times B$ is called a **relation** from A to B.*

Example 1.2.1. Let $A = \{1, 2, 3\}$ and $B = \{1, 2, 3, 4\}$. The Cartesian product of A and B is

$$A \times B = \{(1, 1), (1, 2), (1, 3), (1, 4), (2, 1),$$

$$(2, 2), (2, 3), (2, 4), (3, 1), (3, 2), (3, 3), (3, 4)\}.$$

Some examples of relations from A to B, that is, subsets of $A \times B$, are

$$R_1 = \varnothing,$$

$$R_2 = \{(1, 2)\},$$

$$R_3 = \{(x, y): x = y\} = \{1, 1), (2, 2), (3, 3)\},$$

$$R_4 = \{(x, y): x < y\} = \{(1, 2), (1, 3), (1, 4), (2, 3), (2, 4), (3, 4)\}.$$

The remainder of this section is concerned with a special type of relation which is fundamental to all branches of mathematics; this relation is called a *function*.

Definition 1.2.4. *Let A and B be any two nonempty sets. A **function** F from A into B is a relation from A to B such that*

(a) *for each $x \in A$ there exists $y \in B$ such that $(x, y) \in F$, and*

(b) *if $(x, y) \in F$ and $(x, z) \in F$, then $y = z$.*

Property (a) tells us that, for every $x \in A$, there is a y belonging to B such that the ordered pair (x, y) belongs to the function F, that is, for each $x \in A$ there is an ordered pair with first coordinate x. Property (b) is called the **single-valued property** and says that, for each element x of A, there is only one element y of B such that the ordered pair (x, y) belongs to F.

In Example 1.2.1, the relations R_1 and R_2 are not functions, since they do not satisfy property (a). The relation R_3 is a function; however, the relation R_4 is not a function, since it does not satisfy property (b), that is, it is not *single-valued*.

Synonyms for the word "function" are *transformation*, *map*, and *mapping*.

In general, if a relation F from A to B is a function, then F associates to each element $x \in A$ one and only one element $y \in B$, since each x is the first coordinate of exactly one ordered pair $(x, y) \in F$. For this reason the idea of function is sometimes defined without employing the ordered-pair concept of relation. The following property of functions is often given as the definition of function: *A function from a nonempty set A into a nonempty set B is a rule which associates to each element $x \in A$ exactly one element $y \in B$*. Although this property of functions is sometimes easier for some to understand and apply, it depends on the rather vague meaning of the word "rule." This property is given as sort of a working definition of function to help the student visualize more clearly the meaning of a function.

The **domain** of the function F from A into B is the set A. The domain of a function can be denoted by Dom F. The set $\{y:$ there exists $x \in A$ such that $(x, y) \in F\}$ is called the **range** or **image**. The range of a function F is denoted by Ran F. The domain is the entire set A, and the range is a subset of B. The **values** of a function are the elements of its range. For the function R_3 of Example 1.2.1, the set $A = \{1, 2, 3\}$ is the domain of R_3, the set $\{1, 2, 3\} \subset B$ is the range of R_3, and the values of R_3 are 1, 2, and 3.

If F is a function and $(x, y) \in F$, then x is called the **argument** and y is called the **image** of x under F. The image of x is often denoted by $F(x)$. The argument x is sometimes called the **preimage** of y. If $(x, y) \in F$, then y is the **value** of F corresponding to x.

A student should not merely memorize definitions such as those given above. All definitions should be carefully examined, dissected, and fully assimilated. The motivation for a definition, and the full meaning of it, should be thoroughly understood. To do this, many examples have to be considered in light of the definition. The ideas and concepts contained in a definition should become a part of one's way of thinking.

We shall use the notation $F: A \to B$ to mean that "F is a function from A into B." However, according to the grammatical context, the statement "$F: A \to B$" may be read "F maps A into B" and also "F which maps A into B" or "F mapping A into B," etc. Similarly, the statement "Let $F: A \to B$" may be read "Let F be a mapping of A into B" or "Let F be a function mapping A into B."

We emphasize here that an equation such as $h(x) = x^2 + 3$ does not define a function unless the domain is explicitly specified. Thus the equations

$$f(x) = x^2 + 3 \quad \text{for} \quad 1 \leqslant x \leqslant 2$$

and

$$g(x) = x^2 + 3 \quad \text{for} \quad 0 \leqslant x \leqslant 4$$

define two different functions according to Definition 1.2.4. However, these two functions are related, and it is useful to have terminology to describe this relationship.

Let $f: A \to B$ and $g: C \to D$. If $A \subseteq C$ and $f(x) = g(x)$ for each $x \in A$, then f is said to be the **restriction** of g to A, denoted by $f = g/A$, or g is said to be the **extension** of f to C. It is easily seen that the restriction of a function g to a given subset of its domain is unique, but there are many extensions of a function f to a given set containing its domain. For example, the function g above is one extension of f to the set $I = \{x: 0 \leqslant x \leqslant 4\}$, while the function Φ with domain I defined by

$$\Phi(x) = \begin{cases} x^2 + 3 & \text{for} \quad 1 \leqslant x \leqslant 2, \\ 0 & \text{for} \quad 0 \leqslant x < 1, \\ 1 & \text{for} \quad 2 < x \leqslant 4, \end{cases}$$

is also an extension of f to I. We see that $g \neq \Phi$.

Definition 1.2.5. *Let F be a function from A into B. If the range of F is B, then F is said to be **onto** B, or a function from A **onto** B. If (w, y) and (x, y) belonging to F implies that $w = x$, then F is said to be **one-to-one** (1–1), or a **one-to-one function** from A into B.*

The function R_3 mentioned earlier is 1–1, but it is not onto, since Ran $R_3 \neq B$. An equivalent way of defining a 1–1 function is given in the following theorem. The proof follows from the definition of a 1–1 function and is left as Exercise 15 of Problem Set 1.2.

Theorem 1.2.2. *Let F be a function from A into B. Then F is 1–1 if and only if for $(w, y),(x, z) \in F$, and $w \neq x$, it follows that $y \neq z$.*

Example 1.2.2. Let $A = B$ be the set of real numbers, and let $F = \{(x, x/2): x \in A\}$. Then F is a 1–1 function from A onto B. We use Theorem 1.2.2 to show that F is 1–1. To do this we must show that if (w, y) and (x, z) are any two ordered pairs in F such that $w \neq x$, then $y \neq z$. By the definition of F, we have $y = w/2$ and $z = x/2$. If $w \neq x$, then $w/2 \neq x/2$ and $y \neq z$. Hence, by Theorem 1.2.2, F is 1–1. To show that F is onto B, we need to show that for each $y \in B$ there is an $x \in A$ such that $(x, y) \in F$, that is, for each real number y we must find a real number x such that $x/2 = y$. To do this, let $x = 2y$; then $(2y, y) \in F$, and F is onto B. Therefore F is a 1–1 function from A onto B.

Observe that if $F: A \to B$, then the function $F: A \to \operatorname{Ran} F$ maps A onto $\operatorname{Ran} F$.

In order to show that a function F from A into B is *not* 1–1, it suffices to find two ordered pairs in F with distinct first coordinates but with the same second coordinates, that is, ordered pairs such as (w, y) and (x, y), where $w \neq x$. To show that a function F from A into B is *not* onto B, it is sufficient to find some element $y \in B$ such that the ordered pair $(x, y) \notin F$ for every $x \in A$.

Problem Set 1.2

1. Let $A = \{1, 2, 3, 4\}$, $B = \{a, b, c, d\}$, and $C = \{-3, -2, -1\}$.
   (a) Find $A \times B$.
   (b) Find $B \times A$.
   (c) Is the operation of Cartesian product commutative? That is, does
   $$A \times B = B \times A?$$
   (d) Find $(A \times B) \cup C$ and $(A \cup C) \times (B \cup C)$.
   (e) Find $(A \times B) \cap C$ and $(A \cap C) \times (B \cap C)$.
2. Let A, B, and C be any three nonempty sets. Does $(A \times B) \cap C = (A \cap C) \times (B \cap C)$? If so, prove it; if not, give a counterexample.
3. Let A, B, and C be any three nonempty sets. Show that $(A \cup B) \times C = (A \times C) \cup (B \times C)$. [This rule is called the **Distributive Law.**]
4. Let A and B be nonempty sets. Prove that the following statements are equivalent:
   (I) $A \times B = B \times A$;
   (II) $A \times B \subseteq B \times A$;
   (III) $B \times A \subseteq A \times B$;
   (IV) $A = B$.
5. Let $A = \{1, 2, 3, 4\}$. In each of the following, find all the elements of $A \times A$ that belong to the relation R, where R is defined by
   (a) $R = \{(x, y): y < x\}$,
   (b) $R = \{(x, y): y = x^2\}$.

6. Let $A = \{a, b\}$ and $B = \{1, 2, 3\}$. List all the functions from A into B. Which are 1-1? Which are onto?

7. Let $A = \{a, b, c\}$ and $B = \{1, 2\}$. List all the functions from A into B. Which are 1-1? Which are onto?

8. Let $A = \{a, b, c\}$ and $B = \{1, 2, 3\}$. List all the functions from A into B. Which are 1-1? Which are onto? Why are they the same?

9. If R is the set of real numbers, and if F is that relation in $R \times R$ consisting of all ordered pairs listed below, which are functions?

    (a) $F = \{(x, x - 3): x \in R\}$.

    (b) $F = \{(x, x^2): x \in R\}$.

    (c) $F = \{(x, \cos x): x \in R\}$.

    (d) $F = \{(x, y): x = -y^2\}$.

    Which of the functions are 1-1? Which of the functions are onto?

10. Let Z be the set of integers.

    (a) Give an example of a function from Z to Z that is 1-1 and onto.

    (b) Give an example of a function from Z to Z that is not 1-1 and not onto.

    (c) Give an example of a function from Z to Z that is 1-1, but not onto.

    (d) Give an example of a function from Z to Z that is onto, but not 1-1.

11. Let $g: C \to D$, let $A \subseteq C$, and let $f = g/A$, that is, f is the restriction of g to the set A.

    (a) If g is 1-1, show that f is 1-1.

    (b) Give an example to show that g may be onto D but f need not be onto D.

12. Let $f: A \to B$, let $A \subseteq C$, and let g be an extension of f to C such that $g: C \to B$.

    (a) If f is onto B, show that g is onto B.

    (b) Give an example to show that f may be 1-1 but g need not be 1-1.

13. Let F be a function from A into B. For any set $C \subseteq A$, define $F[C]$ by

$$F[C] = \{y: \text{there exists } x \in C \text{ such that } (x, y) \in F\}.$$

    that is, $F[C]$ is the *set of images of C under F*. Let $D \subseteq A$ and $E \subseteq A$.

    (a) Prove that $F[D \cup E] = F[D] \cup F[E]$.

    (b) Prove that $F[D \cap E] \subseteq F[D] \cap F[E]$.

    (c) Give an example of a function F and sets D and E such that

$$F[D \cap E] \neq F[D] \cap F[E].$$

14. Prove Theorem 1.2.1.

15. Prove Theorem 1.2.2.

1.3. Composite and Inverse Functions

In this section we shall discuss *composition* and *inverse functions*. Both of these concepts are methods of constructing new functions from given functions. Before doing this, however, we need to consider different notations for functions.

Functions, even though they are relations, are very important ın their own right and deserve a special notation. We shall, in general, use lower case letters $f, g, h, \ldots$ to represent functions. For example, if $A = \{1, 2, 3\}$, $B = \{a, b, c, d\}$, and $f = \{(1, a), (2, b), (3, c)\}$, we use the notation

$$f(1) = a, \quad f(2) = b, \quad f(3) = c$$

to specify that a is associated with 1, b is associated with 2, and c is associated with 3, that is, to specify the ordered pairs belonging to f.

If f is a function from a set A to a set B, and if a is any element of A, then the symbol $f(a)$ represents the second coordinate of the *one* ordered pair in the function f that has first coordinate a, that is, $(a, f(a))$ is the only ordered pair in f with a as a first coordinate. For example, if $f = \{(x, x^2): x$ is a real number$\}$, then $f(-1) = 1$, $f(0) = 0$, and $f(\sqrt{2}) = 2$.

Throughout the remainder of this text we shall use R to denote the set of real numbers, unless otherwise specifically stated.

To motivate the next definition, consider the following example.

Example 1.3.1. Let

$$f: R \to R \text{ such that } f(x) = 3x - 1, \text{ and}$$

$$g: R \to R \text{ such that } g(x) = x^2.$$

We can now define the function $h: R \to R$ by $h(x) = g(f(x))$. This function is

$$h: R \to R \text{ such that } h(x) = g(f(x))$$

$$= g(3x - 1) = (3x - 1)^2.$$

The basic idea of this example is *composition of functions* and leads to the following definition.

Definition 1.3.1. *Let* $f: A \to B$ *and* $g: C \to D$. *If* $\operatorname{Ran} f \subseteq C$, *then the* **composite** *of* f *and* g, *denoted by* $g \circ f$, *is the function* $g \circ f: A \to D$ *such that*

$$(g \circ f)(x) = g(f(x))$$

for each $x \in A$.

The operation of forming composite functions is called **composition of functions**. The domain of the composite $g \circ f$ is the domain of f, that is,

$$\operatorname{Dom} g \circ f = \operatorname{Dom} f,$$

while the range of $g \circ f$ is a subset of the range of g, that is,

$$\operatorname{Ran} g \circ f \subseteq \operatorname{Ran} g.$$

The composition of f and g is accomplished by following the function f by the function g.

Example 1.3.2. As an example of the composition of functions, let Z denote the set of integers, and suppose that $f: Z \to Z$ such that $f(x) = 2x$ and $g: Z \to Z$ such that $g(x) = 2x + 1$. Then

$$(f \circ g)(x) = f(g(x)) = f(2x + 1) = 2(2x + 1) = 4x + 2,$$

$$(g \circ f)(x) = g(f(x)) = g(2x) = 2(2x) + 1 = 4x + 1,$$

that is, $(f \circ g): Z \to Z$ such that $(f \circ g)(x) = 4x + 2$, and

$(g \circ f): Z \to Z$ such that $(g \circ f)(x) = 4x + 1$.

It is important to note from Example 1.3.2 that it is possible for both $f \circ g$ and $g \circ f$ to exist but be unequal. It is also possible for only one of the composite functions to exist. For example, if $f: R \to R$ such that $f(x) = -x^2 - 1$, and if $g: R^+ \to R$ such that $g(x) = \sqrt{x}$, where R^+ denotes the set of positive real numbers, then $(f \circ g): R^+ \to R$ such that $(f \circ g)(x) = -x - 1$, but $g \circ f$ does not exist $((\text{Ran } f) \cap (\text{Dom } g) = \varnothing)$. All this can be summarized by the simple statement: *In general, composition of functions is not commutative; that is, in general, $f \circ g \neq g \circ f$.*

We now state and prove one of the most important theorems concerning composition of functions.

Theorem 1.3.1. *If $f: A \to B$, $g: C \to D$, $h: E \to F$, $\text{Ran} f \subseteq C$, and $\text{Ran } g \subseteq E$, then*

$$(h \circ g) \circ f = h \circ (g \circ f),$$

that is, the composition of functions is associative.

Proof. Let $f: A \to B$, $g: C \to D$, $h: E \to F$, $\text{Ran} f \subseteq C$, and $\text{Ran } g \subseteq E$. We first note that $(h \circ g) \circ f: A \to F$ and $h \circ (g \circ f): A \to F$. In order to show that $(h \circ g) \circ f = h \circ (g \circ f)$, we need to show that $[(h \circ g) \circ f](x) = [h \circ (g \circ f)](x)$ for each $x \in A$. Let $x \in A$. Then

$$[(h \circ g) \circ f](x) = (h \circ g)(f(x))$$

$$= h(g(f(x)))$$

$$= h((g \circ f)(x))$$

$$= [h \circ (g \circ f)](x).$$

Thus, $[(h \circ g) \circ f](x) = [h \circ (g \circ f)](x)$ for each $x \in A$, and

$$(h \circ g) \circ f = h \circ (g \circ f).$$

Example 1.3.3. Let $h: R \to R$ such that $h(x) = 3x + 2$, let $g: R \to R$ such that $g(x) = x^2$, and let $f: R \to R$ such that $f(x) = (x - 2)/3$.

Then

$$(g \circ h)(x) = g(h(x)) = g(3x + 2) = (3x + 2)^2,$$
$$(f \circ (g \circ h))(x) = f((g \circ h)(x)) = f([3x + 2]^2)$$
$$= [(3x + 2)^2 - 2]/3,$$
$$(f \circ g)(x) = f(g(x)) = f(x^2) = (x^2 - 2)/3, \text{ and}$$
$$((f \circ g) \circ h)(x) = (f \circ g)(h(x)) = (f \circ g)(3x + 2)$$
$$= [(3x + 2)^2 - 2]/3.$$

Hence $(f \circ (g \circ h))(x) = [(3x + 2)^2 - 2]/3 = ((f \circ g) \circ h)(x)$.

The following two theorems show that the properties of functions being onto and 1–1 are preserved under the operation of composition.

Theorem 1.3.2. *Let f be a function from A onto B, and let g be a function from B onto C. Then g $\circ$ f is a function from A onto C.*

Proof. Let $z \in C$. We must show that there is an $x \in A$ such that $(g \circ f)(x) = z$. Since g is onto C, there exists $y \in B$ such that $z = g(y)$. Since f is onto B, there is $x \in A$ such that $y = f(x)$. By the definition of composition, $(g \circ f)(x) = g(f(x)) = g(y) = z$. Thus $g \circ f$ is onto C.

Theorem 1.3.3. *Let $f: A \to B$, let $g: C \to D$, and let $\text{Ran} f \subseteq C$. If f is 1–1 and g is 1–1, then g $\circ$ f is 1–1.*

The proof of this theorem is left as Exercise 10 of Problem Set 1.3.

Definition 1.3.2. *The function $i_A: A \to A$ such that $i_A(x) = x$ for each $x \in A$ is called the **identity function** on A.*

The function i_A is sometimes denoted more simply by i when there can be no confusion about its domain A.

Example 1.3.4. Let $f: R \to R$ such that $f(x) = 3x + 1$ and let $g: R \to R$ such that $g(x) = (x - 1)/3$. It is easily seen that both f and g are 1–1 functions mapping R onto R. We now find $f \circ g$ and $g \circ f$.

$$(f \circ g)(x) = f[g(x)] = f[(x - 1)/3]$$
$$= 3[(x - 1)/3] + 1 = x.$$
$$(g \circ f)(x) = g[f(x)] = g[3x + 1]$$
$$= ([3x + 1] - 1)/3 = x.$$

Thus $f \circ g = i_R$ and $g \circ f = i_R$.

Any pair of functions f and g, where $f: A \to B$ and $g: B \to A$, such that $f \circ g = i_B$ and $g \circ f = i_A$ are called *inverse functions*, according to the following definition.

Definition 1.3.3. *Let $f: A \to B$ and $g: B \to A$. Then g is said to be an inverse function (or inverse) of f if and only if $g \circ f = i_A$ and $f \circ g = i_B$. That is, $(g \circ f)(x) = x$ for every $x \in A$ and $(f \circ g)(y) = y$ for every $y \in B$.*

Example 1.3.5. Let $A = \{1, 2, 3\}$ and $B = \{a, b, c\}$. If

$$f = \{(1, a), (2, b), (3, c)\},$$

then an inverse of f is

$$g = \{(a, 1), (b, 2), (c, 3)\}.$$

What conditions are necessary for a function f to have an inverse? Theorem 1.3.5 proves that if f is a 1–1 function from A onto B, then f has an inverse. Theorem 1.3.4 states that if f has an inverse, then it is unique, and Theorem 1.3.6 shows that if f is 1–1 and onto, then its inverse is 1–1 and onto.

Theorem 1.3.4. *Let $f: A \to B$. If f has an inverse, then this inverse is unique; that is, if g and h are both inverses of f, then $g = h$.*

Proof. Let g and h be inverse functions of f, that is, $g \circ f = i_A$, $h \circ f = i_A$, $f \circ g = i_B$, and $f \circ h = i_B$, where i_A is the identity function on A and i_B is the identity function on B. We show that $g(y) = h(y)$ for each $y \in B$.

Let $y \in B$. Then $(f \circ g)(y) = y = (f \circ h)(y)$. Applying g to the equality $(f \circ g)(y) = (f \circ h)(y)$ we have that

$$g\big((f \circ g)(y)\big) = g\big((f \circ h)(y)\big) \quad \text{or} \quad [g \circ (f \circ g)](y) = [g \circ (f \circ h)](y).$$

Since the composition of functions is associative (Theorem 1.3.1), this equality may be written as $[(g \circ f) \circ g](y) = [(g \circ f) \circ h](y)$. By the hypothesis at the beginning of the proof, $g \circ f = i_A$, and we have that

$$(i_A \circ g)(y) = (i_A \circ h)(y).$$

Since $i_A \circ g = g$ and $i_A \circ h = h$, it follows that $g(y) = h(y)$. The element y was an arbitrary member of B, and so $g = h$. Therefore, if a function has an inverse, this inverse is unique.

We denote the inverse of f by f^{-1}. Then $f^{-1} \circ f = i_A$ and $f \circ f^{-1} = i_B$, where A is the domain of f and B is the range of f.

Theorem 1.3.5. *Let $f: A \to B$. If f is 1–1 and onto, then f has an inverse.*

Proof. Define the relation g from B into A by

$$g = \{(y, x): (x, y) \in f\}.$$

Then $g \subseteq B \times A$, since $f \subseteq A \times B$. Let $y \in B$. Since f is onto, there exists an $x \in A$ such that $(x, y) \in f$. Thus $(y, x) \in g$ and g is defined for each y of B. If $(y, x_1) \in g$ and $(y, x_2) \in g$, then $(x_1, y) \in f$ and $(x_2, y) \in f$, but since f is 1–1, $x_1 = x_2$. Hence g satisfies the conditions of a function. By the definition of g, $g \circ f = i_A$ and $f \circ g = i_B$. Therefore, g is the inverse of f.

Theorem 1.3.5 tells us that any 1–1 function can be redefined so that it has an inverse. Let $f: A \rightarrow B$ be a 1–1 function. Then $f: A \rightarrow \mathrm{Ran} f$ is 1–1 and onto, and, by Theorem 1.3.5, f has an inverse function $f^{-1}: \mathrm{Ran} f \rightarrow A$.

Theorem 1.3.6. *Let $f: A \rightarrow B$. If f is 1–1 and onto, then its inverse $f^{-1}: B \rightarrow A$ is 1–1 and onto.*

The proof is left as Exercise 11 of Problem Set 1.3.

Problem Set 1.3

1. Let $f: R \rightarrow R$ such that $f(x) = \sin x$, and let $g: R \rightarrow R$ such that $g(x) = 2x + \pi$. Find $g \circ f$ and $f \circ g$. What are the domains of $g \circ f$ and $f \circ g$? Find the range of $g \circ f$ and the range of $f \circ g$.

2. Let $f: R \rightarrow R$ such that $f(x) = 3x^2 + 2x$, and let $g: R \rightarrow R$ such that $g(x) = x - 1$. Find $g \circ f, f \circ g$, the range of $g \circ f$, and the range of $f \circ g$.

3. Let $A = \{a, b, c, d\}$, $B = \{p, q, r, s\}$, $C = \{u, v, w\}$, $f = \{(a, q), (b, r), (c, r), (d, s)\}$, and $g = \{(p, v), (q, w), (r, w), (s, u)\}$. Find $g \circ f$ and the range of $g \circ f$.

4. Let A be the negative real numbers, and let B be the positive real numbers. Define $f: A \rightarrow B$ by $f(x) = 1/(2 - x)$ and $g: B \rightarrow B$ by $g(x) = 1/(1 + x)$. Find $g \circ f$ and the range of $g \circ f$.

5. Let $f: R \rightarrow R$ be defined by $f(x) = x^2 + 1$, and let $g: R \rightarrow R$ such that $g(x) = x^{23}$. Find $f \circ g, g \circ f$, the range of $f \circ g$, and the range of $g \circ f$.

6. Let $f: R \rightarrow R$ such that $f(x) = 3x - 5$. Find f^{-1}.

7. Let $A = \{x: x \geqslant 2\}$ and $B = \{x: x \geqslant -4\}$. Define $f: A \rightarrow B$ by $f(x) = x^2 - 4x$. Show that f is 1–1 and onto, and find f^{-1}.

8. Let A be the nonnegative real numbers, and that $B = \{x: 0 \leqslant x < 1\}$. Define $f: A \rightarrow B$ by $f(x) = x/(1 + x)$. Show that f is 1–1 and onto, and find f^{-1}.

9. Show that the composite of two functions is a function.

10. Prove Theorem 1.3.3.

11. Prove Theorem 1.3.6.

12. Let f be a 1–1 function from A onto B. Show that f^{-1} has an inverse and that $(f^{-1})^{-1} = f$.

13. Let $f: A \rightarrow B$ and $g: B \rightarrow C$. Give examples to show that
   (a) if g is onto C, $g \circ f$ need not be onto C.
   (b) if f is 1–1, $g \circ f$ need not be 1–1.

(c) if f is not onto B, $g \circ f$ may be onto C.

(d) if g is not 1-1, $g \circ f$ may be 1-1.

Prove that

(e) if $g \circ f$ is 1-1, then f is 1-1.

(f) if $g \circ f$ is onto C, then g is onto C.

14. Prove that $i_A: A \to A$ is a 1-1 function from A onto A.

1.4. Countable Sets

It sometimes becomes necessary to compare the number of elements in one set with the number of elements in another set. For example, if a classroom is to be used most efficiently, then each chair in the classroom should be occupied by a student. If each chair is occupied by exactly one student, then the set of chairs and the set of students have the same number of elements. This conclusion can be reached without actually counting the elements of either set, since there is a pairing or a "1-1 correspondence" between the set of chairs and the set of students. The idea of pairing the elements of two sets is now formalized.

Definition 1.4.1. *Let A and B be sets. Then A is said to be **equivalent** to B if and only if there exists a 1-1 function f from A onto B.*

We use the notation "$A \sim B$" to mean "the set A is equivalent to the set B."

In the above example it is easy to see that such a 1-1 onto function exists. In other cases, it may be more difficult to establish the existence of such a function.

In the remainder of this section the set of positive integers or natural numbers plays an important role. The properties of the set of positive integers will be developed rigorously in Chapter Two, but here we will assume that the reader is familiar with the elementary properties of the set of positive integers. Throughout the remainder of the text we shall use N to denote the set of positive integers, unless otherwise specifically stated.

Example 1.4.1. Let $N = \{1, 2, \ldots, n, \ldots\}$ be the set of positive integers. The set $E = \{2n: n \in N\}$ is called the set of positive even integers. We shall show that $N \sim E$. Define $f: N \to E$ by the equation $f(n) = 2n$ for each $n \in N$. Then if $m, n \in N$ and $f(n) = f(m)$, then

$$2n = f(n) = f(m) = 2m,$$

and $n = m$. Hence f is 1-1. Since any positive even integer k can be written as $k = 2m$, where m is a positive integer, we see that $f(m) = 2m = k$. Hence f is onto E. Therefore, f is a 1-1 function from N onto E, and N and E are equivalent.

Theorem 1.4.1. *Let A, B, and C be sets. Then*

(a) $A \sim A$;   *(Reflexive law)*

(b) *If $A \sim B$, then $B \sim A$;*   *(Symmetric law)*

(c) *If $A \sim B$ and $B \sim C$, then $A \sim C$.*   *(Transitive law)*

Proof. (a) The identity function $i_A \colon A \to A$ is a 1–1 function from A onto A. Therefore, $A \sim A$.

(b) If $A \sim B$, there exists a 1–1 function f mapping A onto B. By Theorems 1.3.5 and 1.3.6, f^{-1} is a 1–1 mapping of B onto A. Hence $B \sim A$.

(c) If $A \sim B$ and $B \sim C$, there is a 1–1 function f mapping A onto B, and a 1–1 function g mapping B onto C. Theorems 1.3.2 and 1.3.3 imply that $g \circ f$ is a 1–1 function mapping A onto C. Hence $A \sim C$.

A relation satisfying the three conditions of Theorem 1.4.1 is called an **equivalence relation.**

Definition 1.4.2. *A set A is said to be **finite** if and only if A is empty or there is a positive integer n such that A is equivalent to the set $\{1, 2, \ldots, n\}$. A set A is **infinite** if and only if A is not finite.*

If $A = \varnothing$, then A has no elements. If $A \neq \varnothing$ and A is finite, then there exists a positive integer n such that A is equivalent to the set $\{1, 2, \ldots, n\}$. In this case, we say that the set A contains or has n elements. If the set A is infinite, then we say that A has infinitely many elements.

Definition 1.4.3. *A set A is **denumerable** if and only if A is equivalent to the set of positive integers. A set A is **countable** if and only if it is either finite or denumerable. A set A is **uncountable** if and only if it is not countable.*

We shall see examples of uncountable sets later (in fact, we shall prove that the set of real numbers is uncountable). Examples of finite sets are easily obtained, and so we only give examples of denumerable sets. Theorem 1.4.1 implies that the set of positive integers is itself denumerable, and Theorem 1.4.1 and Example 1.4.1 imply that the set of even positive integers is denumerable.

Theorem 1.4.2. *A nonempty set A is countable if and only if there is a 1–1 function f mapping A into N.*

Proof. Assume that A is a nonempty countable set. Then A is either finite or denumerable. If A is finite, then there exists a positive integer n and a 1–1 function f mapping A onto $\{1, 2, \ldots, n\} \subseteq N$. If A is denumerable, then there exists a 1–1 function f mapping A onto $N \subseteq N$. In either case, there is a 1–1 function f mapping A into N.

Assume that there is a 1–1 function f mapping A into N. Let $\text{Ran} f = B \subseteq N$. If B is finite, A is finite and the proof is complete. Thus assume B is infinite. If $B = N$, then B is denumerable and so A is denumerable. Hence we assume that $B \subset N$. Since $B \neq \varnothing$, B contains at least one element, say b_1. We now consider the set $B \setminus \{b_1\}$ (see Problem 18, Problem Set 1.1, for the definition of the difference of two sets). The set $B \setminus \{b_1\}$ is nonempty, for otherwise the set B would be a finite set. Let b_2 be any element of $B \setminus \{b_1\}$. Continuing in this manner, if the positive integers $b_1, b_2, ..., b_k$ have been chosen, the set $B \setminus \{b_1, b_2, ..., b_k\}$ is a nonempty set (for otherwise B would be a finite set), and we may choose a positive integer b_{k+1} from $B \setminus \{b_1, b_2, ..., b_k\}$. Define the function $g: B \to N$ by $g(b_r) = r$. We need to show that g is a 1–1 function mapping B onto N. By the manner of choosing b_{r+1} from $B \setminus \{b_1, b_2, ..., b_r\}$, it is clear that g is 1–1. If $n \in N$, there is a $b_n \in B$ such that $g(b_n) = n$. Hence g is a 1–1 function from B onto N. Theorems 1.3.2 and 1.3.3 imply that the function $g \circ f$ is a 1–1 mapping of A onto N. Thus A is denumerable. In either case, A is finite or denumerable and hence is countable.

Corollary 1.4.2a. *Any subset of a countable set is countable.*

Proof. If the set A is countable, then there is a 1–1 function f mapping A into N. Let B be any subset of A and define a function $g: B \to N$ as follows:

$$g(a) = f(a) \quad \text{for} \quad a \in B,$$

that is, g is the restriction of f to B. If we show g is 1–1, then we have a 1–1 function mapping B into N, and B is countable. Let $a, b \in B$ such that $a \neq b$. Then $g(a) = f(a) \neq f(b) = g(b)$, since f is 1–1. Hence g is 1–1.

Corollary 1.4.2b. *Any infinite subset of a denumerable set is denumerable.*

Proof. Let B be an infinite subset of a denumerable set A. By Theorem 1.4.2, B is countable and either B is finite or it is infinite. Since B was assumed infinite, B is denumerable

Theorem 1.4.3. *If N is the set of positive integers, then $N \times N$ is denumerable.*

Proof. We define a function f mapping $N \times N$ into N as follows:

$$f((a, b)) = 2^a 3^b \text{ where } (a, b) \in N \times N.$$

If $(a, b), (c, d) \in N \times N$, and if $(a, b) \neq (c, d)$, then either $a \neq c$, $b \neq d$, or both $a \neq c$ and $b \neq d$. In any case

$$f((a, b)) = 2^a 3^b \neq 2^c 3^d = f((c, d)).$$

Hence f is 1–1, and $N \times N$ is countable.

We now need to show that $N \times N$ is actually denumerable. We do this by showing that a subset of $N \times N$ is denumerable, which implies that $N \times N$ is denumerable, since $N \times N$ is countable. Let b be a fixed natural number, and let $A = \{(a, b): a \in N\}$. Since b is a fixed natural number, A is a proper subset of $N \times N$. Define the function g mapping A into N as follows: ·

$$g((a, b)) = a.$$

It is easily verified that g maps A onto N in a 1–1 manner. Hence A, and consequently $N \times N$, are denumerable.

Corollary 1.4.3. *If A and B are nonempty countable sets, then $A \times B$ is countable.*

Proof. Since A and B are countable, there exists a 1–1 function h mapping A into N, and a 1–1 function k mapping B into N. Define a function m mapping $A \times B$ into $N \times N$ by

$$m((a, b)) = (h(a), k(b)), \quad \text{where} \quad (a, b) \in A \times B.$$

We show that the function m is 1–1. Assume that $(a, b), (c, d) \in A \times B$ and that $(a, b) \neq (c, d)$. If $a \neq c$, then $h(a) \neq h(c)$ and so $m((a, b)) = (h(a), k(b)) \neq (h(c), k(d)) = m((c, d))$. If $b \neq d$, then $k(b) \neq k(d)$ and so $m(a, b) = (h(a), k(b)) \neq (h(c), k(d)) = m(c, d)$. Hence m is 1–1.

Let f be the 1–1 function defined in the proof of Theorem 1.4.3 mapping $N \times N$ into N. Then $f \circ m: (A \times B) \to N$, and Theorem 1.3.3 assures us that $f \circ m$ is 1–1. Therefore, $A \times B$ is countable.

The proof of Theorem 1.4.3 actually shows that to establish that a set A is countable, it is only necessary to exhibit a 1–1 function from A into any countable set. Also, to verify that a set is denumerable it is only necessary to establish a 1–1 function from that set onto any denumerable set.

In Section 1.1 union and intersection of a finite number of sets were defined. We shall now extend these definitions to include the union and intersection of infinitely many sets. Let Λ be a set. If for each $\lambda \in \Lambda$ there corresponds a set A_λ, then the collection of sets $\{A_\lambda : \lambda \in \Lambda\}$ is called a **family of sets**; the set Λ is called the **index set** of the family of sets. We now define unions and intersections for families of sets.

Definition 1.4.4. Let $\{A_\lambda : \lambda \in \Lambda\}$ be a family of sets. The set $\{x : x \in A_\lambda$ for at least one $\lambda \in \Lambda\}$ is called the **union** of the family of sets and is denoted by

$$\bigcup \{A_\lambda : \lambda \in \Lambda\} \quad \text{or} \quad \bigcup_{\lambda \in \Lambda} A_\lambda.$$

The set $\{x : x \in A_\lambda$ for every $\lambda \in \Lambda\}$ is called the **intersection** of the family of sets and is denoted by

$$\bigcap \{A_\lambda : \lambda \in \Lambda\} \quad \text{or} \quad \bigcap_{\lambda \in \Lambda} A_\lambda.$$

Using arguments analogous to those in the proof of Theorem 1.1.3, it can easily be verified that DeMorgan's Laws hold for unions and intersections of families of sets. That is,

$$\mathbf{C}\left(\bigcup_{\lambda \in \Lambda} A_\lambda\right) = \bigcap_{\lambda \in \Lambda} \mathbf{C}(A_\lambda),$$

$$\mathbf{C}\left(\bigcap_{\lambda \in \Lambda} A_\lambda\right) = \bigcup_{\lambda \in \Lambda} \mathbf{C}(A_\lambda).$$

The proof is left as Exercise 9 of Problem Set 1.4.

If the index set Λ is finite, then there exists a positive integer n and a 1–1 function f mapping Λ onto the set $\{1, 2, ..., n\}$. Hence for each $\lambda \in \Lambda$ there is a unique $k \in \{1, 2, ..., n\}$ such that $f(\lambda) = k$. Thus we can replace the subscript λ by the corresponding positive integer k and write the finite union and intersection as, respectively,

$$\bigcup_{i=1}^{n} A_i \quad \text{and} \quad \bigcap_{i=1}^{n} A_i.$$

In a similar manner, if the index set Λ is denumerable, then there exists a 1–1 function f mapping Λ onto N. Again for each $\lambda \in \Lambda$ there is a unique $k \in N$ such that $f(\lambda) = k$. We can replace the subscript λ by the corresponding positive integer k and write

$$\bigcup_{i=1}^{\infty} A_i \quad \text{and} \quad \bigcap_{i=1}^{\infty} A_i$$

for the denumerable union and the denumerable intersection, respectively.

If the index set Λ is countable, then it is finite or denumerable. In the case where Λ is finite, we can write the finite union

$$\bigcup_{i=1}^{n} A_i$$

as

$$\bigcup_{i=1}^{\infty} A_i,$$

where we define $A_i = \emptyset$ if $i > n$. In a similar manner, we can write the finite intersection

$$\bigcap_{i=1}^{n} A_i$$

as

$$\bigcap_{i=1}^{\infty} A_i,$$

where we define $A_i = \bigcup_{k=1}^{n} A_k$ if $i > n$. Using these conventions, we can always write countable unions and intersections as, respectively,

$$\bigcup_{i=1}^{\infty} A_i \quad \text{and} \quad \bigcap_{i=1}^{\infty} A_i.$$

Example 1.4.2. Let $A_n = \{x : x \in R \text{ and } x < n\}$ for each $n \in N$. Then

$$\bigcup_{n=1}^{\infty} A_n = R \quad \text{and} \quad \bigcap_{n=1}^{\infty} A_n = \{x : x \in R \text{ and } x < 1\}.$$

Theorem 1.4.4. *If $\{A_n : n \in N\}$ is a countable family of countable sets, then*

$$\bigcup_{n=1}^{\infty} A_n$$

is countable.

We can restate this theorem as: *a countable union of countable sets is countable.*

Proof. Since each set A_n, where $n \in N$, is countable, there is a 1–1 function g_n mapping A_n into N. Define the function h mapping $\bigcup_{n=1}^{\infty} A_n$ into $N \times N$ as follows: Let $x \in \bigcup_{n=1}^{\infty} A_n$. If x belongs to exactly one of the sets A_n, say, A_s, let $h(x) = (s, g_s(x))$. If x belongs to more than one of the sets A_n, let r be the smallest positive integer such that $x \in A_r$. Then let $h(x) = (r, g_r(x))$. We verify that h is 1–1. Let $x, y \in \bigcup_{n=1}^{\infty} A_n$ such that $x \neq y$. Let m and k be the smallest positive integers such that $x \in A_m$ and $y \in A_k$. If $m \neq k$, then

$$h(x) = (m, g_m(x)) \neq (k, g_k(x)) = h(y).$$

If $m = k$, then

$$g_m(x) = g_k(x) \neq g_k(y),$$

since g_k is 1–1. Thus

$$h(x) = (k, g_k(x)) \neq (k, g_k(y)) = h(y).$$

Hence, if $x \neq y$, then $h(x) \neq h(y)$, and h is 1–1. Therefore h is a 1–1 mapping of $\bigcup_{n=1}^{\infty} A_n$ into $N \times N$ and Theorems 1.4.2 and 1.4.3 imply that $\bigcup_{n=1}^{\infty} A_n$ is countable.

Theorem 1.4.4 also implies that the finite union of finite sets is at most countable (it is left to the reader to show that it is actually finite), the finite union of denumerable sets is denumerable, the denumerable union of finite sets is countable, and the denumerable union of denumerable sets is denumerable.

Problem Set 1.4

1. Prove that the set $\{\ldots, -3, -2, -1\}$ is denumerable.
2. Prove that the set $\{\ldots, -3, -2, -1, 0, 1, 2, 3, \ldots\}$ is denumerable.
3. Prove that the set $\{3n: n \in N\}$ is denumerable.
4. Let A be denumerable, and let $B = \{a, b\}$, with $A \cap B = \emptyset$ and $a \neq b$. Show directly that $A \cup B$ is denumerable.
5. Show that a finite union of finite sets is finite.
6. Give examples to show that the denumerable union of finite sets may either be finite or denumerable.
7. Let A be a set, and let B be an infinite subset of A. Show that A is infinite.
8. Prove that the set of rational numbers is denumerable. [*Hint*: Show that there is a 1–1 mapping from the set of rational numbers into the set $N \times N$.]
9. Prove DeMorgan's Law for families of sets.

Chapter Two

THE REAL NUMBER SYSTEM

2.1. Introduction

The structure of mathematical analysis, as it exists today, has as its foundation the *real number system*. All the properties of the real numbers follow by logical inference from five *hypotheses* or *axioms*, called Peano's Postulates. These axioms were first given by the Italian mathematician Giuseppe Peano (1858–1932) in 1889. Beginning with these postulates it is possible to develop, in a rigorous manner, the system of real numbers. The real number system has all the properties needed for our study of advanced calculus.

There are two basic methods of approach that one might employ at this point. The first method would be to begin with Peano's Postulates and develop completely the system of real numbers. This approach would require the five postulates, over 50 definitions, and more than 200 theorems—a task that is much too time consuming here. (See E. Landau, *Foundations of Analysis*, Chelsea Publishing Company, New York, 1951 for a complete discussion of the real number system.) The other basic approach would be to postulate the existence of the real number system, list the desired properties, and advance to the study of functions. This is the approach that we shall use, but first we will introduce part of the algebraic structure of the real number system.

2.2. Groups

Let S be a nonempty set, and let f be a function mapping $S \times S$ into S. Such a function is given a special name.

Definition 2.2.1. *Let S be a nonempty set. A function f mapping $S \times S$ into S is called a **binary operation** on S.*

We now present some examples to illustrate the concept of binary operations.

Example 2.2.1. Let Z denote the set of integers. Let

$$f_1: Z \times Z \to Z \quad \text{such that} \quad f_1(x, y) = x + y,$$

$$f_2: Z \times Z \to Z \quad \text{such that} \quad f_2(x, y) = xy, \quad \text{and}$$

$$f_3: Z \times Z \to Z \quad \text{such that} \quad f_3(x, y) = x - y.$$

Then f_1, f_2, and f_3 are binary operations on the set of integers.

If we replace the set Z by the set Z^+ of positive integers, then f_3 is *not* a binary operation on Z^+, since $f(2, 4) = 2 - 4 = -2 \notin Z^+$, that is, f_3 is *not* a function of $Z^+ \times Z^+$ into Z^+.

In place of the notation for a function f from $S \times S$ into S, it is usually more convenient to use a symbol such as "$+$" or "$\cdot$." For example, in place of $f_1(x, y)$ and $f_2(x, y)$ in Example 2.2.1 we commonly use $x + y$ (addition) and $x \cdot y$ or xy (multiplication).

Some binary operations have special properties. We now define two of these properties and give examples of them.

Definition 2.2.2. *Let $\cdot$ be a binary operation on the set S. Then*

(a) *$\cdot$ is said to be **associative** if and only if*

$$x \cdot (y \cdot z) = (x \cdot y) \cdot z$$

for every $x, y, z \in S$, and

(b) *$\cdot$ is said to be **commutative** if and only if*

$$x \cdot y = y \cdot x$$

for every $x, y \in S$.

Example 2.2.2. The binary operations f_1 and f_2 defined in Example 2.2.1 are both associative and commutative; f_3, however, is neither associative nor commutative, since

$$(3 - 2) - 2 = -1 \neq 3 = 3 - (2 - 2)$$

and
$$3 - 2 = 1 \neq -1 = 2 - 3.$$

In a set with a binary operation, elements with particular properties are given special names. Two such types of elements are now defined.

Definition 2.2.3. *Let* $\cdot$ *be a binary operation on the set S. An element* $e \in S$ *is called an* **identity element** *or an* **identity** *of S with respect to* $\cdot$ *if and only if*

$$x \cdot e = e \cdot x = x \quad \text{for every} \quad x \in S.$$

Example 2.2.3. In the set of integers with the binary operation of addition, the integer 0 is an identity, since $n + 0 = 0 + n = n$ for every integer n. The set of integers under the binary operation of multiplication has the integer 1 as an identity. In each of these cases the operation has only one identity.

The first theorem of this chapter ensures that, if a binary operation on a set has an identity, then it has only one such identity.

Theorem 2.2.1. *Let S be a set with binary operation* $\cdot$. *If S has an identity with respect to the binary operation* $\cdot$, *then it is unique. That is, if e and f are both identities, then* $e = f$.

Proof. Let S be a set with binary operation $\cdot$, and let e and f be identities with respect to $\cdot$. Then

$$e = e \cdot f = f,$$

and the identity is unique.

Definition 2.2.4. *Let* $\cdot$ *be a binary operation on the set S with identity e, and let* $x \in S$. *An element* $x' \in S$ *is called an* **inverse** *of x with respect to* $\cdot$ *if and only if*

$$x \cdot x' = x' \cdot x = e.$$

Example 2.2.4. In the set of integers with the binary operation of addition, for each integer n, $-n$ is an inverse of n, since $n + (-n) = (-n) + n = 0$. In the set of integers under multiplication, the only elements with inverses are 1 and -1, and each of these integers is its own inverse, that is, $1 \cdot 1 = 1$ and $(-1)(-1) = 1$.

Similar to identity elements, if an element has an inverse with respect to an associative binary operation, then the element has exactly one inverse. The proof is left as Exercise 4 of Problem Set 2.2.

Theorem 2.2.2. *Let S be a set with associative binary operation $\cdot$ and identity e. If the element $x \in S$ has an inverse, then that inverse is unique. That is, if x' and x'' are both inverses of x, then $x' = x''$.*

If the element x has an inverse, then it is unique and we denote it by x^{-1}.

Theorem 2.2.3. *Let S be a set with binary operation $\cdot$ and identity e. If $x \in S$ has an inverse, then its inverse x^{-1} has an inverse and $(x^{-1})^{-1} = x$.*

The proof is left as Exercise 5 of Problem Set 2.2.

We are now in a position to define one of the important types of algebraic structures, the *group*.

Definition 2.2.5. *Let S be a set with binary operation $\cdot$. The set S together with the binary operation $\cdot$ is called a **group** if and only if*

(a) *$\cdot$ is associative,*

(b) *S has an identity with respect to $\cdot$, and*

(c) *every element of S has an inverse with respect to $\cdot$.*

We use the notation $(S; \cdot)$ to denote the group whose set is S and whose binary operation is $\cdot$. The binary operation $\cdot$ is referred to as the **operation of the group**, the **group operation**, or the **group multiplication**.

The definition of a group requires that $\cdot$ be a function from $S \times S$ into S, but note that $\cdot$ is actually *onto* S. This follows since for any $x \in S$, $x \cdot e = x$.

A group $(S; \cdot)$ is said to be an **Abelean group** or a **commutative group** if and only if the operation $\cdot$ is commutative. Abelean groups are named in honor of the Norwegian mathematician Neils Henrik Abel (1802–1829), who was one of the early workers in the theory of groups.

If we are considering a group in general, we usually write the operation as multiplication and denote $x \cdot y$ as xy. The identity is denoted by e or by 1. In the case of commutative groups, the operation is often written as addition, and $x \cdot y$ is denoted by $x + y$. In this case 0 is used as the identity.

When we write the group operation as multiplication, the product xx is denoted as x^2, the product xxx is written as x^3, and so on. We also write x^{-1} for the inverse of x, x^{-2} for $x^{-1}x^{-1}$, and so on. For groups whose operation is written additively, we write $x + x$ as $2x$, $x + x + x$ as $3x$, and so on, while the inverse of x is denoted by $-x$.

Example 2.2.5. The proof that the following examples are groups is left as Exercise 6 of Problem Set 2.2.

Let $+$ and $\cdot$ denote respectively the operations of addition and multiplication for the set of integers Z.

(a) Then $(Z; +)$ is a commutative group.

(b) Let

$$A = \{(x, y): x, y \in Z \quad \text{and} \quad y \neq 0\}.$$

Define *addition*, $\oplus$, on A as follows: For $(x,y),(u,v) \in A$, let

$$(x, y) \oplus (u, v) = (xv + yu, yv).$$

Then $(A; \oplus)$ is a commutative group.

(c) Let N denote the set of positive integers, and let

$$B = \{(x, y): x, y \in N\}.$$

Define *multiplication*, $\odot$, on B as follows: For $(x,y),(u,v) \in B$, let

$$(x, y) \odot (u, v) = (xu, yv).$$

Then $(B; \odot)$ is a commutative group.

In the above examples, the sets involved are all infinite. Such groups are called **infinite groups**. If the set of a group is finite, the group is said to be a **finite group**. The following example presents three finite groups. The proof that the following examples are groups is left as Exercise 7 of Problem Set 2.2.

Example 2.2.6. (a) The set $\{1, -1\}$ with the binary operation of ordinary multiplication of integers.

(b) Let $S = \{0, 1, 2\}$, and let *addition* on S be defined by the following table.

+	0	1	2
0	0	1	2
1	1	2	0
2	2	0	1

The set S together with the operation of addition forms a commutative group

(c) Let $S = \{1, i, -1, -i\}$ be a subset of the set of complex numbers, and let the binary operation on S be multiplication of complex numbers. The multiplication table is given below.

·	1	i	-1	$-i$
1	1	i	-1	$-i$
i	i	-1	$-i$	1
-1	-1	$-i$	1	i
$-i$	$-i$	1	i	-1

The set S together with the operation of multiplication forms a commutative group.

In these examples, each group is commutative. Examples can be constructed that are not commutative.

We conclude this section with a discussion about a relation between groups called an *isomorphism*.

Definition 2.2.6. *Let* $(S; \cdot)$ *and* $(T; \odot)$ *be two groups. Let f be a 1–1 function mapping S onto T. Then f is said to be an* **isomorphism** *of* $(S; \cdot)$ *onto* $(T; \odot)$ *if and only if* $f(x \cdot y) = f(x) \odot f(y)$ *for each pair of elements x, y of S. If there is an isomorphism of $(S; \cdot)$ onto $(T; \odot)$, then the groups $(S; \cdot)$ and $(T; \odot)$ are said to be* **isomorphic**.

An isomorphism f is a 1–1 mapping of the set S onto the set T which preserves the operation of multiplication, that is, $f(x \cdot y) = f(x) \odot f(y)$ for each x and y in S. In other words, the image of the product of two elements of S is the same element that would be obtained by mapping the elements of S first and then forming the product of their respective images.

Example 2.2.7. Let $(Z; +)$ be the group with Z the set of integers and $+$ ordinary addition of integers. (Cf. Example 2.2.5(a).)

Let $(T; \cdot)$ be the group with $T = \{2^k : k \in Z\}$ and binary operation $\cdot$ defined by

$$2^m \cdot 2^n = 2^{m+n},$$

where $m, n \in Z$. It is left as an exercise to verify that $(T; \cdot)$ is a group (Exercise 13 of Problem Set 2.2.).

We now show that the groups $(Z; +)$ and $(T; \cdot)$ are isomorphic. We define the function $f: Z \to T$ by the equation

$$f(k) = 2^k$$

for each $k \in Z$. Then it is easily seen that f is a 1–1 mapping of Z

onto T. We only need to show that operations are preserved, that is, that $f(m + n) = f(m) \cdot f(n)$ for each pair of elements m,n of Z. Let m and n be any elements of Z. Then

$$f(m + n) = 2^{m+n} = 2^m \cdot 2^n = f(m) \cdot f(n).$$

Thus f is an isomorphism of the group $(Z; +)$ onto the group $(T; \cdot)$, and the groups $(Z; +)$ and $(T; \cdot)$ are isomorphic.

Example 2.2.8. Let $(T; \oplus)$ be the group with $T = \{0, 1, 2, 3\}$ and binary operation of addition defined as follows:

$\oplus$	0	1	2	3
0	0	1	2	3
1	1	2	3	0
2	2	3	0	1
3	3	0	1	2

It is left as an exercise to verify that $(T; \oplus)$ is a group (Exercise 14 of Problem Set 2.2.).

Let $(S; \cdot)$ be the group defined in Example 2.2.6(c). We shall show that the groups $(S; \cdot)$ and $(T; \oplus)$ are isomorphic.

We define the function $f: S \to T$ by

$$f(1) = 0,$$
$$f(i) = 1,$$
$$f(-1) = 2,$$
$$f(-i) = 3.$$

Then f is a 1–1 function from S onto T. We need only show that operations are preserved under f. If we are to check all possible products of elements in S, we see that there are 16 possible products to check. We shall check the following four products: $1 \cdot i = i$, $(-1) \cdot (-i) = i$, $i \cdot (-i) = 1$, and $(-1) \cdot 1 = -1$. The remaining 12 products are left for the student to check. We have

$$f(1 \cdot i) = \quad f(i) \quad = 1 = 0 \oplus 1 = f(1) \oplus f(i),$$
$$f((-1) \cdot (-i)) = \quad f(i) \quad = 1 = 2 \oplus 3 = f(-1) \oplus f(-i),$$
$$f(i \cdot (-i)) = \quad f(1) \quad = 0 = 1 \oplus 3 = f(i) \oplus f(-i),$$
$$f((-1) \cdot 1) = f(-1) = 2 = 2 \oplus 0 = f(-1) \oplus f(1).$$

Thus f is an isomorphism of $(S; \cdot)$ onto $(T; \oplus)$, and the groups $(S; \cdot)$ and $(T; \oplus)$ are isomorphic.

From Definition 2.2.6 and Examples 2.2.7 and 2.2.8, we see that two groups are isomorphic if their elements can be put into a 1–1 correspondence in such a way that their respective binary operations also correspond.

The existence of an isomorphism between two groups means that the "algebraic properties" or "algebraic structure" of the two groups are identical and, in fact, that they can only be distinguished by the notation used. This is easily seen from the multiplication tables of the isomorphic groups, since isomorphic groups have the "same" multiplication table. In our study of advanced calculus, we shall think of two isomorphic groups as being the same group.

Problem Set 2.2

1. Give an example of a binary operation that is commutative but *not* associative.
2. Give an example of a binary operation that is associative but *not* commutative.
3. Let Z^+ be the set of positive integers, and let $f(x, y) = x^y$.
   (a) Is f a binary operation?
   (b) Is f associative?
   (c) Is f commutative?
4. Prove Theorem 2.2.2.
5. Prove Theorem 2.2.3.
6. Show that the sets in Example 2.2.5 are commutative groups with respect to the given binary operations.
7. Show that the sets in Example 2.2.6 are commutative groups with respect to the given binary operations.
8. With respect to the operation of addition, which of the following sets are groups? Why?
   (a) Z^+, the set of positive integers.
   (b) E, the set of even integers.
   (c) The set of multiples of 5.
   (d) The set $\{0\}$.
   (e) The set $\{1\}$.
   (f) R, the set of real numbers.
9. With respect to the operation of multiplication, which of the sets in Problem 8 are groups? Why?
10. Let $(S; \cdot)$ be a group. Prove that the equation $a \cdot x = b$ has a solution x for all $a, b \in S$. Show that the solution is unique.
11. Let $(S; \cdot)$ be a group. Prove that if $a \cdot x = b \cdot x$ for $a, b, x \in S$, then $a = b$. (This property of groups is called the **cancellation law.)**
12. Let $(S; \cdot)$ be a group. If the inverse of x is denoted by x^{-1}, show that $(xy)^{-1} = y^{-1} x^{-1}$.

13. Show that the set T with the operation defined in Example 2.2.7 is a commutative group.
14. Show that the set T with the operation defined in Example 2.2.8 is a commutative group.
15. Show that the group of integers with respect to addition and the group of even integers with respect to addition are isomorphic.
16. Show that the additive group of integers and the additive group of multiples of 5 are isomorphic.
17. Give an example of a group that is isomorphic to, but not identical with, the group defined in Example 2.2.6(b).
18. Give an example of a group that is *not* isomorphic to the group defined in Example 2.2.6(b).
19. Show that the group defined in Example 2.2.6(b) is *not* isomorphic to the group defined in Example 2.2.6(c).
20. Show that the additive group of integers is *not* isomorphic to the multiplicative group defined in Example 2.2.5(c).
21. Let $(S; \cdot)$ and $(T; \odot)$ be isomorphic groups. Prove that $(S; \cdot)$ is commutative if and only if $(T; \odot)$ is commutative.
22. Let f be an isomorphism from $(S; \cdot)$ onto $(T; \odot)$. Let e be the identity of $(S; \cdot)$, and let i be the identity of $(T; \odot)$. Prove that $f(e) = i$.
23. Let f be an isomorphism from $(S; \cdot)$ onto $(T; \odot)$. Let $x \in S$, and let x^{-1} be its inverse with respect to the binary operation $\cdot$. Prove that $f(x^{-1}) = [f(x)]^{-1}$. [Here $[f(x)]^{-1}$ is the inverse of $f(x)$ with respect to the binary operation $\odot$.]

2.3. Fields

In Section 2.2 we studied an algebraic system with one binary operation, that is, the group. In this section we study an algebraic structure with two binary operations, called a *field*. As we shall see, there are many important mathematical systems satisfying the axioms of a field.

We begin our study of fields by first defining the concept of *distributivity*.

Definition 2.3.1. *Let S be a set with two binary operations $+$ and $\cdot$. The operation $\cdot$ is said to be **distributive** over the operation $+$ if and only if for every $x, y, z \in S$*

$$x \cdot (y + z) = x \cdot y + x \cdot z,$$

$$(x + y) \cdot z = x \cdot z + y \cdot z.$$

Example 2.3.1. Let $S = Z$ be the set of integers, and let f_1, f_2, and f_3 be the operations defined in Example 2.2.1. Then f_2 is distributive over f_1 since $x \cdot (y + z) = x \cdot y + x \cdot z$ and $(x + y) \cdot z = x \cdot z + y \cdot z$

for every $x,y,z \in Z$. Also, f_2 is distributive over f_3 since $x \cdot (y - z) = x \cdot y - x \cdot z$ and $(x - y) \cdot z = x \cdot z - y \cdot z$ for every $x,y,z \in Z$.

Example 2.3.2. Let $S = \{0, 1, 2\}$, and let *addition*, $+$, be defined by the table given in Example 2.2.6(b). Let *multiplication*, $\cdot$, be defined by the table:

$\cdot$	0	1	2
0	0	0	0
1	0	1	2
2	0	2	1

Then multiplication is distributive over addition; that is, the operation $\cdot$ is distributive over the operation $+$. (Cf. Exercise 1 of Problem Set 2.3.)

Example 2.3.3. For the set B defined in Example 2.2.5(c), multiplication is distributive over addition (where addition is defined in Example 2.2.5(b)). Also, the set of real numbers has multiplication distributive over addition.

Definition 2.3.2. *Let S be a set with two binary operations $+$ and $\cdot$, and let 0 be the identity with respect to $+$. Then S with the binary operations $+$ and $\cdot$ is called a field if and only if*

(a) *$(S; +)$ is a commutative group,*

(b) *$(S \setminus \{0\}; \cdot)$ is a commutative group, and*

(c) *the operation $\cdot$ is distributive over the operation $+$.*

The field whose set of elements is S and whose binary operations are $+$ and $\cdot$ is usually denoted by $(S; +, \cdot)$.

In a field the operation $+$ is usually referred to as **addition**, while the operation $\cdot$ is called **multiplication** (although these operations may be quite different from ordinary addition and multiplication). The identity 0 with respect to $+$ is called the **zero element** or **zero** of the field, while the identity 1 with respect to $\cdot$ is called the **identity element** or **unit** of the field. A set S with two binary operations, called addition and multiplication, is a field if and only if the set S is a commutative additive group (commutative group with respect to addition), the nonzero elements of S form a commutative multiplicative group (commutative group with respect to multiplication), and multiplication is distributive over addition.

As examples of fields, we consider the following:

Example 2.3.4. Let $S = \{0, 1, 2\}$. Let addition on S be defined by the table in Example 2.2.6(b), and let multiplication be defined by the table in Example 2.3.2. Then $(S; +, \cdot)$ is a field.

Example 2.3.5. It will be proved in Section 2.6 that the set of rational numbers is a field with respect to the operations of addition and multiplication of real numbers. In Section 2.7 it will be proved that the set of real numbers is a field with respect to the operations of addition and multiplication.

We now develop some properties of fields. Let $(S; +, \cdot)$ be a field. Then $(S; +)$ is a commutative additive group, and hence addition is associative and commutative, that is, $x + (y + z) = (x + y) + z$ and $x + y = y + x$ for every $x, y, z \in S$. The set S has a zero element, 0, and each element x of S has a unique **additive inverse** or **negative**, $-x$, such that $x + (-x) = (-x) + x = 0$. Using these ideas we can now define a new binary operation, called **subtraction**, on S. For any two elements x and y of S, the **difference** of x and y, denoted by $x - y$, is defined by $x - y = x + (-y)$. That is, to **subtract** y from x we add the additive inverse of y to x. In a field we can always solve equations of the type $x + z = y$ for the unknown z. By adding the additive inverse of x to both sides of the equation $x + z = y$, the unique solution is given by $z = y - x$. Another consequence of $(S; +)$ being a commutative group is the **cancellation law of addition**. That is, if $x + z = y + z$, then $x = y$.

In the above paragraph we developed consequences of the field $(S; +, \cdot)$ that follow from the fact that $(S; +)$ is a commutative group. We now obtain some consequences that follow from the fact that $(S \setminus \{0\}; \cdot)$ is a commutative group. Since $(S; +, \cdot)$ is a field, $(S \setminus \{0\}; \cdot)$ is a commutative group, and hence multiplication is associative and commutative, that is, $x \cdot (y \cdot z) = (x \cdot y) \cdot z$ and $x \cdot y = y \cdot x$ for every $x, y, z \in S \setminus \{0\}$. The set $S \setminus \{0\}$ has an identity element 1, and each element x of $S \setminus \{0\}$ has a **multiplicative inverse**, x^{-1}, such that $x \cdot x^{-1} = x^{-1} \cdot x = 1$. Using these ideas we can now define a new binary operation, called **division**, on the set $S \setminus \{0\}$. For any two elements x in S and y in $S \setminus \{0\}$, the **quotient** of x and y, denoted by x/y, is defined by $x/y = x \cdot y^{-1}$. That is, to **divide** x by y, we multiply x by the multiplicative inverse of y. In a field we can always solve equations of the form $x \cdot z = y$ for the unknown z if $x \neq 0$. The unique solution of this equation is $z = y \cdot x^{-1}$, and it is found by multiplying both sides of the equation $x \cdot z = y$ by x^{-1}. Another consequence of $(S \setminus \{0\}; \cdot)$ being a commutative group is the **cancellation law of multiplication**. That is, if $x \cdot z = y \cdot z$ and $z \neq 0$, then $x = y$.

The properties of fields described above depend only on the group structures of a field. The one important property of a field that we have not yet used is the distributive law. We now develop some of the properties of fields that are consequences of the distributive law. We list these as theorems and give their proofs.

Theorem 2.3.1. *Let* $(S; +, \cdot)$ *be a field with zero element* 0. *If* x *is any element of* S, *then* $x \cdot 0 = 0 \cdot x = 0$.

Proof. Let $(S; +, \cdot)$ be a field with additive identity 0, and let x be any element of S. Since 0 is the additive identity of $(S; +, \cdot)$, we can write $0 = 0 + 0$ and $x \cdot 0 = (x \cdot 0) + 0$. It follows from the distributive law and from these two equalities that

$$(x \cdot 0) + 0 = x \cdot 0 = x \cdot (0 + 0) = (x \cdot 0) + (x \cdot 0).$$

Thus $(x \cdot 0) + 0 = (x \cdot 0) + (x \cdot 0)$, and the cancellation law for addition implies that $0 = x \cdot 0$. Since multiplication is commutative, we also have $0 \cdot x = 0$.

In the definition of a field $(S; +, \cdot)$, the associative and commutative laws of multiplication were only required to hold on the set $S \setminus \{0\}$. It follows from Theorem 2.3.1 that the associative and commutative laws for multiplication can be extended to the entire set S.

Theorem 2.3.2. *Let* $(S; +, \cdot)$ *be a field, and let* x *and* y *be any elements of* S. *Then* $x \cdot y = (-x) \cdot (-y)$.

Proof. Let $(S; +, \cdot)$ be a field, and let x and y be any elements of S. Since $-x$ is the negative of x, $x + (-x) = 0$. Multiplying both sides of this equation by $(-y)$ and using Theorem 2.3.1, we obtain

$$[x + (-x)] \cdot (-y) = 0 \cdot (-y) = 0.$$

Applying the distributive law and adding $x \cdot y$ to each side of this equation, we have

$$(x \cdot y) + [x \cdot (-y) + (-x) \cdot (-y)] = (x \cdot y) + 0 = x \cdot y.$$

The associative law of addition allows us to write

$$[x \cdot y + x \cdot (-y)] + (-x) \cdot (-y) = x \cdot y.$$

We use the distributive law and the fact that $-y$ is the negative of y to write this equation as

$$x \cdot [y + (-y)] + (-x) \cdot (-y) = x \cdot 0 + (-x) \cdot (-y) = x \cdot y.$$

From Theorem 2.3.1 and the fact that 0 is the additive identity, we finally obtain

$$(-x) \cdot (-y) = x \cdot y.$$

Corollary 2.3.2. *Let* $(S; +, \cdot)$ *be a field, and let* x *and* y *be any elements of* S. *Then*

$$(-x) \cdot y = x \cdot (-y) = -(x \cdot y).$$

The proof is left as Exercise 6 of Problem Set 2.3.

Theorem 2.3.3. *Let* $(S; +, \cdot)$ *be a field, and let* x *and* y *be any two elements of* S *such that* $x \cdot y = 0$. *Then either* $x = 0$ *or* $y = 0$.

Proof. Let $(S; +, \cdot)$ be a field, and let $x, y \in S$ such that $x \cdot y = 0$. If $x = 0$, then the proof is completed, so we assume that $x \neq 0$ and show that this implies that $y = 0$. From Theorem 2.3.1, $x \cdot 0 = 0$, and thus we have $x \cdot 0 = 0 = x \cdot y$. Since $x \neq 0$, we apply the cancellation law of multiplication to obtain $0 = y$. Therefore either $x = 0$ or $y = 0$.

Corollary 2.3.3. *Let* $(S; +, \cdot)$ *be a field, and let* x *and* y *be any two elements of* S *such that* $x \cdot y \neq 0$. *Then* $x \neq 0$ *and* $y \neq 0$.

We conclude this section with a discussion of *isomorphisms of fields*.

Definition 2.3.3. *Let* $(S; +, \cdot)$ *and* $(T; \oplus, \odot)$ *be two fields. Let* f *be a 1–1 mapping of* S *onto* T. *Then* f *is said to be an* **isomorphism** (*or* **field isomorphism**) *of* $(S; +, \cdot)$ *onto* $(T; \oplus, \odot)$ *if and only if*

(a) f *is an isomorphism of the group* $(S; +)$ *onto the group* $(T; \oplus)$, *and*

(b) f *is an isomorphism of the group* $(S \setminus \{0\}; \cdot)$, *where* 0 *is the zero element of* $(S; +, \cdot)$, *onto* $(T \setminus \{0'\}; \odot)$, *where* $0'$ *is the zero element of* $(T; \oplus, \odot)$.

If there is an isomorphism of $(S; +, \cdot)$ *onto* $(T; \oplus, \odot)$, *then the fields* $(S; +, \cdot)$ *and* $(T; \oplus, \odot)$ *are said to be* **isomorphic.**

An isomorphism f is a 1–1 mapping of the set S onto the set T that preserves the operations of addition and multiplication. This means that $f(x + y) = f(x) \oplus f(y)$ and $f(x \cdot y) = f(x) \odot f(y)$ for each x and y in S. In other words, the image of the sum (product) of two elements of S is the same element that would be obtained by mapping the elements of S first and then forming the sum (product) of their respective images.

Example 2.3.6. Let $(S; +, \cdot)$ be the field defined in Example 2.3.4.

Let $(T; \oplus, \odot)$ be the field with $T = \{a, b, c\}$ and where the operations $\oplus$ and $\odot$ are defined by the tables:

$\oplus$	a	b	c
a	a	b	c
b	b	c	a
c	c	a	b

$\odot$	a	b	c
a	a	a	a
b	a	b	c
c	a	c	b

It is left as an exercise to show that $(T; \oplus, \odot)$ is a field (Exercise 7 of Problem Set 2.3.)

Define $f: S \to T$ as follows:

$$f(0) = a,$$
$$f(1) = b,$$
$$f(2) = c.$$

It is left as an exercise for the student to make the direct computations to show that f is an isomorphism (Exercise 8 of Problem Set 2.3.)

From Definition 2.3.3 and Example 2.3.6 we see that two fields are isomorphic if their elements can be put in a 1–1 correspondence in such a way that their respective binary operations (addition and multiplication) also correspond.

The existence of an isomorphism between two fields means that the "algebraic properties" or "algebraic structure" of the two fields are identical, and, in fact, that they can only be distinguished by the notation used for their elements and operations. This is easily seen from the addition and multiplication tables of the isomorphic fields, since isomorphic fields have the "same" addition and multiplication tables. In our study of advanced calculus, we shall think of two isomorphic fields as being the same field.

Problem Set 2.3

1. Show that multiplication is distributive over addition for the two operations defined in Example 2.3.2.
2. Let $S = \{a + b \sqrt{2} : a, b \text{ are integers}\}$. Let addition and multiplication be the usual addition and multiplication of real numbers. Show that $(S; +)$ is a commutative group. Show that multiplication is distributive over addition. Show that $(S; +, \cdot)$ is not a field.

3. Let S be a nonempty set, and let $\mathscr{P}(S)$ denote the set of all subsets of S. Let $\cap: \mathscr{P}(S) \times \mathscr{P}(S) \to \mathscr{P}(S)$ such that $\cap(X, Y) = X \cap Y$, and let $\cup: \mathscr{P}(S) \times \mathscr{P}(S) \to \mathscr{P}(S)$ such that $\cup(X, Y) = X \cup Y$. Show that $\cap$ and $\cup$ are binary operations on $\mathscr{P}(S)$. Show that $\cap$ is distributive over $\cup$ and that $\cup$ is distributive over $\cap$.

4. Show that $(S; +, \cdot)$ defined in Example 2.3.4 is a field.

5. Let $S = \{0, 1, 2, 3, 4\}$, and let the operations $\oplus$ and $\odot$ be defined by the tables:

$\oplus$	0	1	2	3	4
0	0	1	2	3	4
1	1	2	3	4	0
2	2	3	4	0	1
3	3	4	0	1	2
4	4	0	1	2	3

$\odot$	0	1	2	3	4
0	0	0	0	0	0
1	0	1	2	3	4
2	0	2	4	1	3
3	0	3	1	4	2
4	0	4	3	2	1

Show that $(S; \oplus, \odot)$ is a field.

6. Prove Corollary 2.3.2.

7. Show that $(T; \oplus, \odot)$ defined in Example 2.3.6 is a field.

8. Show that the fields $(S; +, \cdot)$ and $(T; \oplus, \odot)$ defined in Example 2.3.6 are isomorphic.

9. Construct a field which is isomorphic to, but not identical with, the field $(S; \oplus, \odot)$ given in Problem 5.

For Problems 10–22, let $(S; +, \cdot)$ be a field with additive identity 0 and multiplicative identity e.

10. Show that $-0 = 0$.
11. Show that $-(x + y) = -x - y$.
12. Show that $-(x - y) = y - x$.
13. Show that $e^{-1} = e$.
14. Show that $x/y = 0$ if and only if $x = 0$.
15. Show that if $y \neq 0$ and $z \neq 0$, then $x/y = xz/yz$.
16. Show that if $y \neq 0$ and $w \neq 0$, then $x/y + z/w = (xw + yz)/yw$ and $(x/y)(z/w) = xz/yw$.
17. Show that $x(y - z) = xy - xz$.
18. Show that $(x - y) + (y - z) = (x - z)$.
19. Show that $(x - y) - (z - y) = (x - z)$.
20. Show that $(x - y)(z - w) = (xz + yw) - (xw + yz)$.
21. Show that $x - y = z - w$ if and only if $x + w = z + y$.
22. Show that the equation $xy + z = 0$, where $x \neq 0$, has a unique solution for the unknown y.

2.4. Ordered Fields

In Section 2.3 the concept of a field was defined and some elementary properties of fields were discussed. The purpose of this section is to define the notion of *order* and to use it to introduce ordered fields. The ordered field of real numbers provides the basis for our study of advanced calculus.

To simplify the notation we shall use the letter F to denote the field $(F; +, \cdot)$ as well as the set F itself. We shall also use the symbol 0 to denote the additive identity of F, the symbol $-x$ to denote the additive inverse of x, the symbol 1 to denote the multiplicative identity of F, and the symbol x^{-1} or $1/x$ to denote the multiplicative inverse of x.

We begin by defining the concept of a *positive class* in a field.

> **Definition 2.4.1.** *Let* $\mathfrak{p}$ *be a nonempty subset of F. Then* $\mathfrak{p}$ *is called a* **positive class** *of F if and only if*
>
> (a) *if* $x, y \in \mathfrak{p}$, *then the sum* $x + y \in \mathfrak{p}$,
>
> (b) *if* $x, y \in \mathfrak{p}$, *then the product* $xy \in \mathfrak{p}$, *and*
>
> (c) *if* $x \in F$, *then exactly one of the following holds*; $x \in \mathfrak{p}$, $x = 0$, *or* $-x \in \mathfrak{p}$.

Condition (c) implies that if $\mathfrak{p}$ is a positive class of a field F, then the set $\mathfrak{n} = \{-x : x \in \mathfrak{p}\}$ is disjoint from $\mathfrak{p}$. The set $\mathfrak{n}$ is called the **negative class** of F with respect to $\mathfrak{p}$. Note also that the sets $\mathfrak{p}$, $\{0\}$, and $\mathfrak{n}$ are mutually disjoint. Condition (*c*) also implies that $F = \mathfrak{p} \cup \{0\} \cup \mathfrak{n}$. Any element x of F such that $x \in \mathfrak{p}$ is said to be **positive** with respect to $\mathfrak{p}$, while if $x \in \mathfrak{n}$, then x is said to be **negative** with respect to $\mathfrak{p}$.

We are now able to define an *order relation* on F by use of the positive class $\mathfrak{p}$. It is left as an exercise to show that the following definition actually defines a relation on F (Exercise 1 of Problem Set 2.4).

> **Definition 2.4.2.** *Let F be a field with positive class* $\mathfrak{p}$. *Let* $x, y \in F$. *Then x is* **less than** *y (or equivalently, y is* **greater than** *x), denoted by* $x < y$ *(or* $y > x$*), if and only if* $y - x \in \mathfrak{p}$.

We see that $0 < x$ if and only if $x - 0 = x \in \mathfrak{p}$. That is, x is positive if and only if $0 < x$. In a similar manner, x, is negative if and only if $x < 0$.

We shall use the notation $x \leqslant y$ ($y \geqslant x$) to mean that $x < y$ or $x = y$. In addition, if $x < y$ and $y < z$, then we write $x < y < z$ or $z > y > x$. If $x < y$ and $y \leqslant z$, we write $x < y \leqslant z$ or $z \geqslant y > x$.

> **Definition 2.4.3.** *If* $\mathfrak{p}$ *is a positive class in the field F, then F is* **ordered** *by* $\mathfrak{p}$ *(in the sense of Definition 2.4.2), and F together with the ordering defined by* $\mathfrak{p}$ *is called an* **ordered field**.

An ordered field is the ordered pair, $(F; \mathfrak{p})$, where F denotes the field and $\mathfrak{p}$ denotes the particular positive class of F that defines the order on F. When there is no possibility of confusion as to the positive class being considered, then we denote by F the ordered field $(F; \mathfrak{p})$.

We now consider the following examples of ordered fields.

Example 2.4.1. Let Q denote the set of rational numbers, that is, the set of all quotients of the form p/q where p and q are integers and $q \neq 0$.

In Section 2.6 it will be proved that $(Q; +, \cdot)$, where $+$ and $\cdot$ are ordinary addition and multiplication, is a field. Let $\mathfrak{p}$ be the set of all quotients p/q such that p and q are positive integers. Using the ordinary properties of rational numbers, it is easily verified that $\mathfrak{p}$ is a positive class for the field $(Q; +, \cdot)$.

Example 2.4.2. For the field of real numbers R, with ordinary addition and multiplication, the set of all positive real numbers (that is, the set of all real numbers x such that $x > 0$) is a positive class for the field R.

Example 2.4.3. Let $(F; \oplus, \odot)$ be the field with $F = \{0, 1\}$ and whose binary operations are defined by the tables:

$\oplus$	0	1
0	0	1
1	1	0

$\odot$	0	1
0	0	0
1	0	1

If $\mathfrak{p}_1 = \{0\}$, then $\mathfrak{p}_1$ satisfies properties (a) and (b) of Definition 2.4.1 but does not satisfy property (c). If $\mathfrak{p}_2 = \{1\}$, then $\mathfrak{p}_2$ satisfies property (b) but does not satisfy properties (a) and (c). Furthermore, if $\mathfrak{p}_3 = F = \{0, 1\}$, then $\mathfrak{p}_3$ does not satisfy property (c). Thus the field $(F; \oplus, \odot)$ does not have a positive class.

Theorem 2.4.1. (*Law of Trichotomy*) *Let F be an ordered field, and let $x, y \in F$. Then exactly one of the following holds*: $x < y$, $x = y$, *or* $y < x$.

Proof. By definition 2.4.1(c), exactly one of the following holds: $x - y \in \mathfrak{p}$, $x - y = 0$, $y - x \in \mathfrak{p}$. Hence $y < x$, $x = y$, or $x < y$.

Theorem 2.4.2. (*Transitive Law*) *Let F be an ordered field, and let $x, y, z \in F$. If $x < y$ and $y < z$, then $x < z$.*

Proof. Since $x < y$ and $y < z$, we have that $y - x \in \mathfrak{p}$ and $z - y \in \mathfrak{p}$. By Definition 2.4.1(a) and the field property $z - x = (z - y) + (y - x)$ (cf. Exercise 18 of Problem Set 2.3), we have $z - x = (z - y) + (y - x) \in \mathfrak{p}$ and $x < z$.

Theorem 2.4.3. *Let F be an ordered field. Then for $x,y,z,w \in F$ the following properties hold*:

(a) *If $x < y$, then $x + z < y + z$ and $x - z < y - z$.*

(b) *If $0 < x$, then $0 < 1/x$.*

(c) *If $x < 0$, then $1/x < 0$.*

(d) *If $x < y$ and $0 < z$, then $xz < yz$ and $x/z < y/z$.*

(e) *If $x < y$ and $z < 0$, then $yz < xz$ and $y/z < x/z$.*

(f) *If $x < y$ and $z < w$, then $x + z < y + w$.*

(g) *If $0 < x < y$ and $0 < z < w$, then $0 < xz < yw$.*

(h) *If $0 < x$ and $0 < y$, then $0 < x + y$ and $0 < xy$.*

(i) *If $x < 0$ and $y < 0$, then $x + y < 0$ and $0 < xy$.*

(j) *If $0 < x$ and $y < 0$, then $xy < 0$.*

(k) *If $x \in F$, then $0 \leqslant x^2$. (If $x \neq 0$, then $0 < x^2$.)*

(l) *If $0 < x < y$, then $0 < 1/y < 1/x$.*

The proof of this theorem is similar to the proofs of Theorems 2.4.1 and 2.4.2 and is left as Exercises 2–14 of Problem Set 2.4.

Corollary 2.4.3. *Let F be an ordered field. Then $0 < 1$; that is, the additive identity is less than the multiplicative identity.*

Theorem 2.4.4. (*Antisymmetric Law*) *Let F be an ordered field, and let $x,y \in F$. If $x \leqslant y$ and $y \leqslant x$, then $x = y$.*

Proof. Since $x \leqslant y$, then $y - x \in \mathfrak{p} \cup \{0\}$, and since $y \leqslant x$, then $x - y \in \mathfrak{p} \cup \{0\}$. Now, by the field property $-(x - y) = y - x$ (cf. Exercise 12 of Problem Set 2.3), we have

$$-(x - y) = y - x \in \mathfrak{n} \cup \{0\}.$$

But then $y - x \in (\mathfrak{p} \cup \{0\}) \cap (\mathfrak{n} \cup \{0\}) = \{0\}$. Hence $y - x = 0$, or $x = y$.

We conclude this section with a short discussion of *absolute value.*

Definition 2.4.4. *Let F be an ordered field, and let $x \in F$. Then the* **absolute value** *of x, denoted by $|x|$, is defined by*

$$|x| = \begin{cases} x & \text{if } x \geqslant 0, \\ -x & \text{if } x < 0. \end{cases}$$

We may think of the absolute value of an element x as the *distance* between x and the additive identity or *origin* 0. The absolute value $|x - y|$ may be considered as the *distance* between the two elements x and y.

We now list some of the properties of absolute value.

Theorem 2.4.5. *Let F be an ordered field, and let $x,y \in F$. Then*

(a) $|x| \geqslant 0$, *and* $|x| = 0$ *if and only if* $x = 0$,

(b) $|-x| = |x|$,

(c) $|xy| = |x| |y|$,

(d) $|x|^2 = |x^2|$,

(e) $|x| \leqslant y$ *if and only if* $-y \leqslant x \leqslant y$,

(f) $|x| < y$ *if and only if* $-y < x < y$, *and*

(g) $-|x| \leqslant x \leqslant |x|$.

These seven properties are direct consequences of the definition of absolute value and are left for the reader to verify (Exercises 22–28 of Problem Set 2.4).

Theorem 2.4.6. (*Triangle Inequality*) *Let F be an ordered field, and let $x,y \in F$. Then*

$$|x + y| \leqslant |x| + |y|.$$

Proof. Property (e) of Theorem 2.4.5 implies that

$$-|x| \leqslant x \leqslant |x| \quad \text{and} \quad -|y| \leqslant y \leqslant |y|.$$

By Theorem 2.4.3(f), we may add these two inequalities, to get

$$-(|x| + |y|) \leqslant x + y \leqslant |x| + |y|.$$

Property (e) of Theorem 2.4.5 now implies that

$$|x + y| \leqslant |x| + |y|.$$

Corollary 2.4.6. *Let F be an ordered field, and let $x,y \in F$. Then*

(a) $|x - y| \leqslant |x| + |y|$,

(b) $||x| - |y|| \leqslant |x + y|$, *and*

(c) $||x| - |y|| \leqslant |x - y|$.

The proof of this corollary is left as Exercise 29 of Problem Set 2.4.

Problem Set 2.4

1. Show that Definition 2.4.2 defines a relation on F.

For Exercises 2–14, let F be an ordered field, and let $x,y,z,w \in F$.

2. If $x < y$, show that $x + z < y + z$ and $x - z < y - z$.
3. If $x > 0$, show that $1/x > 0$.
4. If $x < 0$, show that $1/x < 0$.
5. If $x < y$ and $z > 0$, show that $xz < yz$ and $x/z < y/z$.
6. If $x < y$ and $z < 0$, show that $xz > yz$ and $x/z > y/z$.
7. If $x < y$ and $z < w$, show that $x + z < y + w$.
8. If $0 < x < y$ and $0 < z < w$, show that $0 < xz < yw$.
9. If $x > 0$ and $y > 0$, show that $xy > 0$ and $x + y > 0$.
10. If $x < 0$ and $y < 0$, show that $x + y < 0$ and $xy > 0$.
11. If $x > 0$ and $y < 0$, show that $xy < 0$.
12. Show that $x^2 \geqslant 0$.
13. If $x \neq 0$, show that $x^2 > 0$.
14. If $y > x > 0$, show that $1/x > 1/y > 0$.
15. The **arithmetic mean** or **average** of two elements x and y in an ordered field is defined by
$$(x + y)/2 = (x + y) \cdot 2^{-1}.$$

Show that if $x < y$, then $x < (x + y)/2 < y$.
16. Using Problem 15, show that an ordered field has no smallest positive element (that is, a positive element x_0 such that $x_0 \leqslant x$ for every positive element x).
17. Can an order be defined on the field $(S; +, \cdot)$ given in Example 2.3.4? Why?
18. Can an order be defined on the field $(S; \oplus, \odot)$ given in Problem 5 of Problem Set 2.3? Why?
19. Show that an ordered field has no largest element (that is, a positive element x_0 such that $x \leqslant x_0$ for every element x).
20. Let x and y be elements of an ordered field. If $xy > 0$, prove that either $x > 0$ and $y > 0$ or $x < 0$ and $y < 0$.
21. Let x and y be elements of an ordered field. If $xy < 0$, prove that either $x < 0$ and $0 < y$, or $0 < x$ and $y < 0$.

For Exercises 22–28, let F be an ordered field and let $x,y \in F$.

22. Show that $|x| \geqslant 0$.
23. Show that $|x| = 0$ if and only if $x = 0$.
24. Show that $|x| = |-x|$.
25. Show that $|xy| = |x||y|$.
26. Show that $|x|^2 = |x^2|$.
27. If $y \geqslant 0$, show that $|x| \leqslant y$ if and only if $-y \leqslant x \leqslant y$.
28. Show that $-|x| \leqslant x \leqslant |x|$.

29. Prove Corollary 2.4.6.

30. Show that the equation $x^2 + 1 = 0$ has no root in F.

For Exercises 31–36, let F be an ordered field and let $x, y \in F$.

31. If $x \geqslant 0, y \geqslant 0$, show that $x < y$ if and only if $x^2 < y^2$.

32. If $x \leqslant 0, y \leqslant 0$, show that $x < y$ if and only if $y^2 < x^2$.

33. Show that $x + x = 0$ implies that $x = 0$.

34. Show that $x + x + x = 0$ implies that $x = 0$.

35. Show that $x^2 + y^2 \geqslant 0$, and that $x^2 + y^2 > 0$ unless $x = y = 0$.

36. Let $x < \varepsilon$ for every positive element $\varepsilon \in F$. Show that $x \leqslant 0$. [*Hint*: If $0 < x$ one could choose $\varepsilon = x$.]

37. Show that Theorem 2.4.2 holds if $<$ is replaced by $\leqslant$.

38. What parts of Theorem 2.4.3 hold if $<$ is replaced by $\leqslant$?

39. Let w, x, y, z be elements of an ordered field such that $y < w < z$ and $y < x < z$. Show that $|x - w| < z - y$. If $y \leqslant w \leqslant z$ and $y \leqslant x \leqslant z$, show that $|x - w| \leqslant z - y$.

40. Let x, y be elements of an ordered field. The **maximum** of the elements x and y, denoted by $\max\{x, y\}$, is defined by

$$\max\{x, y\} = \begin{cases} x & \text{if } y \leqslant x, \\ y & \text{if } x < y. \end{cases}$$

Show that $\max\{x, -x\} = |x|$.

41. Let x, y be elements of an ordered field. Show that

$$\max\{x, y\} = (x + y + |x - y|)/2.$$

42. Let x, y be elements of an ordered field. The **minimum** of the elements x and y, denoted by $\min\{x, y\}$, is defined by

$$\min\{x, y\} = \begin{cases} x & \text{if } x \leqslant y, \\ y & \text{if } y < x. \end{cases}$$

Show that $\min\{x, -x\} = -|x|$.

43. Let x, y be elements of an ordered field. Show that

$$\min\{x, y\} = (x + y - |x - y|)/2.$$

44. Let x, y be elements of an ordered field. Show that

$$-\max\{x, y\} = \min\{-x, -y\}.$$

2.5. The Natural Numbers and Mathematical Induction

Let F be an ordered field. We shall refer to the elements of F as *numbers*. In this section we consider the ordered field F and a subset of F called the *natural numbers*. In a very imprecise manner, we may think of defining the natural numbers one after another in the following way: 1 is the multiplicative identity of F, $2 \equiv 1 + 1$, $3 \equiv 2 + 1$, $4 \equiv 3 + 1$, and so forth. Since Corollary 2.4.3 implies that $0 < 1$, Theorem 2.4.2 implies that we have a collection of positive numbers, called *natural numbers*, such that

$$1 < 2 < 3 < 4 < \dots .$$

Since this is a very rough method of defining the natural numbers, we shall now proceed to give a precise definition of the natural numbers. We begin by defining an *inductive set*.

> **Definition 2.5.1.** *Let F be an ordered field. A subset S of F is called an* **inductive set** *if and only if it has the following two properties*:
>
> (a) *The multiplicative identity is in S; that is, $1 \in S$.*
>
> (b) *If the number $x \in S$, then the number $x + 1 \in S$; that is, $x \in S$ implies $x + 1 \in S$.*

We now present some examples of inductive sets. Let F be an ordered field and $\mathfrak{p}$ be the positive class of F.

> **Example 2.5.1.** The set F is itself an inductive set. This follows from the fact that the multiplicative identity 1 is an element of F, and, for any x in F, $x + 1$ is in F.

> **Example 2.5.2.** The set $\mathfrak{p}$ is an inductive set since $1 \in \mathfrak{p}$, and if $x \in \mathfrak{p}$ then $x < x + 1 \in \mathfrak{p}$.

> **Example 2.5.3.** The set $\{x : x \in F \text{ and } 1 \leqslant x\}$ is an inductive set. The proof is left to the reader (Exercise 1 of Problem Set 2.5).

With the concept of an inductive set in mind, we are now ready to give a precise definition of a natural number and of the set of natural numbers.

> **Definition 2.5.2.** *A number n of F is called a* **natural number** *if and only if n is an element of every inductive set of F.*

We denote the set of natural numbers by N. Since each natural number belongs to each inductive set of F, the set of natural numbers N is a subset of every inductive set of F. We shall show that N itself is an inductive set.

> **Theorem 2.5.1.** *The set of natural numbers N is an inductive set.*

> *Proof.* In order to show that N is an inductive set, we must show that $1 \in N$ and that if $n \in N$ then $n + 1$ also belongs to N. Part (a) of Definition 2.5.1 implies that 1 belongs to each inductive set, and thus $1 \in N$ by the definition of a natural number. We now assume n is a natural number. Then n belongs to each inductive set S of F.

But if $n \in S$, then Part (b) of Definition 2.5.1 implies that $n + 1$ also belongs to S. Hence if n belongs to every inductive set, $n + 1$ also belongs to every inductive set, and so $n + 1$ is a natural number. Therefore, the set of natural numbers N is an inductive set.

We now show that no proper subset of natural numbers is an inductive set. That is, if S is a set of natural numbers such that (a) $1 \in S$, and (b) if $n \in S$ then $(n + 1) \in S$, then $S = N$.

Theorem 2.5.2. (*Axiom of Induction*) *If S is a subset of the set of natural numbers and S is an inductive set, then $S = N$.*

Proof. By assumption $S \subseteq N$. Definition 2.5.2. implies that $N \subseteq S$ since each natural number belongs to every inductive set. Therefore $S = N$.

We continue by establishing one of the important tools for proving theorems in mathematical analysis, the *Principle of Mathematical Induction*.

Theorem 2.5.3. (*The Principle of Mathematical Induction*) *If for each natural number n there is associated a proposition $P(n)$ such that $P(1)$ is true and such that $P(k + 1)$ is true whenever $P(k)$ is true, where k is a natural number, then $P(n)$ is true for every natural number n.*

Proof. Let

$$S = \{n : n \in N \quad \text{and} \quad P(n) \text{ is true}\}.$$

Then $S \subseteq N$, and S is an inductive set since $1 \in S$, and if $n \in S$ then $n + 1 \in S$. Thus Theorem 2.5.2 implies that $S = N$. That is, the proposition $P(n)$ is true for each natural number n.

As pointed out above, one of the important uses of the Principle of Mathematical Induction is for proving theorems that involve propositions or statements about natural numbers. A second important use of the Principle of Mathematical Induction is as a method of defining concepts or statements involving the natural numbers. If we wish to define some concept $D(n)$ for each natural number n, we can define $D(1)$ and then show how to define $D(n + 1)$ in terms of $D(n)$; this is sufficient for defining $D(n)$ for all natural numbers n, as we now verify. Let

$$S = \{n : n \in N \quad \text{and} \quad D(n) \quad \text{is defined}\}.$$

Then $S \subseteq N$ and $1 \in S$. If $n \in S$, then $D(n)$ is defined. Since $D(n + 1)$ is defined in terms of $D(n)$, we have that $D(n + 1)$ is defined. Thus $n + 1 \in S$. Hence S is an inductive set and Theorem 2.5.2 implies $S = N$. That is, $D(n)$ is defined for each natural number n. Definitions given by induction are called **recursive definitions.** We now give an example of a recursive definition.

Example 2.5.4. Let $S_1 = 1$ and $S_{n+1} = S_n + (n + 1)$ for each natural number n. By the method mentioned above, we have now defined S_n for every natural number n: $S_1 = 1$, $S_2 = S_1 + 2 = 3$, $S_3 = S_2 + 3 = 6$, ...

In the above example we used the Principle of Mathematical Induction to define the number S_n for each natural number n. We now proceed to *proof by the Principle of Mathematical Induction*.

Assume that for each natural number n, $T(n)$ is a statement that is either true or false. Let

$$S = \{n : n \in N \quad \text{and} \quad T(n) \quad \text{is true}\}.$$

Suppose that we can show that $T(1)$ is true, and suppose that, whenever the statement $T(n)$ is true for the natural number n, we are able to show that the statement $T(n + 1)$ is also true. Then we see that $S \subseteq N$ such that $1 \in S$ and, whenever the natural number $n \in S$, the natural number $n + 1$ also belongs to S. Thus S is an inductive set and $S = N$. That is, the statement $T(n)$ is true for each natural number n. This method is called **proof by Mathematical Induction**.

We illustrate this procedure as follows.

Example 2.5.5. Let S_n be defined in Example 2.5.4 for each natural number n. We shall show that $S_n = n(n + 1)/2$ for each natural number n. Since

$$S_1 = 1 = 1 \cdot (1 + 1)/2,$$

we have that the statement $S_n = n(n + 1)/2$ is true for 1. Assume that the statement $S_n = n(n + 1)/2$ is true for the natural number k, that is, $S_k = k(k + 1)/2$. Then

$$S_{k+1} = S_k + (k + 1) = k(k + 1)/2 + (k + 1).$$

Using the distributive law of the field F, we may write

$$S_{k+1} = (k + 1)(k/2 + 1).$$

Now by the field property $x/y + z/w = (xw + yz)/yw$ (cf. Exercise 1f of Problem Set 2.3), we may write $S_{k+1} = (k + 1)(k + 2)/2$. Finally, we can use the associative law and the fact that the number $2 = 1 + 1$ to write $S_{k+1} = (k + 1)[(k + 1) + 1]/2$. Hence the statement $S_n = n(n + 1)/2$ is true for $k + 1$. The Principle of Mathematical Induction implies that the statement $S_n = n(n + 1)/2$ is true for all natural numbers n.

Other examples of the Principle of Mathematical Induction appear in the proofs of some of the properties of natural numbers.

We will now list some of the properties of the natural numbers. First remember that the set of natural numbers is a subset of the field F. Each natural number is positive since by Definition 2.5.2 each natural number is an element of each inductive set of F, and Example 2.5.2 assures us that p is an inductive set. That is, $N \subset p$, and so each number in N is positive. Example 2.5.3 implies that the set

$$S = \{x : x \in F \quad \text{and} \quad 1 \leqslant x\}$$

is an inductive set. Since Definition 2.5.2 implies that $N \subseteq S$, we see that for each natural number $n, n \geqslant 1$.

The remaining properties of the natural numbers will be given as theorems. The proofs of these involve the Principle of Mathematical Induction.

Theorem 2.5.4. *Let m and n be natural numbers. Then $m + n$ is a natural number. That is, N is closed with respect to addition.*

Proof. Let m be any natural number, and let

$$S_m = \{k : k \in N \quad \text{and} \quad m + k \in N\},$$

that is, S_m is the set of all natural numbers k such that $m + k$ is a natural number. Then $S_m \subseteq N$. Since m is a natural number, Theorem 2.5.1 implies that $m + 1$ is a natural number. Thus $1 \in S_m$. Let $n \in S_m$, then, by hypothesis, both n and $m + n$ are natural numbers. By the associative law of addition for the field F, we may write $m + (n + 1) = (m + n) + 1$. Since $m + n$ is a natural number, Theorem 2.5.1 implies that $(m + n) + 1$ is a natural number. Thus $m + (n + 1)$ is a natural number, and the Principle of Mathematical Induction implies $S_m = N$. That is, for any natural number $n, m + n$ is a natural number. Since m was an arbitrary natural number, for any natural numbers m and $n, m + n$ is also a natural number.

We now show that in special cases subtraction in the set of natural numbers is allowed.

Theorem 2.5.5. *Let m and n be natural numbers with $m < n$. Then $n - m$ is a natural number.*

Proof. We shall use the Principle of Mathematical Induction to prove this theorem. Let S be the set of all natural numbers m such that for every natural number n greater than m the number $n - m$ is a natural number.

We first show that $1 \in S$. Assume that there exists a natural number t, greater than 1, such that $t - 1$ is not a natural number. Let $T = N \setminus \{t\}$, that is, let T be the set of all natural numbers except t.

If we can show that T is an inductive set, then, since the set of natural numbers N is a subset of every inductive set, we have

$$N \subseteq T = N \setminus \{t\},$$

which is a contradiction.

Since t is greater than 1, we have $1 \in T$. Let n be any natural number of T. In order to show that T is inductive, we need to show $n + 1$ also belongs to T. Since n is a natural number, it follows that $n + 1$ is also a natural number. Assume $n + 1 \notin T$. Then, by the definition of T, $n + 1 = t$. It now follows that $n = t - 1$. But this is a contradiction since n is a natural number and $t - 1$ is not a natural number. Thus $(n + 1) \in T$ and T is an inductive set. Definition 2.5.2 implies that

$$N \subseteq T = N \setminus \{t\},$$

which is a contradiction. Hence, for each natural number n greater than 1, the number $n - 1$ is also a natural number and $1 \in S$.

We now assume that the natural number $m \in S$ and show that $m + 1$ also belongs to S. Since $m \in S$ for any natural number n greater than m, we have that the number $n - m$ is a natural number. Let k be any natural number greater than $m + 1$. Since m is a natural number we have that $k > m + 1 > 1$ and, since $1 \in S$, $k - 1$ is a natural number. But $k - 1 > (m + 1) - 1 = m + (1 - 1) = m$. The induction hypothesis now implies that $(k - 1) - m$ is a natural number. We may write the natural number $(k - 1) - m$ as $k - (1 + m)$ or $k - (m + 1)$, and so the number $k - (m + 1)$ is a natural number. Hence $(m + 1) \in S$. Therefore the Principle of Mathematical Induction implies that $S = N$.

Theorem 2.5.6. *Let m and n be natural numbers. Then mn is a natural number. That is, N is closed with respect to multiplication.*

The proof is left as an exercise (Exercise 11 of Problem Set 2.5).

Theorem 2.5.7. *Let m be a natural number. Then there exists no natural number n such that $m < n < m + 1$. That is, there is no natural number between the natural numbers m and $m + 1$.*

Proof. Let m be a natural number and let

$$S_{m+1} = \{k : k \in N \text{ and } k \leqslant m + 1\} \cup \{x : x \in F \text{ and } m + 1 < x\}.$$

Then S_{m+1} is an inductive set containing no numbers between m and $m + 1$. (Cf. Exercise 3 of Problem Set 2.5.) Since the set N is contained in every inductive set, $N \subseteq S_{m+1}$. But, since S_{m+1} has no numbers between m and $m + 1$, N can have no natural numbers between m and $m + 1$.

Corollary 2.5.7. *If m and n are natural numbers such that $m \leqslant n \leqslant m + 1$, then either $m = n$ or $n = m + 1$. If $m \leqslant n < m + 1$, then $m = n$. If $m < n \leqslant m + 1$, then $n = m + 1$.*

We conclude this section with a discussion about *well-ordered sets*. We begin by defining the concept of a *least element* of a set.

Definition 2.5.3. *Let S be a subset of F. Then the set S is said to have a least (smallest) element x_0 if and only if $x_0 \in S$ and, for each $x \in S$, $x_0 \leqslant x$.*

Definition 2.5.4. *Let S be a subset of F. Then the set S is said to have a greatest (largest) element x_0 if and only if $x_0 \in S$ and, for each $x \in S$, $x \leqslant x_0$.*

Example 2.5.6. In the set of natural numbers N, the number 1 is a least element. Since, for every $n \in N$, $n + 1$ also is an element of N, the set of natural numbers does *not* have a greatest element.

Example 2.5.7. In the set $\{-7, -4, -2, 1, 3, 5\}$, the number -7 is a least element and the number 5 is a greatest element.

Example 2.5.8. The positive class p of F has no least element or greatest element. (Cf. Exercises 16 and 19 of Problem Set 2.4.)

Notice in Examples 2.5.6–2.5.8 that the sets with a least element have exactly one least element. The same remark can be made about sets having a greatest element. The following two theorems ensure that least elements and greatest elements are unique.

Theorem 2.5.8. *If $S \subset F$ and S has a least element, then it is unique. That is, if x_0 and y_0 are least elements of S, then $x_0 = y_0$.*

Proof. Since x_0 is a least element of S, $x_0 \leqslant y_0$. Since y_0 is a least element of S, $y_0 \leqslant x_0$. Theorem 2.4.4 implies that $x_0 = y_0$.

Theorem 2.5.9. *If $S \subset F$ and S has a greatest element, then it is unique. That is, if x_0 and y_0 are greatest elements of S, then $x_0 = y_0$.*

The proof is left as Exercise 20 of Problem Set 2.5.

Definition 2.5.5. *Let S be a nonempty subset of F. Then S is said to be well ordered if and only if every nonempty subset of S has a least element. If the set S is well ordered, then S is called a well-ordered set.*

There are many examples of well-ordered sets. In fact, any finite subset of F is a well-ordered set. The next theorem proves that the set of natural numbers is well-ordered.

Theorem 2.5.10. *The set of natural numbers is a well-ordered set.*

Proof. Let T be a nonempty subset of the set of natural numbers N. We need to show that T has a least element. Define the set S by

$$S = \{n : n \in N \quad \text{and} \quad n \leqslant t \quad \text{for each} \quad t \in T\}.$$

Let t be an element of T. Then $t < t + 1$ and $(t + 1) \notin S$. That is, there is a natural number $t + 1$ which does not belong to S, and so $S \neq N$. The number $1 \in S$ since $1 \leqslant t$ for all $t \in T$. Now since $1 \in S$ and some natural number does not belong to S, there must exist a natural number $s \in S$ such that $(s + 1) \notin S$ (for otherwise, by the Axiom of Induction, Theorem 2.5.2, S would be the set of natural numbers). Since $(s + 1) \notin S$, there exists $t \in T$ such that $t < (s + 1)$. But, since $s \in S$, $s \leqslant t$. Hence $s \leqslant t < (s + 1)$, and Corollary 2.5.7 implies that $s = t$. Therefore, $s \in T$ and $s \leqslant t$ for each $t \in T$. Hence s is the least element of T and the set of natural numbers is well ordered.

Corollary 2.5.10. *Any nonempty subset of the set of natural numbers is well ordered.*

In the following theorem we shall use the symbol S_n to denote the set of all natural numbers less than the natural number n, that is, $S_n = \{m : m \in N$ and $m < n\}$. We note that $S_1 = \emptyset$. We shall now use the fact that the set of natural numbers is well ordered (Theorem 2.5.10) to prove a *Second Principle of Mathematical Induction*.

Theorem 2.5.11. (*The Second Principle of Mathematical Induction*) *Let S be a subset of N, and assume that, for each natural number n, the inclusion $S_n \subseteq S$ implies that $n \in S$. Then $S = N$.*

Proof. Since $S_1 = \emptyset$, the hypothesis of this theorem implies that $1 \in S$. Let $T = N \setminus S$. If we can show that $T = \emptyset$, then the proof is finished. Assume that $T \neq \emptyset$. Then Theorem 2.5.10 implies that T has a least element, say t. Since $1 \in S$, $t > 1$. If the natural number n is less than t, then $n \in S$ since t is the least element of T. Then $S_t \subseteq S$, and the hypothesis implies that $t \in S$. Since the sets S and T are disjoint, this contradicts the fact that t is the least element of T. Therefore, T has no least element, and T must be empty. Hence $S = N$.

In order to give an example of the Second Principle of Mathematical Induction, we must first define the "nth power" of an element of a field. We use the Principle of Mathematical Induction for this definition.

Definition 2.5.6. *Let x be any element of F. Define*

(a) $x^1 = x$, *and*

(b) $x^{n+1} = x^n x^1$ *for any natural number n.*

By the Principle of Mathematical Induction, x^n is then defined for any natural number n.

Theorem 2.5.12. *Let x and y be any elements of F, and let n and m be any two natural numbers. Then*

(a) $x^m x^n = x^{m+n}$,

(b) $(x^m)^n = x^{mn}$,

(c) $(xy)^n = x^n y^n$, *and*

(d) $(x/y)^n = x^n/y^n$.

We shall prove part (a) of this theorem and leave the proof of the remaining parts as Exercise 21 of Problem Set 2.5.

Proof. Part (a). Let m be any natural number, and let

$$S_m = \{n : n \in N \quad \text{and} \quad x^m x^n = x^{m+n}\}.$$

Since the Definition 2.5.6 implies that $x^{m+1} = x^m x^1$, the natural number $1 \in S_m$. Assume that the natural number n belongs to S_m, that is, $x^m x^n = x^{m+n}$. We need to show that $n + 1 \in S_m$, that is, that $x^{m+(n+1)} = x^m x^{n+1}$. Using the associative law of addition, we may write $x^{m+(n+1)} = x^{(m+n)+1}$. By Definition 2.5.6, we have $x^{(m+n)+1} = x^{m+n} x^1$. Since $n \in S_m$, we obtain $x^{m+n} = x^m x^n$ or $x^{m+n} x^1 = (x^m x^n) x^1$. Using the associative law of multiplication, we may write $(x^m x^n) x^1 = x^m(x^n x^1)$. Applying Definition 2.5.6 once again, we have $x^m(x^n x^1) = x^m x^{n+1}$. Putting all these equalities together, we have $x^{m+(n+1)} = x^m x^{n+1}$. Thus the natural number $n + 1$ also belongs to S_m. By the Principle of Mathematical Induction, we have $S_m = N$. That is, $x^{m+n} = x^m x^n$ for every pair of natural numbers m and n.

Example 2.5.7. Let $x_1 = 1$, $x_2 = 2$, and $x_{n+2} = x_{n+1} + x_n$ for all natural numbers $n \geqslant 1$. Then by the Principle of Mathematical Induction (Theorem 2.5.3), x_n is defined for each natural number n.

We shall use the Second Principle of Mathematical Induction (Theorem 2.5.11) to show that

$$4^n x_n < 7^n$$

for every natural number n. Let $S = \{n : n \in N \text{ and } 4^n x_n < 7^n\}$. Then we need to show that $S = N$. Assume that the set

$$\{m : m \in N \quad \text{and} \quad m < (n + 2) \in N\} \subseteq S,$$

that is, $4^m x_m < 7^m$ for each natural number $m < (n + 2)$. We now demonstrate that $(n + 2) \in S$. Since $n \in S$ and $(n + 1) \in S$, we obtain

$$\begin{aligned}
4^{n+2} x_{n+2} &= 4^{n+2}(x_{n+1} + x_n) \\
&= 4(4^{n+1} x_{n+1} + 4 \cdot 4^n x_n) \\
&< 4(7^{n+1} + 4 \cdot 7^n) \\
&= 4(7 + 4) \cdot 7^n \\
&= 4(11) \cdot 7^n \\
&< 7^2 \cdot 7^n = 7^{n+2}.
\end{aligned}$$

Thus

$$4^{n+2} x_{n+2} < 7^{n+2}$$

and $(n + 2) \in S$. The Second Principle of Mathematical Induction then implies that $S = N$. Therefore,

$$4^n x_n < 7^n$$

for every natural number n.

Problem Set 2.5

1. Show that the set $S_1 = \{1\} \cup \{x : x \in F \text{ and } 1 < x\}$ is an inductive set. That is, show that the set given in Example 2.5.3 is an inductive set.
2. Show that the set $S_2 = \{1, 2\} \cup \{x : x \in F \text{ and } 2 < x\}$ is an inductive set.
3. Let m be a natural number. Let

$$S_m = \{k : k \in N \text{ and } k \leqslant m\} \cup \{x : x \in F \text{ and } m < x\}.$$

Show that S_m is an inductive set.
4. Inductively define s_n as follows: $s_1 = 1$ and $s_{n+1} = s_n + (n + 1)^2$ for each natural number n. Show that

$$s_n = n(n + 1)(2n + 1)/6$$

for each natural number n.
5. Inductively define s_n as follows: $s_1 = 1$ and $s_{n+1} = s_n + (n + 1)^3$ for each natural number n. Show that

$$s_n = n^2(n + 1)^2/4$$

for each natural number n.

6. Inductively define s_n as follows: $s_1 = 1$ and $s_{n+1} = s_n + (n + 1)^4$ for each natural number n. Show that

$$s_n = n(n + 1)(2n + 1)(3n^2 + 3n - 1)/30$$

for each natural number n.

7. Inductively define s_n as follows: $s_1 = 1/2$ and $s_{n+1} = s_n + 1/(n + 1)(n + 2)$ for each natural number n. Show that

$$s_n = n/(n + 1)$$

for each natural number n.

8. Let $a, r \in F$ with $r \neq 1$. Inductively define s_n as follows: $s_1 = ar$ and $s_{n+1} = s_n + ar^{n+1}$ for each natural number n. Show that

$$s_n = ar(r^n - 1)/(r - 1)$$

for each natural number n.

9. Let $c \in F$ such that $0 < c \leqslant 1$. Inductively define s_n as follows: $s_1 = c/2$ and $s_{n+1} = (s_n^2 + c)/2$ for each natural number n. Show that $s_n < s_{n+1}$ for each natural number n.

10. Show that $n < 2^n$ for each natural number n.

11. Prove Theorem 2.5.6.

12. Prove that the sum of any n natural numbers is a natural number.

13. Prove that the product of any n natural numbers is a natural number.

14. Prove that the sum of any n positive numbers is positive.

15. Prove that the product of any n positive numbers is positive.

16. If $- x_1 - x_2 - \ldots - x_n$ is defined to be $(- x_1) + (- x_2) + \ldots + (- x_n)$, show that $-(x_1 + x_2 + \ldots + x_n) = - x_1 - x_2 \ldots - x_n$.

17. If $x_1, x_2, \ldots, x_n$ are nonzero, show that $x_1 x_2 \ldots x_n \neq 0$.

18. If $x_1, x_2, \ldots, x_n$ are nonzero, show that

$$(x_1 x_2 \ldots x_n)^{-1} = x_1^{-1} x_2^{-1} \ldots x_n^{-1}.$$

19. If $x_1 < x_2, x_2 < x_3, \ldots, x_n < x_{n+1}$, show that $x_1 < x_{n+1}$.

20. Prove Theorem 2.5.9.

21. Prove the remaining parts of Theorem 2.5.12.

22. Inductively define s_n as follows: $s_1 = 1, s_2 = 3$, and $s_{n+2} = 3s_{n+1} - 2s_n$ for each natural number n. Show that

$$s_n = 2^n - 1$$

for each natural number n.

23. Let $S \subset F$. The number $x_0 \in F$ is said to be an **upper bound** of S if and only if $x \leqslant x_0$ for each $x \in S$. If S is a nonempty subset of N which has an upper bound, show that S has a greatest element.

24. Prove that a nonempty subset S of N which has a greatest element is finite.

25. Let $S \subseteq N$. Let k be the least natural number belonging to S. If for each $n \geqslant k$, $n \in S$ implies that $(n + 1) \in S$, show that S is the set of all natural numbers greater than or equal to k. [This is commonly called the **Generalized Axiom of Induction.**]

26. Show that $2^n < n!$ ($n! = 1 \cdot 2 \cdot \ldots \cdot n$) for every $n \geqslant 4$.

27. Show that $n^3 < 2^n$ for every $n > 9$.

28. Let $x_1, x_2, \ldots, x_n$ be natural numbers each greater than 1. Show that $x_1 x_2 \ldots x_n > n$ for each natural number n.

29. A natural number p is called **composite** if and only if there exsits natural numbers $m > 1$ and $n > 1$ such that $p = mn$. A natural number p is **prime** if and only if $p > 1$ and p is not composite. Use Problem 28 to show that any natural number $p > 1$ is either a prime or the product of primes.

30. Let x and y be natural numbers. Show that there exists a natural number n such that $nx > y$.

31. **(Fundamental Theorem of Euclid.)** If x and y are natural numbers, show that there exists unique numbers n and r, each of which is either 0 or a natural number, where $r < x$ such that

$$y = nx + r.$$

[*Hint*: Let $(n + 1)$ be the smallest natural number such that $(n + 1)x > y$.]

32. The **sigma summation notation** is defined by

$$\sum_{k=m}^{n} f(k) = f(m) + f(m + 1) + \ldots + f(n),$$

where m and n are natural numbers with $n \geqslant m$. Show that

$$\sum_{k=m}^{n} \left(f(k) + g(k) \right) = \sum_{k=m}^{n} f(k) + \sum_{k=m}^{n} g(k).$$

33. Show that

$$\sum_{k=m}^{n} cf(k) = c \sum_{k=m}^{n} f(k),$$

where c is a real number.

34. Show that

$$\sum_{k=1}^{n} [f(k + 1) - f(k)] = f(n + 1) - f(1).$$

35. Using Problem 34 with $f(n) = n^3$, show that

$$\sum_{k=1}^{n} (2k + 1) = n^2 + 2n.$$

36. Using Problem 35, show that

$$\sum_{k=1}^{n} k = n(n + 1)/2.$$

37. Using Problems 34–36 with $f(n) = n^3$, show that

$$\sum_{k=1}^{n} k^2 = n(n + 1)(2n + 1)/6.$$

38. Using Problems 34–37 with $f(n) = n^4$, show that

$$\sum_{k=1}^{n} k^3 = n^2(n + 1)^2/4.$$

39. Using Problems 34–38 with $f(n) = n^5$, show that

$$\sum_{k=1}^{n} k^4 = n(n + 1)(2n + 1)(3n^2 + 3n - 1)/30.$$

40. Using Problem 34 with $f(n) = -1/n$, show that

$$\sum_{k=1}^{n} 1/k(k + 1) = n/(n + 1).$$

41. Using Problem 34 with $f(n) = -1/n^2$, show that

$$\sum_{k=1}^{n} (2k + 1)/k^2(k + 1)^2 = n(n + 2)/(n + 1)^2.$$

42. Using Problem 34 with $f(n) = -1/(3n - 2)$, show that

$$\sum_{k=1}^{n} 1/(3k - 2)(3k + 1) = n/(3n + 1).$$

43. Let $a,b \in F$. Using Problem 34 with $f(n) = -1/(an + b)$, show that

$$\sum_{k=1}^{n} 1/(ak + b)(ak + a + b) = n/(a + b)(an + a + b).$$

44. Find the fallacy in the following "proof" that any two natural numbers are equal:
Let $S = \{n : n \in N$ and if $a,b \in N$ such that $\max(a, b) = n$ then $a = b\}$.
Then $S \subseteq N$ and $1 \in S$. Suppose that $k \in S$, that is, if $a,b \in N$ such that $\max(a, b) = k$, then $a = b$. We wish to show that $(k + 1) \in S$, that is, if $a,b \in N$ such that $\max(a, b) = k + 1$, then $a = b$. Let a and b be any two natural numbers such that $\max(a, b) = k + 1$. Consider the two numbers

$$c = a - 1, \text{ and}$$
$$d = b - 1;$$

then $\max(c, d) = k$. Hence $c = d$ since $k \in S$. By the Principle of Mathematical
Induction, $S = N$.

Now if a and b are any two natural numbers, denote $\max(a, b)$ by r. Since
$r \in N$, $a = b$.

2.6. The Integers and Rational Numbers

In Section 2.5 the natural numbers were defined and some of their
basic properties were proved. If m and n are natural numbers, it is not always
possible to find a natural number x such that $m + x = n$. Indeed, this is the
case whenever $n \leqslant m$. Another way of stating this limitation is that the set of
natural numbers is not closed under subtraction, that is, the difference of two
natural numbers need not be a natural number. The next step in developing
the properties of the real numbers will be to find the smallest subset (containing the natural numbers) of our field F that is closed under subtraction.
This subset of F is called the *set of integers*.

Let A be any subset of the field F. Let

$$-A = \{a : a \in F \quad \text{and} \quad -a \in A\}.$$

The set $-A$ is the set of all additive inverses of elements of A.

Definition 2.6.1. *Let x be an element of F. Then x is said to be an*
integer if and only if $x \in N$, $x = 0$, or $-x \in N$. The set

$$Z = N \cup \{0\} \cup (-N)$$

*is called the **set of integers**.*

The set of integers is composed of all the natural numbers, the additive
identity 0, and the set of all additive inverses of natural numbers. If x is an
integer and $x \in N$, then x is called a **positive integer**. For the remainder of this
book we shall use the names "natural number" and "positive integer"
interchangeably. If x is an integer and $x \in -N$ (or $-x \in N$), then x is called
a **negative integer**. The set $-N$ is called the **set of negative integers**.

Since the set of integers is a subset of our field F, the set of integers has an
additive binary operation and a multiplicative binary operation defined on it.
Since the operations are associative and commutative on F, they are also
associative and commutative on Z. The additive identity 0 of F is also the
additive identity for Z, while the multiplicative identity 1 of F is also the
multiplicative identity of Z. In order to show that $(Z; +)$ is a commutative
group, we need only show that the sum of two integers is an integer and that

each integer has an additive inverse. The proof of these two properties is left as Exercises 1 and 2 of Problem Set 2.6. Thus we have the following theorem:

Theorem 2.6.1. *The set of integers is a commutative group with respect to addition.*

Since $(Z; +)$ is an additive group, we now have all the properties associated with an additive group; for example, the cancellation law holds, there is a binary operation called subtraction defined by $m - n = m + (-n)$, and equations of the type $m + x = n$ can be solved and have the unique solution $x = n - m$.

As pointed out above, the set of integers has a multiplicative operation defined on it. This operation is associative and commutative. It is easily verified that the product of two integers is an integer. (Cf. Exercise 3 of Problem Set 2.6).

In Section 2.5, the law of exponents was defined for natural numbers and some of the properties were proved. We now extend the law of exponents (Theorem 2.5.12) where the exponents are natural numbers to a law of exponents where the exponents are integers.

Definition 2.6.2. *Let x be a nonzero element of F. Then $x^0 \equiv 1$. If $x \neq 0$ and n is a negative integer, then $x^n \equiv 1/x^{-n}$.*

Example 2.6.1. $5^0 = 1$, $3^{-2} = 1/3^2 = 1/9$, and $(4/5)^{-3} = 1/(4/5)^3 = 1/(4^3/5^3) = 5^3/4^3 = 125/64$.

Theorem 2.6.2. *Let x and y be any two nonzero elements of the ordered field F, and let m and n be any two integers of F. Then*

(a) $x^m x^n = x^{m+n}$,

(b) $x^m/x^n = x^{m-n}$,

(c) $(x^m)^n = x^{mn}$,

(d) $(xy)^n = x^n y^n$, *and*

(e) $(x/y)^n = x^n/y^n$.

We shall prove part (a) and leave the proof of the remaining parts as Exercises 7–10 of Problem Set 2.6.

Proof. Part (a). We consider a number of cases, depending on whether the integers m and n are positive, zero, or negative.

CASE 1. Either $m = 0$, or $n = 0$, or both $m = 0$ and $n = 0$. We assume $n = 0$. Then $x^m x^n = x^m x^0 = x^m \cdot 1 = x^m = x^{m+0} = x^{m+n}$.

CASE 2. Both m and n are positive. Theorem 2.5.12(a) implies that $x^m x^n = x^{m+n}$.

CASE 3. Both m and n are negative. Let $p = -m$ and $q = -n$. Then p and q are positive, and so $x^m x^n = (1/x^p)(1/x^q) = 1/(x^p x^q)$. Theorem 2.5.12(a) implies that $1/(x^p x^q) = 1/x^{p+q}$. By the definition of p and q,

$$1/x^{p+q} = 1/x^{-m-n} = 1/x^{-(m+n)} = x^{m+n}.$$

Thus $x^m x^n = x^{m+n}$.

CASE 4. One of the integers m and n is positive while the other is negative. Without loss of generality, assume that m is positive and n is negative. Let $q = -n$.

If $m > q$, then $m - q$ and q are positive integers whose sum is m. Theorem 2.5.12(a) implies that $x^m = x^{(m-q)+q} = x^{m-q} x^q$. Dividing by x^q, we get $x^m/x^q = x^{m-q}$. But

$$m - q = m + n, \quad \text{and} \quad x^m/x^q = x^m/x^{-n} = x^m x^n.$$

Thus $x^m x^n = x^{m+n}$.

If $m < q$, then $q - m$ and m are positive integers whose sum is q. Theorem 2.5.12(a) implies that $x^{q-m} x^m = x^q$. Dividing by x^m, we get $x^{q-m} = x^q/x^m$. Now

$$x^{m+n} = x^{m-q} = 1/x^{-(m-q)} = 1/x^{q-m} = 1/(x^q/x^m) =$$

$$1/(x^q x^{-m}) = (1/x^q)(1/x^{-m}) = (1/x^{-m})(1/x^q) = x^m x^{-q} = x^m x^n.$$

Thus $x^m x^n = x^{m+n}$.

The set of integers has many properties that the set of natural numbers does not have, for example, additive inverses. The set of integers, however, is not a field, since 1 and -1 are the only integers with integral multiplicative inverses. For example, the multiplicative inverse of 2 is a $1/2$, which is not an integer. Another way of stating this limitation is that the set of integers is not closed under division, that is, the quotient of two integers need not be an integer. We next consider a subset (containing the set of integers) of our ordered field F that is closed under division. This subset of F is called the *set of rational numbers*.

Definition 2.6.3. *Let $x \in F$. Then x is called a **rational number** if and only if there exist integers m and n, where $n \neq 0$, such that $x = m/n = mn^{-1}$. The set $Q = \{m/n : m \in Z$ and $n \in Z \setminus \{0\}\}$ is called the **set of rational numbers**. All elements of $F \setminus Q$ are called **irrational numbers**.*

Since the set of rational numbers is a subset of our field F, the set of rational numbers has an additive binary operation and a multiplicative binary operation defined on it. Since these operations are associative and commutative on F, they are also associative and commutative on Q. The additive identity 0 of F is also the additive identity of Q, since we can write $0 = 0/1 = 0 \cdot 1^{-1}$, and the multiplicative identity 1 of F is also the multiplicative identity of Q, since $1 = 1/1 = 1 \cdot 1^{-1}$. Let x and y be two rational numbers. Then there are integers p, q, r, s, with $q \neq 0$ and $s \neq 0$, such that $x = p/q$ and $y = r/s$. Since $ps + qr$ and qs, with $qs \neq 0$, are integers, and since $x + y = p/q + r/s = (ps + qr)/qs$ (cf. Exercise 16 of Problem Set 2.3), the sum of two rational numbers is a rational number. That is, the set of rational numbers is closed under addition. If x is a rational number, then there exists integers p and q, with $q \neq 0$, such that $x = p/q$. But then $-x = -1 \cdot x = (-1/1)(p/q) = (-p)/q$, and thus $-x$ is also a rational number. Hence each rational number has an additive inverse, and the set $(Q; +)$ is a commutative group. In a similar manner the system $(Q \setminus \{0\}; \cdot)$ can be shown to be a commutative group. (Cf. Exercise 20 and 21 of Problem set 2.6.) Since Q is a subset of the ordered field F, Q is ordered and multiplication is distributive over addition. Combining these properties with the results that $(Q; +)$ and $(Q \setminus \{0\}; \cdot)$ are commutative groups, we have the following theorem.

Theorem 2.6.3. *The set of rational numbers Q is an ordered field under the operation of F.*

Since $(Q; +, \cdot)$ is an ordered field, we have all the usual field and order properties for the set of rational numbers; for example, the additive and multiplicative cancellation laws, subtraction and division, and any equation of the form $ax + b = c$, with $a \neq 0$, has the unique solution $x = (c - b)/a$.

We now define two properties associated with ordered fields, namely the *Archimedean property* and *denseness*. The Archimedean property is named for Archimedes (287–212 B.C.), the great mathematician of antiquity, who first discovered the property.

Definition 2.6.4. *An ordered field F is said to be **Archimedean** or to have the **Archimedean property** if and only if for each $x \in \mathfrak{p}$, the positive class of F, there is a natural number n such that $x < n$.*

We may also state this property as follows: *F is Archimedean if and only if for each positive number x of F there is a natural number n such that $n - x$ is positive, that is, $n - x \in \mathfrak{p}$.* We shall first prove some properties about Archimedean fields and then show that the ordered field of rational numbers is Archimedean.

Theorem 2.6.4. *The ordered field F is Archimedean if and only if for each pair of positive elements x and y of F there is a natural number n such that $x < ny$.*

Proof. Let F be an Archimedean field, and let x and y be positive elements of F. Then x/y is a positive element of F, and Definition 2.6.4 implies that there is a natural number n such that $x/y < n$. Thus $x < ny$.

Assume for each pair of positive elements x and y of F there is a natural number n such that $x < ny$. Let $y = 1$. Then for each positive element x of F there is a natural number n such that $x < n \cdot 1 = n$. Thus F is Archimedean.

The proofs of the following corollaries to Theorem 2.6.4 are left as Exercises 26 and 27 of Problem Set 2.6.

Corollary 2.6.4a. *Let F be an Archimedean field. If x is a positive element of F, then there is a natural number n such that $0 < 1/n < x$.*

Corollary 2.6.4b. *Let F be an Archimedean field. If x is any element of F, then there are integers m and n such that $m < x < n$.*

With the aid of these results we can now establish an important result between elements of an Archimedean field and the integers of the Archimedean field.

Theorem 2.6.5. *Let F be an Archimedean field. If x is any element of F, then there is a unique integer n of F such that $n \leqslant x < n + 1$. This integer is denoted by $n = [x]$.*

Proof. We first show that such an integer exists and then show that it is unique.

Corollary 2.6.4b implies that there exist integers r and s such that $r < x < s$. Let

$$T = \{t : t \in N \quad \text{and} \quad x < r + t\}.$$

Then $T \subseteq N$. The natural number $t_0 = s - r$ belongs to T since $r < x < s = r + (s - r) = r + t_0$, and thus $T \neq \emptyset$. Since any nonempty subset of the set of natural numbers is well ordered (cf. Corollary 2.5.10), the set T has a least element, say p. Let $n = r + p - 1$. If $p = 1$, then $n = r \leqslant x < r + 1 = n + 1$. If $p \geqslant 2$, then $p - 1$ is a natural number and $n = r + (p - 1) \leqslant x$ since p is the least natural number such that $x < r + p$. Hence

$$n = r + (p - 1) \leqslant x < r + p = n + 1.$$

Assume that there are distinct integers m and n with $m < n$ such that $m \leqslant x < m + 1$ and $n \leqslant x < n + 1$. Then $m < n \leqslant x < m + 1$ and $0 < n - m < 1$. Since $m < n$, the number $n - m$ is a natural number (positive integer). This is a contradiction because there are no natural numbers less than 1. This completes the proof.

Using the properties of an Archimedean field F, it is now possible to show that the ordered field of rational numbers of F is also Archimedean. (Cf. Exercise 28 of Problem Set 2.6.)

Theorem 2.6.6. *Let F be an Archimedean field and let Q be its ordered field of rational numbers. Then Q is Archimedean.*

We now define the concept of *density* in an ordered field.

Definition 2.6.5. *Let F be an ordered field, and let $S \subseteq F$. Then S is said to be **dense** in F if and only if between any two elements of F there is an element of S. That is, if x and y are elements of F with $x < y$, then there is $s \in S$ such that $x < s < y$.*

The set of rational numbers of an Archimedean field F are dense in F.

Theorem 2.6.7. *Let F be an Archimedean field, and let Q be its set of rational numbers. Then Q is dense in F.*

Proof. Let x and y, $x < y$, be any two elements of F. Corollary 2.6.4a implies that there is a natural number q such that $0 < 1/q < (y - x)/2$. Having found the natural number q, we need only find an integer p such that $x < p/q < y$. Let $p = [xq + 1]$. Using Theorem 2.6.5, we have

$$xq < xq + 1 \leqslant p < xq + 2.$$

Dividing this inequality by q, and using the fact that

$$0 < 1/q < (y - x)/2,$$

we have

$$x < x + 1/q \leqslant p/q < x + 2/q$$
$$< x + 2(y - x)/2 = x + (y - x) = y.$$

Hence we have found a rational number p/q such that $x < p/q < y$. Therefore the set of rational numbers Q is dense in the field F.

Theorem 2.6.6 assures us that the field of rational numbers is an Archimedean field. Thus an Archimedean field need not have irrational numbers. But if an ordered field has at least one irrational number η, then it has infinitely many irrational numbers since $n \eta \in F$ for each $n \in N$. Also, if η is an irrational number of F, then either η or $-\eta$ is positive, and so F has positive irrational numbers. It is left for the reader to show that if F has an irrational number, then F has arbitrarily small irrational numbers, F has arbitrarily large irrational numbers, and the set of irrational numbers of F is dense in F. (Cf. Exercise 30, 31, and 32 of Problem Set 2.6.)

Problem Set 2.6

1. Show that the sum of any two integers is an integer.
2. Show that each integer has an additive inverse.
3. Show that the product of any two integers is an integer.
4. Show that $1^n = 1$ for each integer n.
5. Show that $(-1)^{2n} = 1$ for each integer n.
6. Show that $(-1)^{2n+1} = -1$ for each integer n.

For Exercises 7–10. let x and y be any two nonzero elements of the ordered field F, and let m and n be any two integers of F.

7. Show that $x^m/x^n = x^{m-n}$.
8. Show that $(x^m)^n = x^{m \cdot n}$.
9. Show that $(xy)^n = x^n y^n$.
10. Show that $(x/y)^n = x^n/y^n$.
11. Let S be a nonempty subset of Z. The number $x_0 \in F$ is said to be a **lower bound** of S if and only if $x_0 \leqslant x$ for each $x \in S$. Show that if S has a lower bound, then S has a least element.
12. Let S be a nonempty subset of Z. Show that if S has an upper bound, then S has a greatest element. (Cf. Problem 23 of Problem Set 2.5.)
13. Let S be a subset of Z. Show that if S has an upper bound and a lower bound, then S is finite.
14. Let $a > 1$, and let m and n be positive integers. Show that $m < n$ if and only if $a^m < a^n$.
15. Let $0 < a < 1$, and let m and n be positive integers. Show that $m < n$ if and only if $a^n < a^m$.
16. Let $0 < a$ and $a \neq 1$, and let m and n be integers. Show that $m = n$ if and only if $a^m = a^n$.
17. Let $0 < a$ and $0 < b$, and let n be a nonzero integer. Show that $a = b$ if and only if $a^n = b^n$.
18. Let m be any integer. Show that there does *not* exist an integer n such that $m - 1 < n < m$.
19. Let n be an integer, and let m be a positive integer. Show that there exists unique integers q and r with $0 \leqslant r < m$ such that
$$n = qm + r.$$
20. Show that the product of two rational numbers is a rational number.
21. Show that each nonzero rational number has a multiplicative inverse.
22. Let m and n be integers. Then m is said to be a **factor** of n if and only if there exists an integer k such that $n = km$. Let m, n, and p be integers. Show that if m is a factor of n and n is a factor of p, then m is a factor of p.
23. Let $r = m/n$ be a rational number. Show that r can be written as the quotient of two integers whose only common factors are $+1$ and -1. [*Hint*: Use induction.]

24. Let $x, y \in F$ with $y > 0$. If $x^2 = y$, then x is said to be a **square root** of y. If y has square roots, then the positive one is called the **square root** of y and is denoted by $\sqrt{y}$. Show that $\sqrt{2}$ is *not* a rational number. [*Hint*: Assume that $\sqrt{2} = p/q$, where p and q are integers whose only common factors are $+1$ and -1. Show this leads to a contradiction.]

25. Let p be a prime natural number. Show that $\sqrt{p}$ is *not* a rational number.

26. Prove Corollary 2.6.4a.

27. Prove Corollary 2.6.4b.

28. Prove Theorem 2.6.6.

29. Show that an ordered field F is Archimedean if and only if for each positive element x of F there is a natrual number n such that

$$0 < 1/2^n < x.$$

30. Let F be an ordered field. If η an is irrational number of F, show that $r\eta$ is an irrational number of F for every rational number $r \neq 0$. Show that $r + \eta$ is an irrational number of F for every rational number r.

31. Let F be an Archimedean field containing a positive irrational number η. Show that if x is a positive element of F, then there exists a natural number n such that $0 < \eta/n < x$.

32. Let F be an Archimedean field containing an irrational number η. Show that the set of all irrational numbers of F is dense in F. [*Hint*: Show that if $x, y \in F$ with $x < y$, then there is a rational number r such that the irrational number $r + \eta$ satisfies the relation $x < r + \eta < y$.]

33. Let p and q be rational numbers with $0 < q$. Show that there exist a unique natural number n and a unique rational number r with $0 \leqslant r < q$ such that

$$p = nq + r.$$

34. Show that the set of positive rational numbers Q^+ is a unique positive class for the field of rational numbers; that is, show that if $\mathfrak{p}$ is any positive class for the field of rational numbers, then $Q^+ = \mathfrak{p}$. [*Hint*: If n is a positive integer and $-1/n \in \mathfrak{p}$, then $(-1/n)^2 = 1/n^2 \in \mathfrak{p}$, and therefore $n/n^2 = 1/n \in \mathfrak{p}$.]

2.7. The Real Numbers

In the previous section we discussed the properties of the set of rational numbers as an ordered field. We now introduce the *completeness property* which distinguishes the ordered field of real numbers from other ordered fields. We begin our discussion of the completeness property by defining the concepts of *upper bounds, lower bounds, least upper bounds,* and *greatest lower bounds.*

Definition 2.7.1. *Let F be an ordered field, and let S be a nonempty subset of F. An element $x \in F$ is called an **upper bound** (**lower bound**)* *of S if and only if $y \leqslant x$ $(x \leqslant y)$ for each $y \in S$. The set S is said to be **bounded** if and only if S has both an upper bound and a lower bound.*

Example 2.7.1.(a) Let F be an ordered field. The set of natural numbers of F has no upper bounds, but any number less than or equal to 1 is a lower bound.

(b) The set of rational numbers of F have neither an upper bound nor a lower bound.

(c) The set $S = \{x : x \in F \text{ and } 0 < x < 1\}$ has 0 as a lower bound and 1 as an upper bound. Also any number less than zero is a lower bound while any number greater than one is an upper bound.

From the above example, it is clear that a set may have many upper bounds and many lower bounds. For finite sets of an ordered field one of the upper bounds is contained in the set (the greatest element) and one of the lower bounds is contained in the set (the least element). For the set S defined in Example 2.7.1(c), no upper bound and no lower bound are contained in the set S and so the set S has no greatest element and no least element. But for sets of this type there are elements in F comparable to the greatest element and the least element; these elements are called the *least upper bound* and the *greatest lower bound*.

Definition 2.7.2. *Let F be an ordered field, and let S be a nonempty subset of F. An element $x \in F$ is called a **least upper bound** (or **supremum**) of S if and only if (a) x is an upper bound of S and (b) if y is any upper bound of S, then $x \leqslant y$. An element $x \in F$ is called a **greatest lower bound** (or **infimum**) of S if and only if (a) x is a lower bound of S and (b) if y is any lower bound of S, then $y \leqslant x$.*

The following theorem assures that if a set has a least upper bound (greatest lower bound), then it is unique.

Theorem 2.7.1. *A set S has at most one least upper bound (greatest lower bound).*

The proof of this theorem is similar to the proof of Theorem 2.5.8 and is left as Exercise 3 of Problem Set 2.7.

The least upper bound of the set S is denoted by lub S or sup S, while the greatest lower bound of the set S is denoted by glb S or inf S.

Example 2.7.2.(a) Let F be an ordered field, and let N be the set of natural numbers. Since the set of natural numbers does not have an upper bound, sup N does not exist. For the set N, inf $N = 1$.

(b) For the set S defined in Example 2.7.1(c), sup $S = 1$ and inf $S = 0$.

We are now ready to define the *completeness property*.

> **Definition 2.7.3.** *An ordered field F is said to be **complete** (or to have the **completeness property**) if and only if every nonempty subset of F that has an upper bound has a least upper bound in F.*

> **Example 2.7.3.** Let Q denote the ordered field of rational numbers. Let $S = \{r : r \in Q \text{ and } r^2 < 2\}$. Then $S \neq \varnothing$ since $1 = 1^2 < 2$, and so $1 \in S$. Also S has 2 as an upper bound. The proof of Theorem 2.7.4 shows that the least upper bound of S is $\sqrt{2}$, and $\sqrt{2} \notin Q$. (Cf. Exercise 24 of Problem Set 2.6.) Hence the ordered field or rational numbers is not complete.

> **Postulate 2.7.1.** *There exists a complete ordered field.*

We shall assume for the remainder of this book that there is a complete ordered field. We will call this complete ordered field the **real number system** and denote it by R. Elements of R will be called **real numbers**. The properties of the real number system are given throughout this chapter.

We have introduced the real number system axiomatically in that we assumed that it satisfies certain properties. The question may arise as to whether a complete ordered field actually exists, and, if it does exist, whether it is unique. As pointed out in Section 2.1, beginning with Peano's Postulates, one can construct the system of real numbers. It is also possible to show that the system of real numbers is unique in the sense of isomorphism.

We shall say no more about the existence or uniqueness of the real number system but proceed to prove some more results about it. The completeness axiom assures us that every nonempty subset of R which has an upper bound has a least upper bound. From this property it follows that every nonempty subset of R that has a lower bound has a greatest lower bound.

> **Theorem 2.7.2.** *Let S be any nonempty subset of R that has a lower bound. Then S has a greatest lower bound.*

The proof follows from the completeness property and is left as Exercise 5 of Problem Set 2.7.

It was pointed out in Theorem 2.6.6 that the ordered field of rational numbers is an Archimedean field. We now show that the real number system is an Archimedean field.

> **Theorem 2.7.3.** *The real number system is an Archimedean field.*

Proof. Assume that the real number system is not Archimedean. Then there exists a positive real number x such that $n < x$ for each natural number n. Thus the set of natural numbers N has an upper bound, and the completeness property of R implies that N has a least upper bound y. But then $(n + 1) < y$ for each natural number n, and so $n < y - 1$ for each natural number n. Hence $y - 1$ is an upper bound of N that is smaller than the least upper bound of N. This is a contradiction, and so R is an Archimedean field.

We conclude this section by showing that the real number system contains irrational numbers.

Theorem 2.7.4. *There exists a real number whose square is 2. That is, $\sqrt{2}$ is a real number.*

Proof. Let $S = \{x : x \in R$ and $x^2 < 2\}$. Then $S \neq \emptyset$ since $1 \in S$. Also, for each $x \in S$, $x^2 < 2 < 4 = 2^2$, and thus $x < 2$. (Cf. Exercise 14 of Problem Set 2.6.) Hence 2 is an upper bound of S. The completeness property of R then implies that S has a least upper bound, say y.

We now need to show that $y^2 = 2$.

Assume that $y^2 < 2$, that is, $y \in S$. Let $h = (2 - y^2)/(2y + 2)$. Then $0 < h < 2$. Let $z = y + h$. Then $y < z$ and

$$\begin{aligned} z^2 &= (y + h)^2 = y^2 + 2hy + h^2 = y^2 + (2y + h) \\ &= y^2 + (2y + h)(2 - y^2)/(2y + 2) \\ &< y^2 + (2y + 2)(2 - y^2)/(2y + 2) \\ &= y^2 + (2 - y^2) = 2. \end{aligned}$$

Hence $z^2 < 2$ and $z \in S$. But since $y < z$, we have an element of S that is greater than the least upper bound of S. This is a contradiction and so $y^2 \geqslant 2$.

Assume that $2 < y^2$. Let $k = (y^2 - 2)/(2y + 2)$. Then $0 < k < y$. Let $w = y - k$. Then $w < y$ and

$$\begin{aligned} w^2 &= (y - k)^2 = y^2 - 2yk + k^2 = y^2 - (2y - k)k \\ &= y^2 - (2y - k)(y^2 - 2)/(2y + 2) \\ &> y^2 - (2y + 2)(y^2 - 2)/(2y + 2) \\ &= y^2 - (y^2 - 2) = 2. \end{aligned}$$

Thus $2 < w^2$, and so $x < w$ for every $x \in S$. (Cf. Exercise 15 of Problem Set 2.6.) Hence w is an upper bound of S and w is less than the least upper bound of S. This is a contradiction, and so $y^2 \leqslant 2$.

Combining the above results gives $y^2 = 2$, or $y = \sqrt{2}$, and since $y \in R$ we have $\sqrt{2} \in R$.

Problem Set 2.7

1. Let S denote the sets of numbers described below. Determine sup S and inf S if they exist.

   (a) $\{3, 5, 7, 9, 11, 13\}$.

   (b) $\{(1/n) - 1 : n \in N\}$.

   (c) $\{(-1)^n(2-4/n^2) : n \in N\}$.

   (d) $\{x : x \in Q \text{ and } 1/(1-x) \leqslant 1 + 2x\}$.

   (e) $\{x : x \in Q \text{ and } x^2 - 2x < 8\}$.

2. Let $S \subset R$. If S has a greatest element x, show that $x = \sup S$.

3. Prove Theorem 2.7.1.

4. Give an example of a bounded set of real numbers that does not have a greatest element or a least element.

5. Prove Theorem 2.7.2.

6. Show that any finite set of real numbers has a least upper bound and a greatest lower bound.

7. Let x be any real number. Show that the set $S = R \setminus \{x\}$ is *not* complete, that is, find a nonempty subset of S which has an upper bound but *not* a least upper bound.

8. Let S be a nonempty set of real numbers. Prove that $x \in R$ is the least upper bound of S if and only if for each $\varepsilon > 0$ (a) each element s of S satisfies the inequality $s < x + \varepsilon$, and (b) at least one element s of S satisfies the inequality $s > x - \varepsilon$.

9. Let S be a nonempty set of real numbers. Prove that $x \in R$ is the greatest lower bound of S if and only if for each $\varepsilon > 0$ (a) each element s of S satisfies the inequality $s > x - \varepsilon$, and (b) at least one element s of S satisfies the inequality $s < x + \varepsilon$.

10. Let S and T be nonempty sets of real numbers. If $S \subseteq T$ and T is bounded, show that S is bounded.

11. Let S and T be nonempty sets of real numbers. If $S \subseteq T$ and T is bounded, show that sup $S \leqslant$ sup T and inf $S \geqslant$ inf T.

12. Let S be a nonempty set of real numbers. Let $-S = \{-x : x \in S\}$. If S is bounded, show that

$$- (\sup S) = \inf(-S)$$

and

$$- (\inf S) = \sup(-S).$$

13. A real number x is said to be **nonnegative** if and only if $x = 0$ or $x > 0$. If x is a nonnegative real number and $x < 1/n$ for each natural number n, show that $x = 0$.

14. Let n be a natural number, and let r be a positive real number. Prove that there exists a unique positive real number s such that $s^n = r$. [*Hint*: Let $S = \{x: 0 < x$ and $x^n < r\}$.]

15. Show that the set of positive real numbers R^+ is a unique positive class for the field of real numbers, that is, if $\mathfrak{p}$ is any positive class for R, then $\mathfrak{p} = R^+$. [*Hints*: If $0 < x \in R$, then either $\sqrt{x} \in \mathfrak{p}$ or $-\sqrt{x} \in \mathfrak{p}$, hence $x \in \mathfrak{p}$; 0 is not a number of $\mathfrak{p}$; if $x \in \mathfrak{p}$ and if $x < 0$, then $-x > 0$ and so $-x \in \mathfrak{p}$.]

16. Let $T = \{r + s\sqrt{2}: r,s \in Q\}$. Show that T is a field with respect to the operation of the field of real numbers R.

17. Show that there is not a unique positive class for the field T defined in Exercise 16; that is, show that T is an ordered field in two ways. [*Hint*: Let $\mathfrak{p}$ be a positive class of T. Define $\mathfrak{p}'$ by

$$\mathfrak{p}' = \{r + s\sqrt{2}: r - s\sqrt{2} \in \mathfrak{p}\}.]$$

18. If a field has a positive class with square roots, then show that its positive class is unique. [*Hint*: Let $\mathfrak{p}$ be a positive class with square roots, and let $\mathfrak{p}'$ be a positive class. If $x \in \mathfrak{p}$, then either $\sqrt{x}$ or $-\sqrt{x}$ is in $\mathfrak{p}'$, and so $x \in \mathfrak{p}'$. If $y \in \mathfrak{p}'$ and $y \notin \mathfrak{p}$, then $-y \in \mathfrak{p}$ and, by the above, $-y \in \mathfrak{p}'$.]

19. Use Exercise 18 to show that if a field has a positive class with respect to which it is complete, then the positive class is unique.

Chapter Three

TOPOLOGY OF THE REAL NUMBER SYSTEM

3.1. Sequences

In real analysis one studies functions whose domain and range are sets of real numbers. These functions are called *real functions of a real variable* or *real-valued functions of a real variable*. Since our main interest is the study of real functions of a real variable, we shall use the word *function* to mean a real function of a real variable and the notation $f(x)$ to indicate that f is a function of the real variable x.

This section is devoted to the study of a particular kind of function—the sequence. A thorough understanding of sequences is a valuable tool in understanding the material in real analysis.

> **Definition 3.1.1.** *A* **sequence** *is a function f whose domain is the set of positive integers. If n is a positive integer, then $f(n)$ is called the* **nth term** *of the sequence.*

If f is a sequence, then we write $f(n) = f_n$ for each positive integer n and write $f = \{f_n\}$ or $f = \{f_n\}_{n=1}^{+\infty}$. If $m \neq n$, then f_m and f_n are considered to be different terms of the sequence f even though f_m and f_n may be equal. It is sometimes convenient to describe a sequence by giving a formula for its nth term. For example, the sequence whose nth term is n^2 can be written as $\{n^2\}$ or $\{n^2\}_{n=1}^{+\infty}$. The third term of this sequence is 9 and the fifth term is 25. We often denote a sequence by writing $f_1, f_2, f_3, \ldots$, or, in the case of the example given above, $1, 4, 9, \ldots, n^2, \ldots$. We frequently use symbols other than f to denote a sequence, for example, $\{a_n\}$ or $\{s_n\}$.

Example 3.1.1.(a) $\{1/n^2\} = 1, 1/4, 1/9, \dots$.

(b) $\{(-1)^n\} = -1, 1, -1, \dots$.

From the sequence $\{1/n^2\}$ given in Example 3.1.1(a), we see that if n is taken to be a large positive integer, then the nth term of the sequence is small. That is, as n becomes very large, $1/n^2$ becomes very small or very close to zero. For the sequence $\{(-1)^n\}$, we notice that for n even, the nth term $(-1)^n$ is very close to 1 (actually equal to 1), while if n is odd, the nth term $(-1)^n$ is very near -1 (actually equal to -1). In Example 3.1.1(a), the sequence behaves nicely in the sense that the terms approach some fixed real number (in this case zero) as n becomes large. In this case, we say that the sequence *converges to* 0. In Example 3.1.1(b) the terms do not approach a fixed real number as n becomes large (actually in this case the terms are alternately $+1$ and -1). This sequence is said to *diverge*. With this introduction, we define *convergence of a sequence*.

> **Definition 3.1.2.** *A sequence $\{a_n\}$ is said to* **converge** *to the real number A if and only if for each $\varepsilon > 0$ there exists a positive real number $N = N(\varepsilon) > 0$ such that $|a_n - A| < \varepsilon$ for all $n > N$. The sequence $\{a_n\}$ is said to be* **convergent** *if and only if there is a real number A such that $\{a_n\}$ converges to A. The sequence $\{a_n\}$ is said to be* **divergent** *or to* **diverge** *if and only if it is not convergent.*

Intuitively, we think of a sequence $\{a_n\}$ being convergent if and only if there is some real number A such that a_n is "near" A when n is large, that is,

$$|a_n - A|$$

is small when n is large.

If the sequence $\{a_n\}$ converges to A, we say that the sequence $\{a_n\}$ has **limit** A and we denote this by $\{a_n\} \to A$ or $\lim_n a_n = A$. Theorem 3.1.1 will show that if a sequence has a limit, then the limit is unique. Thus, the number A is called the **limit** of the sequence.

In working with sequences, we must decide if the sequence is convergent (has a limit) or is divergent. If we think the sequence has a limit, say, A, we check the definition to see if A is actually the limit of the sequence. If A satisfies the conditions of the definition, then we have found the limit, and the sequence converges to that limit and the sequence is convergent. If A does not satisfy the conditions for the limit of the sequence, we must show that some other real number is the limit or prove that the sequence does not have a limit, that is, the sequence is divergent. We present two examples to illustrate this process.

Example 3.1.2. Consider the sequence $\{(-1)^n/n\}$. If n is a large positive integer, then $1/n$ is small, that is, near zero. For example, if $n = 10{,}000$ then $(-1)^n/n = .0001$, which is very near zero. Since $(-1)^n/n$ is near zero for large n, it is natural to consider 0 as the possible limit of the sequence $\{(-1)^n/n\}$. Let $\varepsilon > 0$. We see that

$$\left| \frac{(-1)^n}{n} - 0 \right| = \frac{1}{n}.$$

Thus if $n > N = 1/\varepsilon$, we have

$$\left| \frac{(-1)^n}{n} - 0 \right| = \frac{1}{n} < \frac{1}{N} = \varepsilon.$$

Then 0 satisfies the conditions of Definition 3.1.2, and the sequence $\{(-1)^n/n\}$ is convergent and has the limit 0.

Example 3.1.3. As a second example we consider the sequence $\{a^n\}$, where $0 < a < 1$. We first choose a possible limit for this sequence. Since $0 < a < 1$, we have $0 < a^n < 1$ and $a^{n+1} = a^n \cdot a < a^n \cdot 1 = a^n$. Thus all the terms of the sequence are between 0 and 1 and are becoming smaller as n becomes larger. Hence we select zero as a reasonable limit for the sequence $\{a^n\}$. In order to prove that 0 is the limit of this sequence, we make use of the binomal theorem. Let $b = 1/a$. Then $b > 1$ and we can write $b = 1 + h$, where $h > 0$. From the binomal theorem, it follows that

$$b^n = (1 + h)^n > nh,$$

for each positive integer n. We see that

$$|a^n - 0| = a^n = 1/b^n = 1/(1 + h)^n < 1/hn.$$

If ε is any positive real number and N is a positive real number such that $N > 1/h\varepsilon$, then for each positive integer $n > N$, we have $1/hn < 1/hN$. Thus, for the given ε, we have

$$|a^n - 0| < 1/hn < 1/hN < \varepsilon$$

whenever $n > N$. Hence, for each $\varepsilon > 0$, we have found a suitable positive number N such that

$$|a^n - 0| < \varepsilon$$

whenever $n > N$, and so the sequence $\{a^n\}$ converges to 0.

Examples 3.1.2 and 3.1.3 demonstrate the procedure for proving that a sequence is convergent. Before giving examples of divergent series, we shall define some terms related to sequences and prove some results about convergent sequences.

Sequences that converge to zero have a special name; they are called **null sequences**. The sequences in Examples 3.1.2 and 3.1.3 are null sequences.

Our first theorem about sequences deals with the limit of a sequence; it establishes that the limit of a convergent sequence is unique.

Theorem 3.1.1. *If the sequence $\{a_n\}$ converges to A, then the sequence $\{a_n\}$ cannot also converge to a limit distinct from A. That is, if $a_n \to A$ and $a_n \to B$, then $A = B$.*

Proof. Suppose that $\{a_n\} \to A$, $\{a_n\} \to B$, and $A \neq B$. Let $\varepsilon > 0$. Since $\{a_n\} \to A$, there exists a positive number N_A such that $|A - a_n| < \varepsilon/2$ when $n > N_A$. Since $\{a_n\} \to B$, there exists a positive number N_B such that $|B - a_n| < \varepsilon/2$ when $n > N_B$. Let $N = \max (N_A, N_B)$. Then for $n > N$, we have that $n > N_A$ and $n > N_B$. Thus $|A - a_n| < \varepsilon/2$, $|B - a_n| < \varepsilon/2$, and so the Triangle Inequality (Theorem 2.4.6) implies that

$$|B - A| = |B - a_n + a_n - A| \leqslant |B - a_n| + |a_n - A|$$
$$= |B - a_n| + |A - a_n| < \varepsilon/2 + \varepsilon/2 = \varepsilon,$$

when $n > N$. Hence for each $\varepsilon > 0$ we have $|B - A| < \varepsilon$ and so $|B - A| = 0$, or $B = A$. (Cf. Exercise 36 of Problem Set 2.4.) This is a contradiction, and so the limit of a convergent sequence is unique.

Definition 3.1.3. *The sequence $\{a_n\}$ is said to be **increasing** (or **nondecreasing**) if and only if $a_m \leqslant a_n$ whenever $m < n$. The sequence $\{a_n\}$ is said to be **strictly increasing** if and only if $a_m < a_n$ whenever $m < n$.*

Definition 3.1.4. *The sequence $\{a_n\}$ is said to be **decreasing** (or **nonincreasing**) if and only if $a_m \leqslant a_n$ whenever $n < m$. The sequence $\{a_n\}$ is said to be **strictly decreasing** if and only if $a_m < a_n$ whenever $n < m$.*

If a sequence is either increasing or decreasing, then it is said to be **monotone**.

Theorem 3.1.2. *The sequence $\{a_n\}$ is increasing (decreasing) if and only if $a_n \leqslant a_{n+1}$ $(a_n \geqslant a_{n+1})$ for each positive integer n. The sequence $\{a_n\}$ is strictly increasing (strictly decreasing) if and only if $a_n < a_{n+1}$ $(a_n > a_{n+1})$ for each positive integer n.*

The proof is left as Exercise 15 of Problem Set 3.1.

We see that the sequences $\{1/n\}$ and $\{a^n\}$, $0 < a < 1$, given in Examples 3.1.2 and 3.1.3, respectively, are strictly decreasing.

As in the case of subsets of sets, we now introduce the concept of a *subsequence* of a sequence.

> **Definition 3.1.5.** *Let $\{a_n\}$ be a sequence, and let $\{n_k\}$ be a sequence of positive integers such that $n_k < n_{k+1}$ for each k, that is, $\{n_k\} = \{n_k\}_{k=1}^{+\infty}$ is a strictly increasing sequence. Then the sequence $\{a_{n_k}\} = \{a_{n_k}\}_{k=1}^{+\infty}$ is called a **subsequence** of the sequence $\{a_n\}$.*

> **Example 3.1.4.** To give an example of a subsequence, we consider the sequence $\{1/n\}$. If we let $n_k = 2k$ for each positive integer k, the corresponding subsequence of $\{1/n\}$ is $\{1/2k\} = \{1/2n\}$. Furthermore, if we let $\{n_k\}$ be any strictly increasing sequence of positive integers, then the sequence $\{1/n_k\}$ is a subsequence of the sequence $\{1/n\}$.

If $\{a_n\}$ is a sequence, then $\{a_n\}$ is trivially a subsequence of itself.

> **Theorem 3.1.3.** *If the sequence $\{a_n\}$ converges to A, then every subsequence of the sequence $\{a_n\}$ also converges to A.*

> *Proof.* Assume the sequence $\{a_n\}$ converges to A and let $\{a_{n_k}\}$ be any subsequence of the sequence $\{a_n\}$. Since the sequence $\{n_k\}$ is strictly increasing, $n_k \geq k$ for each positive integer k. The fact that the sequence $\{a_n\} \to A$ implies that for each $\varepsilon > 0$ there exists a positive number N such that $|A - a_n| < \varepsilon$ whenever $n > N$. Thus, whenever $n_k \geq k > N$, we have that $|A - a_{n_k}| < \varepsilon$. Hence the subsequence $\{a_{n_k}\}$ converges to A.

Theorem 3.1.3 is a result dealing with convergent sequences. We use this theorem to produce two results about divergent sequences.

> **Corollary 3.1.3a.** *If the sequence $\{a_n\}$ has two subsequences which converge to different limits, then the sequence $\{a_n\}$ is divergent.*

> **Corollary 3.1.3b.** *If the sequence $\{a_n\}$ has a subsequence which is divergent, then the sequence $\{a_n\}$ is divergent.*

We now present three examples of divergent sequences.

> **Example 3.1.5.** We consider the sequence $\{a_n\}$ defined by the equation $a_n = (-1)^n$ for each positive integer n. Then the subsequence $\{a_{2n}\}$ is defined by $a_{2n} = 1$ for each positive integer n, and the subsequence $\{a_{2n-1}\}$ is defined by $a_{2n-1} = -1$ for each positive integer. It is left as an exercise to show that $\{a_{2n}\} \to 1$ and $\{a_{2n-1}\} \to -1$. Hence Corollary 3.1.3a implies that the sequence $\{(-1)^n\}$ diverges.

Example 3.1.6. Let $a_n = \sin(n\pi/4)$ for each positive integer n, and let $n_k = 4k - 2$ for each positive integer k. Corollary 3.1.3b implies that the sequence $\{a_n\} = \{\sin(n\pi/4)\}$ is divergent, since this sequence has the divergent subsequence

$$\{a_{n_k}\} = \{\sin[(4k - 2)\pi/4]\} = \{(-1)^k\}.$$

Example 3.1.7. The sequence $\{n\}$ is a divergent sequence since for any real number A, if $n > 2A$, then $|A - n| > |A|$. This sequence is said to diverge to $+\infty$. (The symbol "$+\infty$" is read *plus infinity*.)

Example 3.1.7 leads to the following definition.

Definition 3.1.6. *The sequence $\{a_n\}$ is said to **diverge** to $+\infty$ ($-\infty$) or to have **limit** $+\infty$ ($-\infty$) if and only if to each real number M there is a positive real number $N = N(M)$ such that*

$$a_n > M \ (a_n < M)$$

for each positive integer $n > N$.

If the sequence $\{a_n\}$ diverges to $+\infty$ ($-\infty$), we usually denote this by the symbols $\{a_n\} \to +\infty$ or $\lim_n a_n = +\infty$ ($\{a_n\} \to -\infty$ or $\lim_n a_n = -\infty$). Using this notation, for the sequence $\{n\}$ of Example 3.1.7 we write $\{n\} \to +\infty$ or $\lim_n n = +\infty$.

Definition 3.1.7. *If the sequence $\{a_n\}$ diverges but does not diverge to plus infinity and does not diverge to minus infinity, then the sequence $\{a_n\}$ is said to **oscillate** or to be an **oscillating sequence**.*

The sequences $\{(-1)^n\}$ and $\{\sin(n\pi/4)\}$ given in Examples 3.1.5 and 3.1.6, respectively, are oscillating sequences. We point out that the term *oscillate* does not mean that *the terms go up and down*. The terms of the sequence $\{(-1)^n/n\}$ *go up and down*, but this sequence does not oscillate since it is a convergent sequence. The term oscillate is only used for a certain type of divergent sequence. Roughly speaking, a sequence oscillates if the terms *go up and down too much*. It follows from the above definitions that for any sequence to diverge, either it oscillates or it diverges to plus infinity or minus infinity.

The following theorem tells us that the convergence and divergence of a sequence is unaltered by a change of a finite number of its terms.

Theorem 3.1.4. *Let $\{a_n\}$ and $\{b_n\}$ be two sequences. Let S and T be two positive integers such that $a_{S+n} = b_{T+n}$ for each positive integer n.*

The sequence $\{a_n\}$ converges if and only if the sequence $\{b_n\}$ converges. If $\{a_n\} \to A$, then $\{b_n\} \to A$.

The proof is left as Exercise 16 of Problem Set 3.1.

Example 3.1.8. We consider the two sequences $\{(-1)^n/n\}$ and $\{a_n\}$, where $a_n = n^2$ if $n \leqslant 100$ and $a_n = (-1)^n/n$ if $n > 100$. For $n > 100$, the terms of the two sequences are identical and Theorem 3.1.4 implies that both sequences either converge to the same limit or diverge. In Example 3.1.2 it was proved that the sequence $\{(-1)^n/n\}$ converges to 0; thus the sequence $\{a_n\}$ also converges to 0.

Theorem 3.1.5. *If $a_n = A$ for each positive integer n, then $\{a_n\}$ converges to A.*

Proof. Let $\varepsilon > 0$, then for each positive integer n,

$$|a_n - A| = 0 < \varepsilon.$$

Thus, $\{a_n\} \to A$.

Corollary 3.1.5. *If there is a positive real number N such that $a_n = A$ for each $n > N$, then $\{a_n\}$ converges to A.*

We now develop a method for determining the convergence of a sequence without direct application of the definition of convergence. We begin by defining some special types of sequences.

Definition 3.1.8. *The sequence $\{a_n\}$ is said to be **bounded** if and only if there exists a real number M such that $|a_n| \leqslant M$ for each n.*

Example 3.1.9. The sequences in Examples 3.1.1–3.1.6 are all bounded. For the sequence $\{(-1)^n/n\}$ given in Example 3.1.2, we see that $|(-1)^n/n| \leqslant 1$ for each positive integer n.

The following theorem gives a necessary condition for a sequence to converge.

Theorem 3.1.6. *If the sequence $\{a_n\}$ converges, then it is bounded.*

Proof. Since $\{a_n\}$ is convergent, there is a real number A such that $\lim_n a_n = A$. Let $\varepsilon = 1$. Then there is a positive integer N such that $|A - a_n| < 1$ when $n > N$. From this we have $|a_n| < |A| + 1$ for $n > N$. Let

$$M = \max(|a_1|, |a_2|, ..., |a_N|, |A| + 1).$$

Then $|a_n| \leqslant M$ for each positive integer n and $\{a_n\}$ is bounded.

We see that in Examples 3.1.2 and 3.1.3 the sequences $\{1/n\}$ and $\{a^n\}$, $0 < a < 1$, are strictly decreasing and bounded. Likewise, the constant sequence $\{a\}$ is decreasing (or increasing) and bounded. The property common to these three sequences (other than decreasing) is that each converges. As the following theorem indicates, being increasing (or decreasing) and bounded is enough to guarantee convergence of a sequence.

Theorem 3.1.7. *If $\{a_n\}$ is a monotone bounded sequence, then it is convergent.*

Proof. We consider the case where $\{a_n\}$ is increasing; the decreasing case follows in an analogous manner. In order to prove this result, we use the Completeness Property of the real numbers (Definition 2.7.3). Consider the set $S = \{a_n : n = 1, 2, ...\}$. Since the sequence $\{a_n\}$ is bounded, we know there is a real number M such that $|a_n| \leqslant M$ for each n. Hence the set S has an upper bound and, by the Completeness Property, has a least upper bound, say, A. To show that A is the limit of the sequence $\{a_n\}$, we first note that, since A is is an upper bound of S, $a_n \leqslant A$ for each n. Assume that $\varepsilon > 0$ is given and that, for each positive integer n, $A - a_n \geqslant \varepsilon$. Then $A - \varepsilon \geqslant a_n$, and $A - \varepsilon$ is an upper bound less than the least upper bound. Since this assumption leads to a contradiction, there must be some positive integer N such that $A - a_N < \varepsilon$. Then, if $n > N$, $a_n \geqslant a_N$ and $A - a_n \leqslant A - a_N < \varepsilon$. Therefore, for $n > N$, $0 \leqslant A - a_n = |A - a_n| < \varepsilon$ and the sequence $\{a_n\}$ converges to A.

The sequence $\{a^n\}$, $0 < a < 1$, given in Example 3.1.3 is an example of a bounded monotone sequence. This sequence is strictly decreasing, and $|a^n| < 1$ for each positive integer n. It was proved in Example 3.1.3 that the sequence $\{a^n\}$, $0 < a < 1$, converges to zero. An alternate proof that this sequence is convergent follows directly from Theorem 3.1.7.

Problem Set 3.1

For Exercises 1–6, determine whether the given sequence is convergent or divergent. If the sequence converges, find its limit.

1. $\{2 - 1/n\}$.
2. $\{(n^2 + n)/(n^2 - n)\}$.
3. $\{1/\sqrt{n}\}$.
4. $\{n^2/(n + 1)\}$.
5. $\{\sqrt{n+1} - \sqrt{n}\}$.
6. $\{(1 + h)^n\}$, $h > 0$.

7. Show that the sequence $\{(\log n)/n\}$, where $\log n$ denotes $\log_e n$, converges to zero.

For Exercises 8–13, find the limit of the given sequence and prove that the sequence converges to that limit.

8. $\{1/2^n\}$.

9. $\{1/n\}$.

10. $\{1 - 1/n\}$.

11. $\{(-1)^n/n^2\}$.

12. $\{1 + (-1)^n/\sqrt{n}\}$.

13. $\{n/(1 + n^2)\}$.

14. Prove that the sequence $\{a_n\}$ converges to zero if and only if the sequence $\{|a_n|\}$ converges to zero.

15. Prove Theorem 3.1.2.

16. Prove Theorem 3.1.4.

17. Prove that if the sequence $\{a_n\}$ converges to A, then the sequence $\{|a_n|\}$ converges to $|A|$. Is the converse true? Give an example of a divergent sequence $\{a_n\}$ such that the sequence $\{|a_n|\}$ is convergent.

18. Prove that if the sequence $\{a_n\}$ converges to A, then the sequence $\{-a_n\}$ converges to $-A$.

19. Show that the sequence $\{\sqrt{n}\}$ diverges to $+\infty$.

20. If $a > 1$, show that the sequence $\{a^n\}$ diverges to $+\infty$.

21. If the sequence $\{a_n\}$ diverges to $+\infty$, prove that any subsequence of the sequence $\{a_n\}$ also diverges to $+\infty$.

22. Let $a_1 = \sqrt{2}$ and $a_n = \sqrt{2 + a_{n-1}}$ for $n > 1$. Show that the sequence $\{a_n\}$ converges, and find its limit.

23. Let $a > 0$, $a_1 > 0$, and $a_n = (a_{n-1} + a)^{\frac{1}{2}}$ for $n > 1$. Show that the sequence $\{a_n\}$ converges, and find its limit.

24. Let $\{a_n\}$ be a sequence with $a_n \geq 0$ for each positive integer n. If the sequence $\{a_n\}$ converges to A, show that $A \geq 0$.

25. Let $\{a_n\}$ be a sequence with $a_n \leq M$ for each positive integer n. If the sequence $\{a_n\}$ converges to A, show that $A \leq M$.

26. Let $\{a_n\}$ be a sequence such that for every $\varepsilon > 0$,

$$|a_n - A| < \varepsilon$$

when $n > N$, where N does *not* depend on ε. Prove that all but a finite number of the terms of the sequence $\{a_n\}$ are equal to A.

27. Let P be a polynomial of degree three, that is,

$$P(x) = ax^3 + bx^2 + cx + d,$$

where $a, b, c, d, x \in R$. Show that

$$\lim_n \frac{P(n + 1)}{P(n)} = 1.$$

28. Let $\{a_n\}$ be a sequence such that

$$\lim_n a_{2n} = A \text{ and } \lim_n a_{2n-1} = A.$$

Prove that the sequence $\{a_n\}$ converges to A.

3.2. Operations with Sequences

In this section we define new sequences in terms of given sequences and establish their convergence by checking the convergence of the original sequences. We first define the *sum* and the *product* of two sequences. If $\{a_n\}$ and $\{b_n\}$ are two sequences, then the sequence $\{a_n + b_n\}$ is called the **sum sequence** and the sequence $\{a_n b_n\}$ is called the **product sequence** of the sequences $\{a_n\}$ and $\{b_n\}$.

Theorem 3.2.1. *If $\{a_n\}$ converges to A and $\{b_n\}$ converges to B, then $\{a_n + b_n\}$ converges to $A + B$.*

Proof. Assume that $\varepsilon > 0$ is given. Since $\{a_n\} \to A$, there exists N_A such that $|A - a_n| < \varepsilon/2$ when $n > N_A$. Likewise, there exists N_B such that $|B - b_n| < \varepsilon/2$ when $n > N_B$. Let $N = \max(N_A, N_B)$. Then for $n > N$, we have $|A - a_n| < \varepsilon/2$ and $|B - b_n| < \varepsilon/2$, and so

$$
\begin{aligned}
|(A + B) - (a_n + b_n)| &= |(A - a_n) + (B - b_n)| \\
&\leqslant |A - a_n| + |B - b_n| \\
&< \varepsilon/2 + \varepsilon/2 = \varepsilon.
\end{aligned}
$$

Therefore, $\{a_n + b_n\}$ converges to $A + B$.

Example 3.2.1. In Example 3.1.2 and 3.1.3 it was proved that the sequence $\{(-1)^n/n\}$ converges to zero and that the sequence $\{a^n\}$, $0 < a < 1$, converges to zero. Theorem 3.2.1 now implies that the sequence $\{(-1)^n/n + a^n\}$, $0 < a < 1$, also converges to zero.

Corollary 3.2.1. *If $\{a_n\}$ converges to A and $\{b_n\}$ converges to B, then the difference sequence $\{a_n - b_n\}$ converges to $A - B$.*

Proof. This follows from Theorem 3.2.1 and Exercise 18 of Problem Set 3.1.

Theorem 3.2.2. *If $\{a_n\}$ converges to A and $\{b_n\}$ converges to B, then $\{a_n b_n\}$ converges to AB.*

Proof. In order to show that $\{a_n b_n\} \to AB$, given $\varepsilon > 0$, we need to find a positive integer N such that $|AB - a_n b_n| < \varepsilon$ when $n > N$. If we consider $|AB - a_n b_n|$, we can write it as follows,

$$
\begin{aligned}
|AB - a_n b_n| &= |AB - a_n B + a_n B - a_n b_n| \\
&= |B(A - a_n) + a_n(B - b_n)| \\
&\leqslant |B||A - a_n| + |a_n||B - b_n|
\end{aligned}
$$

We can now make $|A - a_n|$ and $|B - b_n|$ small since $\{a_n\} \to A$ and $\{b_n\} \to B$. The number $|B|$ presents no problem since it is a fixed number. The numbers $|a_n|$ depend on the positive integer n, but this case causes no trouble since by Theorem 3.1.6 the sequence is bounded, say, $|a_n| \leqslant M$ for each n, where $M > 0$. With this introduction to the proof, we see what steps are necessary in order to finish the proof.

Assume that $\varepsilon > 0$. Since $\{a_n\} \to A$, there is some N_A such that $|A - a_n| < \varepsilon/(|B| + M)$ when $n > N_A$. Similarly, there is some N_B such that $|B - b_n| < \varepsilon/(|B| + M)$ when $n > N_B$. Let $N = \max (N_A, N_B)$. Then for $n > N$, we have $|A - a_n| < \varepsilon/(|B| + M)$ and $|B - b_n| < \varepsilon/(|B| + M)$, and, from above,

$$
\begin{aligned}
|AB - a_n b_n| &\leqslant |B||A - a_n| + |a_n||B - b_n| \\
&< \frac{|B|\varepsilon}{(|B| + M)} + \frac{M\varepsilon}{(|B| + M)} = \varepsilon.
\end{aligned}
$$

Therefore, $\{a_n b_n\}$ converges to AB.

Example 3.2.2. Consider the sequence $\{(2n^2 - n - 1)/(n^2 - n)\}$. For each positive integer $n > 1$ we can write

$$
\frac{2n^2 + n - 1}{n^2 - n} = \left(\frac{n^2 + n}{n^2 - n}\right)\left(2 - \frac{1}{n}\right).
$$

Thus the sequence $\{(2n^2 + n - 1)/(n^2 - n)\}$ is the product sequence of the sequences $\{(n^2 + n)/(n^2 - n)\}$ and $\{2 - 1/n\}$. The sequence $\{(n^2 + n)/(n^2 - n)\}$ converges to 1 (cf. Exercise 2 of Problem Set 3.1), while the sequence $\{2 - 1/n\}$ converges to 2 (cf. Exercise 1 of Problem Set 3.1). Therefore, Theorem 3.2.2 implies that the sequence $\{(2n^2 + n - 1)/(n^2 - n)\}$ converges to 2, the product of 1 and 2.

Corollary 3.2.2. *If $\{a_n\}$ converges to A and b is any real number, then $\{ba_n\}$ converges to bA.*

Note that Theorems 3.2.1 and 3.2.2 are only for the sum and the product of *convergent* sequences. If $\{a_n\}$ and $\{b_n\}$ are two divergent sequences, then the sequences $\{a_n + b_n\}$ and $\{a_n b_n\}$ may either converge or diverge. The construction of such examples is left as Exercises 7 and 8 of Problem Set 3.2.

If $\{a_n\}$ and $\{b_n\}$, $b_n \neq 0$ for each n, are sequences, the sequence $\{a_n/b_n\}$ is called the **quotient sequence** of the sequences $\{a_n\}$ and $\{b_n\}$. Before proving results about quotient sequences, we need the following *lemma* (a lemma is an auxiliary theorem used in proving other theorems).

Lemma 3.2.1. *If $\{a_n\}$ converges to $A \neq 0$, then there exists N such that $|a_n| > |A|/2$ for $n > N$.*

Proof. Let $\varepsilon = |A|/2$. Since $\{a_n\} \to A$, there exists N such that $|A - a_n| < \varepsilon = |A|/2$ when $n > N$. Hence

$$|a_n| = |-a_n| = |-A + A - a_n| \geqslant |-A| - |A - a_n|$$
$$\geqslant |A| - |A|/2 = |A|/2$$

when $n > N$.

Using the above lemma, we prove the following theorem about quotient sequences.

Theorem 3.2.3. *If $\{a_n\}$ converges to A and $\{b_n\}$ converges to B, where $b_n \neq 0$ for each n and $B \neq 0$, then $\{a_n/b_n\}$ converges to A/B.*

Proof. As in the proof of Theorem 3.2.2, we first indicate the inequality needed. For all large values of n, we would like to make

$$|A/B - a_n/b_n| < \varepsilon.$$

Now

$$|A/B - a_n/b_n| = |(Ab_n - a_n B)/Bb_n|$$

$$= \frac{|Ab_n - AB + AB - a_n B|}{|Bb_n|}$$

$$\leqslant \frac{|A||b_n - B| + |B||A - a_n|}{|B||b_n|}$$

$$\leqslant \frac{|A||b_n - B|}{|B||b_n|} + \frac{|A - a_n|}{|b_n|}$$

$$= \frac{1}{|b_n|} \left\{ \frac{|A||b_n - B|}{|B|} + |A - a_n| \right\}.$$

Hence,

$$|A/B - a_n/b_n| \leqslant \frac{1}{|B||b_n|} (|A||b_n - B| + |B||A - a_n|).$$

If $\varepsilon > 0$ is given, then, by Lemma 3.2.1, there is N_1 such that $|b_n| > |B|/2$ for $n > N_1$. Since $\{a_n\}$ converges to A, there is N_A such that $|A - a_n| < \varepsilon |B|^2/2(|A| + |B|)$ when $n > N_A$. Also, there is N_B such that $|B - b_n| < \varepsilon |B|^2/2(|A| + |B|)$ when $n > N_B$. Let

$$N = \max (N_1, N_A, N_B).$$

Then for $n > N$, we get

$$|A/B - a_n/b_n| < \frac{2}{|B|^2} \frac{|A| \varepsilon |B|^2 + |B| \varepsilon |B|^2}{2(|A| + |B|)} = \varepsilon.$$

Therefore, $\{a_n/b_n\}$ converges to A/B.

Example 3.2.3. As an example of quotient sequences, consider the sequence $\{1/(2n - 1)\}$. For each positive integer we can write

$$\frac{1}{2n - 1} = \frac{1/n}{2 - 1/n}.$$

Thus the sequence $\{1/(2n - 1)\}$ is the quotient sequence of the sequences $\{1/n\}$ and $\{2 - 1/n\}$. Since the sequence $\{1/n\}$ converges to 0 (cf. Exercise 9 of Problem Set 3.1) and the sequence $\{2 - 1/n\}$ converges to 2 (cf. Exercise 1 of Problem Set 3.1), Theorem 3.2.3 implies that the quotient sequence $\{1/(2n - 1)\}$ converges to $0/2 = 0$.

Problem Set 3.2

1. Give an example to show that if $\{a_n\}$ is a sequence such that $a_n > a_{n+1} > 0$ for all n, then $\{a_n\}$ need not converge to 0.
2. Let $\{a_n\} \to A$ and $\{b_n\} \to B$. If $a_n \geqslant b_n$ for all n, prove that $A \geqslant B$. Show that it is possible to have $a_n > b_n$ for all n, but $A = B$.
3. Let $\{a_n\} \to 0$, where $a_n > 0$ for all n. Let $\{b_n\}$ be a sequence such that $|b_n| < a_n$, for all n. Prove that $\{b_n\} \to 0$.
4. Assume $\{a_n\}$ and $\{b_n\}$ both converge to A. If, for each n, $a_n \leqslant c_n \leqslant b_n$, prove that $\{c_n\}$ converges to A.
5. If $\{a_n\}$ is a null sequence with $a_n \geqslant 0$ for each n, prove that $\{\sqrt{a_n}\}$ is a null sequence.
6. If $\{a_n\}$ converges to $A > 0$ with $a_n \geqslant 0$ for each n, prove that $\{\sqrt{a_n}\}$ converges to $\sqrt{A}$. [*Hint*: $\sqrt{a_n} - \sqrt{A} = (a_n - A)/(\sqrt{a_n} + \sqrt{A})$.]
7. Give an example of two divergent sequences whose sum sequence converges.
8. Give an example of two divergent sequences whose product sequence converges.
9. Assume $\{a_n\}$ and $\{b_n\}$ are two sequences such that $\{a_n\}$ and $\{a_n + b_n\}$ both converge. Prove that $\{b_n\}$ converges.

10. Assume that $\{a_n\}$ and $\{b_n\}$ are two sequences such that $\{a_n\}$ converges to $A \neq 0$ and $\{a_n b_n\}$ converges. Prove that $\{b_n\}$ converges.
11. Let $\{a_n\}$ be a sequence, and define

$$\sigma_n = \frac{a_1 + a_2 + \ldots + a_n}{n}$$

for each n. If $\{a_n\}$ converges to 0, prove that $\{\sigma_n\}$ converges to 0. [*Hint*: Let M be a positive integer. For $n > M$ write

$$\sigma_n = \frac{1}{n}(a_1 + a_2 + \ldots + a_M) + \frac{1}{n}(a_{M+1} + a_{M+2} + \ldots + a_n).$$

For $\varepsilon > 0$, choose M such that $|a_m| < \varepsilon/2$ when $m > M$. With M fixed, choose $N > M$ such that $|a_1 + a_2 + \ldots + a_M|/N < \varepsilon/2$. Let $n > N$ and consider each part of σ_n separately.]
12. Using the notation of Exercise 11, show that if $\{a_n\} \to A$, then $\{\sigma_n\} \to A$.
13. Using the notation of Exercise 11, show that if $\{a_n\} \to +\infty\,(-\infty)$, then $\{\sigma_n\} \to +\infty\,(-\infty)$.
14. Using the notation of Exercise 11, give an example of a sequence $\{a_n\}$ such that $\{a_n\}$ diverges, but $\{\sigma_n\}$ converges.

3.3. Nested Intervals

An important type of a set of real numbers is the *interval*. An **interval** I is a set of real numbers such that, for each $x, y \in I$, any real number between x and y also belongs to I.

If I is a bounded interval, that is, the interval I as a set is bounded, then, by the Completeness Property, I has a greatest lower bound a and a least upper bound b. The greatest lower bound a is called the **left endpoint** of I, and the least upper bound b is called the **right endpoint** of I. If both endpoints belong to I, then we say that I is **closed** and denote this by $I = [a, b]$. If neither of the endpoints belong to I, then we say I is **open** and denote it by $I = (a, b)$. If only one of the endpoints belongs to I we say I is **half open** and denote it by either $I = [a, b)$ or $I = (a, b]$, depending on which endpoint belongs to I. Then for any interval I with endpoints $a < b$, we have

$$[a, b] = \{x : x \in R \text{ and } a \leqslant x \leqslant b\},$$

$$(a, b) = \{x : x \in R \text{ and } a < x < b\},$$

$$(a, b] = \{x : x \in R \text{ and } a < x \leqslant b\},$$

$$[a, b) = \{x : x \in R \text{ and } a \leqslant x < b\}.$$

For any interval I with endpoints $a < b$, the **length** of I, denoted by $|I|$, is the nonnegative real number

$$|I| = b - a.$$

If the interval I is not bounded, that is, the interval I as a set is unbounded, then I is called an **infinite interval**. If the infinite interval I is bounded below, the Completeness Property implies that I has a greatest lower bound a. If $a \in I$, then I is denoted by $[a, +\infty)$, and, if $a \notin I$, then I is denoted by $(a, +\infty)$. Similarly, if I is bounded above, I has a least upper bound b. If $b \in I$, then I is denoted by $(-\infty, b]$, and, if $b \notin I$, then I is denoted by $(-\infty, b)$. Again the real numbers a and b are called **endpoints** of these infinite intervals. An infinite interval can have at most one endpoint. If an infinite interval contains an endpoint, it is said to be a **closed infinite interval**; otherwise it is called an **open infinite interval**. If an infinite interval I does not have an endpoint, then it is the set of all real numbers and is denoted by R or by $(-\infty, +\infty)$. There is no length assigned to infinite intervals.

If for each positive integer n there is an interval I_n, then this set of intervals is called a **sequence of intervals** and is denoted by $\{I_n\}$. A sequence of intervals $\{I_n\}$ is said to be **nested** if and only if $I_{n+1} \subseteq I_n$ for each n and the sequence of lengths $\{|I_n|\}$ is a null sequence, that is, each interval of the sequence is contained in its predecessor and $\lim_n |I_n| = 0$.

The following theorem, called the **Nested Interval Theorem**, assures us that the intersection of a nested sequence of closed nonempty intervals is nonempty.

Theorem 3.3.1. *If $\{I_n\}$ is a nested sequence of closed nonempty intervals, then $\bigcap_{n=1}^{\infty} I_n$ contains exactly one real number (or point).*

Proof. Let $I_n = [a_n, b_n]$ for each n. Since $\{I_n\}$ is nested,

$$a_n \leqslant a_{n+1} < b_{n+1} \leqslant b_n$$

and $\{b_n - a_n\} \to 0$.

If $m \leqslant n$, then $a_n < b_n \leqslant b_m$, and, if $n < m$, then $a_n \leqslant a_m < b_m$. In any case, $a_n < b_m$ for every n and m.

Since the sequence $\{a_n\}$ is bounded and increasing, it converges, say to a. We note $a_n \leqslant a$ for every n. The real number $a \leqslant b_m$ for every m. For assume that for some $m_0, a > b_{m_0}$. Let $\varepsilon = (a - b_{m_0})/2$. Since $\{a_n\} \to a$, then, for some n_0, we have $a - a_{n_0} < (a - b_{m_0})/2$. It follows that $b_{m_0} < a_{n_0}$, which is a contradiction. Hence $a \leqslant b_m$ for each m. Since the sequence $\{b_n\}$ is bounded and decreasing, it converges, say to b. In a manner similar to that above, $a_n \leqslant b$ for each n. But now $|b - a| \leqslant b_n - a_n$ for each n, and since $\{b_n - a_n\}$ is a null sequence, we get $b = a$. Hence there is at least one real number b in $\bigcap_{n=1}^{\infty} I_n$.

Let c be any real number, $b \neq c$. Since $\{b_n - a_n\}$ is a null sequence, there is an N such that $|b - c| > b_N - a_N$. Hence $c \notin I_N$ and $c \notin \bigcap_{n=1}^{\infty} I_n$. Therefore, there is exactly one real number $b \in \bigcap_{n=1}^{\infty} I_n$.

Example 3.3.1. As an example of a nested sequence of closed nonempty intervals, we consider the sequence $I_n = [0, 1/n]$ for each positive integer n. Then, for each positive integer n, the interval I_n is closed and nonempty, and $I_{n+1} \subseteq I_n$. Since $|I_n| = 1/n$ for each positive integer n, we have $\lim_n |I_n| = 0$. Hence the sequence $\{I_n\}$ is a nested sequence of closed nonempty intervals. We see that the only real number common to all the intervals is the point 0.

A sequence of nested open intervals may be defined in a similar manner. The following example indicates that the intersection of a sequence of nested open intervals may be either a *singleton set* or the empty set.

Example 3.3.2. Let $I_n = (-1/n, 1/n)$ and $J_n = (0, 1/n)$ for each positive integer n. Then $\{I_n\}$ and $\{J_n\}$ are sequences of nested open intervals. But

$$\bigcap_{n=1}^{\infty} I_n = \{0\} \quad \text{and} \quad \bigcap_{n=1}^{\infty} J_n = \varnothing.$$

We apply the Nested Interval Theorem to prove that the set of real numbers is not countable.

Theorem 3.3.2. *The set R of real numbers is not countable.*

Proof. Let S be any countable subset of R. Since S may be put in a one-to-one correspondence with the set of positive integers, the elements of S may be labeled

$$a_1, a_2, \ldots, a_n, \ldots.$$

Let I_1 be a closed interval with $|I_1| < 1$ such that $a_1 \notin I_1$. Let $I_2 \subset I_1$ be a closed interval with $|I_2| < 1/2$ such that $a_2 \notin I_2$. Let $I_3 \subset I_2$ be a closed interval with $|I_3| < 1/3$ such that $a_3 \notin I_3$. By induction, we define a sequence of closed intervals $\{I_n\}$ such that $I_{n+1} \subset I_n$, $|I_n| < 1/n$, and $a_n \notin I_n$ for each positive integer n. By Theorem 3.3.2, there is a unique real number A such that $A \in \bigcap_{n=1}^{\infty} I_n$.

Then $A \neq a_n$ for each n, since $A \in I_n$ and $a_n \notin I_n$. Since R is a subset of itself and for every countable subset S of R there is a real number A not belonging to S, we conclude that R cannot be countable.

We now introduce two concepts that are important in analysis, that is, the ideas of *neighborhood* and *limit point*.

> **Definition 3.3.1.** *Let x be a real number. Then a set T of real numbers is called a **neighborhood** of x if and only if there is an open interval I such that $x \in I \subseteq T$.*

From the definition of neighborhood, we see that any open interval containing x is a neighborhood of x, and that any closed interval containing x, not as an endpoint, is a neighborhood of x. For our purposes, we shall mainly be interested in neighborhoods that are open intervals.

> **Definition 3.3.2.** *Let S be a set of real numbers. Then a real number x is a **limit point** of S if and only if each neighborhood of x contains infinitely many points of S.*

The term *accumulation point* or *cluster point* is sometimes used instead of limit point. For x to be a limit point of a set of real numbers S, for each neighborhood T of x, $T \cap S$ must be an infinite set.

If x is a limit point of a set S, then every neighborhood of x contains at least one point of S other than x. If x is not a limit point of S, then there is some neighborhood of x that contains no points of S, with the possible exception of x itself. Thus x *is a limit point of a set S if and only if every neighborhood of x contains a point of S other than x.*

> **Example 3.3.3.** The set of integers is an example of an infinite set without a limit point. The set $\{1, 1/2, 1/3, ..., 1/n, ...\}$ has 0 as its only limit point. For the set of rational numbers every real number is a limit point.

If $x \in S$ and x is not a limit point of S, then x is called an **isolated point** of S.

From the definition of limit point, it is clear that any finite set has no limit points, that is, every point of a finite set is an isolated point.

The following important theorem is named for the Austrian mathematician B. Bolzano (1781–1848) and the German mathematician K. W. T. Weierstrass (1815–1897).

> **Theorem 3.3.3.** *(Bolzano–Weierstrass Theorem) Every bounded infinite set has a limit point.*

Proof. Let S be a bounded infinite set. Since S is bounded, there is a closed interval $I_0 = [a_0, b_0]$ such that $S \subseteq I_0$, and $|I_0| = b_0 - a_0$. Let

$$I_{1,1} = [a_0, (a_0 + b_0)/2], \quad I_{1,2} = [(a_0 + b_0)/2, b_0].$$

Then at least one of the sets $I_{1,1} \cap S$ or $I_{1,2} \cap S$ is infinite. Without loss of generality, we let $I_1 = [a_1, b_1]$ be such a set. The length of I_1 is

$$|I_1| = b_1 - a_1 = (b_0 - a_0)/2.$$

Let

$$I_{2,1} = [a_1, (a_1 + b_1)/2], \quad I_{2,2} = [(a_1 + b_1)/2, b_1].$$

Then at least one of the sets $I_{2,1} \cap S$ or $I_{2,2} \cap S$ is infinite. Let $I_2 = [a_2, b_2]$ be such a set. The length of I_2 is

$$|I_2| = b_2 - a_2 = (b_0 - a_0)/2^2.$$

By induction, we define a sequence of closed intervals $\{I_n\}$ such that $I_n \cap S$ is infinite, $|I_n| = (b_0 - a_0)/2^n$, and $I_{n+1} \subset I_n$ for each n. Then $\{I_n\}$ is a nested sequence of closed intervals and, by Theorem 3.2.2, there is a unique point x belonging to the intersection $\bigcap_{n=1}^{\infty} I_n$.

We now show that x is a limit point of S. Let T be any neighborhood of x. Then there is an open interval $I = (x - \varepsilon, x + \varepsilon) \subseteq T$. (Cf. Exercise 5 of Problem Set 3.3.) Let n be a positive integer such that $|I_n| = (b_0 - a_0)/2^n < \varepsilon$. Thus $S \cap I_n \subseteq S \cap T$, and, since $S \cap I_n$ is infinite, $S \cap T$ is infinite. Hence x is a limit point of S.

Problem Set 3.3

1. Prove that the intersection of a sequence of bounded closed intervals is either the empty set, a single point, or a closed interval. Give examples to show that each of these cases exist.
2. Prove that the intersection of a sequence of bounded open intervals is either the empty set, a single point, a closed interval, an open interval, or a half open interval. Give examples to show that each of these cases exist.
3. Let x and y be two distinct real numbers, Prove that there is a neighborhood T_x of x and a neighborhood T_y of y such that $T_x \cap T_y = \varnothing$.
4. Let x, y, and z be real numbers with $y \neq z$. Prove that there is a neighborhood of x which excludes either y or z.
5. Let x be a real number, and let T be a neighborhood of x. Prove that there is an $\varepsilon > 0$ such that $(x - \varepsilon, x + \varepsilon) \subseteq T$.
6. Prove that every bounded infinite set has a maximum limit point and a minimum limit point.
7. Let $\{a_n\}$ be a sequence which converges to a. If for every real number N there is an $n > N$ such that $a \neq a_n$, prove that a is a limit point of the set $\{a_n : n = 1, 2, \ldots\}$.
8. Let S be a set of real numbers. Prove that there are only countably many isolated points of S.

3.4. Cauchy Sequences

We now return to sequences to obtain a condition that is equivalent to convergence; this condition is called the *Cauchy condition*. The concept is due to the French mathematician A. L. Cauchy (1789–1857).

Definition 3.4.1. *A sequence $\{a_n\}$ is said to be a **Cauchy sequence** (or to satisfy the **Cauchy condition**) if and only if for each $\varepsilon > 0$ there is a positive number $N = N(\varepsilon)$ such that*

$$|a_n - a_m| < \varepsilon$$

when $n, m > N$.

We continue by establishing two theorems concerning Cauchy sequences; the first gives an equivalent condition, and the second gives a necessary condition for a sequence to be a Cauchy sequence.

Theorem 3.4.1. *A sequence $\{a_n\}$ is a Cauchy sequence if and only if, for each $\varepsilon > 0$, there is a closed interval I of length $|I| < \varepsilon$ and a positive number $N = N(\varepsilon)$ such that $a_n \in I$ for each $n > N$.*

We shall prove the "only if" part of this theorem; the "if" part is left as Exercise 4 of Problem Set 3.4.

Proof. If the sequence $\{a_n\}$ is a Cauchy sequence, then for each $\varepsilon > 0$ there is a positive number N such that

$$|a_n - a_m| < \varepsilon/3$$

when $n, m > N$. Let $m = N + 1$, then

$$|a_n - a_{N+1}| < \varepsilon/3 \quad \text{or} \quad -\varepsilon/3 < a_n - a_{N+1} < \varepsilon/3$$

for each $n > N$. This implies that

$$a_{N+1} - \varepsilon/3 < a_n < a_{N+1} + \varepsilon/3$$

or

$$a_n \in [a_{N+1} - \varepsilon/3, a_{N+1} + \varepsilon/3].$$

Let $I = [a_{N+1} - \varepsilon/3, \ a_{N+1} + \varepsilon/3]$. Then $|I| < \varepsilon$ and $a_n \in I$ for $n > N$. This completes the proof.

Theorem 3.4.2. *If $\{a_n\}$ is a Cauchy sequence, then it is bounded.*

The proof follows by an argument similar to the proof of Theorem 3.1.6 and is left as Exercise 5 of Problem Set 3.4.

We can now prove that the convergence condition and the Cauchy condition are equivalent; this equivalence is called the **Cauchy criterion.**

Theorem 3.4.3. *A sequence converges if and only if it is a Cauchy sequence.*

Proof. Assume that $\{a_n\}$ is a convergent sequence that converges to A. Let $\varepsilon > 0$. Then there is a positive number N such that

$$|A - a_n| < \varepsilon/2$$

for each $n > N$. If $n, m > N$, then

$$\begin{aligned}
|a_m - a_n| &= |a_m - A + A - a_n| \\
&\leqslant |a_m - A| + |A - a_n| \\
&= |A - a_m| + |A - a_n| \\
&< \varepsilon/2 + \varepsilon/2 = \varepsilon.
\end{aligned}$$

Therefore, $\{a_n\}$ is a Cauchy sequence.

Assume that $\{a_n\}$ is a Cauchy sequence. We need to verify that $\{a_n\}$ is convergent. Let $\varepsilon_1 = 1$; by Theorem 3.4.1, there is a closed interval I_1 of length $|I_1| < 1$ and an N_1 such that $a_n \in I_1$ for each $n > N_1$. Let $\varepsilon_2 = 1/2$. Then, by Theorem 3.4.1, there is a closed interval J_2 of length $|J_2| < 1/2$ and an M_2 such that $a_n \in J_2$ for each $n > M_2$. Let $I_2 = I_1 \cap J_2$ and $N_2 = \max(N_1, M_2)$. Then $N_2 \geqslant N_1$, $I_2 \subseteq I_1$, $|I_2| < 1/2$, and $a_n \in I_2$ for each $n > N_2$. Let $\varepsilon_3 = 1/3$. Then, by Theorem 3.4.1, there is a closed interval J_3 of length $|J_3| < 1/3$ and M_3 such that $a_n \in J_3$ for each $n > M_3$. Let $I_3 = I_2 \cap J_2$ and $N_3 = \max(N_2, M_3)$. Then $N_3 \geqslant N_2$, $I_3 \subseteq I_2$, $|I_3| < 1/3$, and $a_n \in I_3$ for each $n > N_3$. By induction we define a sequence of closed intervals $\{I_n\}$ and a sequence of positive integers $\{N_n\}$ such that $I_{n+1} \subseteq I_n$, $N_{n+1} \geqslant N_n$, and $|I_n| < 1/n$ for each n, and $a_k \in I_n$ for each $k > N_n$. Then $\{I_n\}$ is a nested sequence of closed intervals and by Theorem 3.3.1, there is a unique real number A belonging to the intersection $\bigcap_{n=1}^{\infty} I_n$.

To complete the proof, we need only show that $\{a_n\} \to A$. Assume that $\varepsilon > 0$ is given. Since $|I_n| < 1/n$ there is an n such that $|I_n| < 1/n < \varepsilon$. Since A also belongs to I_n, we have $|A - a_k| < \varepsilon$ for each $k > N_n$. Therefore, $\{a_n\}$ converges to A.

Observe that the Cauchy criterion allows one to prove that a sequence is convergent without finding its limit; we shall see that this is advantageous for many problems.

We next show that every bounded sequence has a convergent subsequence.

Theorem 3.4.4. *Every bounded sequence has a convergent subsequence.*

Proof. We again use the idea of nested intervals to construct such a subsequence.

Assume that $\{c_n\}$ is a bounded sequence. Then there is a closed interval $I_0 = [a_0, b_0]$ such that $c_n \in I_0$ for each n, and $|I_0| = (b_0 - a_0)$. Let

$$I_{1,1} = [a_0, (a_0 + b_0)/2],$$
$$I_{1,2} = [(a_0 + b_0)/2, b_0].$$

At least one of these closed intervals contains c_n for infinitely many values of n. Denote such an interval by $I_1 = [a_1, b_1]$. Then $|I_1| = b_1 - a_1 = (b_0 - a_0)/2$. Let

$$I_{2,1} = [a_1, (a_1 + b_1)/2],$$
$$I_{2,2} = [(a_1 + b_1)/2, b_1].$$

At least one of these closed intervals contains c_n for infinitely many values of n. Denote such an interval by $I_2 = [a_2, b_2]$. Then $|I_2| = b_2 - a_2 = (b_0 - a_0)/2^2$. Continuing the process we define by induction a sequence of closed intervals $\{I_n\}$ such that I_n contains c_k for infinitely many positive integers k, $|I_n| = (b_0 - a_0)/2^n$, and $I_{n+1} \subset I_n$ for each n. Then $\{I_n\}$ is a nested sequence of closed intervals and, by Theorem 3.3.1, there is a unique real number A belonging to the intersection $\bigcap_{n=1}^{\infty} I_n$.

To complete the proof, we need only construct a subsequence of $\{c_n\}$ that converges to A. Let $c_{n_1} \in I_1$. Since I_2 contains c_n for infinitely many values of n, there is an $n_2 > n_1$ such that $c_{n_2} \in I_2$. Since I_3 contains c_n for infinitely many values of n, there is an $n_3 > n_2$ such that $c_{n_3} \in I_3$. By induction we define a sequence of positive integers $\{n_k\}$ such that

$$n_1 < n_2 < \ldots < n_k < \ldots$$

and $c_{n_k} \in I_k$, $k = 1, 2, \ldots$. We now need to show that the subsequence $\{c_{n_k}\}$ converges to A. Assume that $\varepsilon > 0$ is given. Let N be such that $(b_0 - a_0)/2^n < \varepsilon$ when $n > N$. Then for $k > N$, we have $|I_{n_k}| < \varepsilon$ and, since A and c_{n_k} both belong to I_{n_k}, we see that $|A - c_{n_k}| < \varepsilon$. Hence the subsequence $\{c_{n_k}\}$ converges to A.

Theorem 3.4.4 implies that every bounded sequence has a convergent subsequence; this leads to the following definition.

Definition 3.4.2. *Let $\{a_n\}$ be a sequence. The number a is called a* **limit point** *of $\{a_n\}$ if and only if there is a subsequence of $\{a_n\}$ converging to a.*

Theorem 3.4.5. *Let $\{a_n\}$ be a sequence. The point a is a limit point of $\{a_n\}$ if and only if for each pair of positive numbers ε and N there is a $k > N$ such that*

$$|a_k - a| < \varepsilon.$$

Proof. Let $\{a_n\}$ be a sequence. Assume that a is a limit point of $\{a_n\}$. Let ε and N be any pair of positive numbers. Since a is a limit point of $\{a_n\}$, there exists a subsequence $\{a_{n_k}\}$ converging to a. Thus there exists a positive number M such that $k > M$ implies that

$$|a_{n_k} - a| < \varepsilon.$$

For $k > \max{(N, M)}$, we have

$$|a_{n_k} - a| < \varepsilon.$$

Since $\{a_{n_k}\}$ is a subsequence of $\{a_n\}$, we have $n_k \geqslant k > N$. Hence for the pair of positive numbers ε and N, $n_k > N$ and

$$|a_{n_k} - a| < \varepsilon.$$

Assume that for each pair of positive numbers ε and N there is a $k > N$ such that

$$|a_k - a| < \varepsilon.$$

For each positive integer n, let $\varepsilon_n = 1/n$ and $N_n = n$. Then for each n there is a positive integer k_n such that $k_n > N_n = n$ and

$$|a_{k_n} - a| < \varepsilon_n = 1/n.$$

Thus the sequence $\{a_{k_n}\}$ is a subsequence of $\{a_n\}$ converging to a. Hence a is a limit point of the sequence $\{a_n\}$.

Corollary 3.4.5. *Let $\{a_n\}$ be a sequence. The point a is a limit point of $\{a_n\}$ if and only if for each $\varepsilon > 0$*

$$|a_n - a| < \varepsilon$$

for infinitely many values of n.

The following theorem points out the relationship between the limit points of a sequence and the convergence of the sequence. The proof is left as Exercise 12 of Problem Set 3.4.

Theorem 3.4.6. *A bounded sequence converges if and only if it has exactly one limit point.*

Any bounded sequence has at least one limit point, one if the sequence is convergent and more otherwise. Two of the limit points of a sequence have special names—the *limit superior* and the *limit inferior*; the remainder of this section deals with these special limit points.

Let $\{a_n\}$ be a bounded sequence. For each positive integer n, let A_n be the set

$$A_n = \{a_n, a_{n+1}, ...\}.$$

Since the sequence $\{a_n\}$ is bounded, each set A_n is bounded and the Completeness Property ensures that each set A has a least upper bound and a greatest lower bound. Let

$$\lambda_n = \sup A_n \quad \text{and} \quad \gamma_n = \inf A_n.$$

Then the sequence $\{\lambda_n\}$ is a decreasing sequence and the sequence $\{\gamma_n\}$ is an increasing sequence. (Cf. Exercise 22 of Problem Set 3.4.) Since $\lambda_n \geqslant \gamma_n$ for each positive integer n, it follows that the sequences $\{\lambda_n\}$ and $\{\gamma_n\}$ both converge. The limits of these sequences are called the *limit superior* and the *limit inferior*.

Definition 3.4.3. *The **limit superior** of the sequence $\{a_n\}$, denoted by*

$$\lim_n \sup a_n \quad \text{or} \quad \overline{\lim_n} \, a_n,$$

is the $\lim_n \lambda_n$. *The **limit inferior** of the sequence $\{a_n\}$, denoted by*

$$\lim_n \inf a_n \quad \text{or} \quad \underline{\lim_n} \, a_n,$$

is the $\lim_n \gamma_n$.

The limit superior is sometimes referred to as the **upper limit** or **greatest limit**, while other names for the limit inferior are **lower limit** and **least limit**.

Example 3.4.1. Consider the sequence $\{(-1)^n (1 + 1/n)\}$. The limit points of this sequence are 1 and -1. The set A_n is

$$A_n = \{(-1)^n (1 + 1/n), (-1)^{n+1} (1 + 1/(n + 1)), ...\}.$$

Then

$$\lambda_n = \sup A_n = \begin{cases} (-1)^n (1 + 1/n) & n \text{ even} \\ (-1)^{n+1} (1 + 1/(n + 1)) & n \text{ odd} \end{cases}$$

and

$$\gamma_n = \inf A_n = \begin{cases} (-1)^n (1 + 1/n) & n \text{ odd} \\ (-1)^{n+1} (1 + 1/(n + 1)) & n \text{ even}. \end{cases}$$

Then

$$\lim_n \sup \, (-1)^n \, (1 + 1/n) = \lim_n \lambda_n = 1$$

and

$$\lim_n \inf \, (-1)^n \, (1 + 1/n) = \lim_n \gamma_n = -1.$$

Theorem 3.4.7. *Let $\{a_n\}$ be a bounded sequence. Let S be the set of all limit points of $\{a_n\}$. Then*

$$\lim_n \sup a_n = \sup S,$$
$$\lim_n \inf a_n = \inf S.$$

We shall prove the statement for the limit superior; the statement for the limit inferior follows in an analogous manner and is left as Exercise 27 of Problem Set 3.4.

Proof. Let $\{a_n\}$ be a bounded sequence, and let S be the set of all limit points of $\{a_n\}$. Let

$$\Lambda = \lim_n \sup a_n = \lim_n \lambda_n$$

(λ_n was defined in the paragraph preceding Definition 3.4.3) and let

$$L = \sup S.$$

We first show that $L \leqslant \Lambda$. For each $\varepsilon > 0$ there exists a positive number N such that $n > N$ implies that

$$|\Lambda - \lambda_n| < \varepsilon.$$

Then, for $n > N$, we have $\lambda_n < \Lambda + \varepsilon$. Since

$$\lambda_n = \sup \{a_n, a_{n+1}, ...\},$$

$a_n < \Lambda + \varepsilon$ for all $n > N$. Hence any convergent subsequence of $\{a_n\}$ must have its limit less than or equal to Λ and $\sup S = L \leqslant \Lambda$.

We now show that $\Lambda \leqslant L$. Since $L = \sup S$, for each $\varepsilon > 0$, there is a limit point a of $\{a_n\}$ in S such that $a > L - \varepsilon/2$. Since a is a limit point of $\{a_n\}$, there exists a subsequence $\{a_{n_k}\}$ of $\{a_n\}$ converging to a, and so there exists a positive number K such that $k > K$ implies that

$$|a_{n_k} - a| < \varepsilon/2.$$

Hence, for $k > K$,

$$a_{n_k} > a - \varepsilon/2 > L - \varepsilon.$$

Thus $\lambda_{n_k} \geqslant L - \varepsilon$ for $k > K$. Since $\{\lambda_{n_k}\}$ is a subsequence of the convergent sequence $\{\lambda_n\}$, we get

$$\Lambda = \lim_n \lambda_n = \lim_k \lambda_{n_k} \geqslant L.$$

Therefore $\Lambda = L$, which completes the proof.

One of the important properties of the limit superior and limit inferior is given in the following theorem. Roughly, it says that for each $\varepsilon > 0$ all but a finite number of the a_n's lie between $\lim_n \inf a_n - \varepsilon$ and $\lim_n \sup a_n + \varepsilon$.

Theorem 3.4.8. *Let $\{a_n\}$ be a bounded sequence and $\varepsilon > 0$. Then there is a positive number $N = N(\varepsilon)$ such that*

$$a_n < \lim_n \sup a_n + \varepsilon$$

and

$$a_n > \lim_n \inf a_n - \varepsilon$$

for all $n > N$.

We prove the first statement and leave the second as Exercise 28 of Problem Set 3.4.

Proof. Assume, to the contrary, that the first statement is not true. That is, assume that for some $\varepsilon > 0$ there is a subsequence $\{a_{n_k}\}$ of the sequence $\{a_n\}$ such that

$$a_{n_k} > \lim_n \sup a_n + \varepsilon$$

for each positive integer k. Then the subsequence $\{a_{n_k}\}$ has a convergent subsequence that converges to some real number $a \geqslant \lim_n \sup a_n + \varepsilon$. Then $a \in S$ and $a \geqslant \sup S + \varepsilon$. This is a contradiction and the desired result follows.

We now establish a relation between the limit of a sequence and the limit inferior and limit superior of a sequence.

Theorem 3.4.9. *Let $\{a_n\}$ be a bounded sequence. Then $\{a_n\}$ is convergent if and only if*

$$\lim_n \sup a_n = \lim_n \inf a_n.$$

Proof. Assume that the sequence $\{a_n\}$ converges to a. Then the set S of limit points is $\{a\}$ and $\sup S = \inf S = a$.

Assume that $\lim_n \sup a_n = \lim_n \inf a_n$. Let $\varepsilon > 0$. Then, by Theorem 3.4.2, there is a positive number N such that

$$\lim_n \inf a_n - \varepsilon < a_n < \lim_n \sup a_n + \varepsilon$$

for each $n > N$. Since $\lim_n \sup a_n = \lim_n \inf a_n$, we can rewrite the above statement as

$$|a_n - \lim_n \sup a_n| < \varepsilon$$

for each $n > N$. Hence $\{a_n\}$ converges to $\lim_n \sup a_n$.

Problem Set 3.4

1. Without the aid of Theorem 3.4.3, prove that the sequence $\{1/n\}$ is a Cauchy sequence.
2. Without the aid of Theorem 3.4.3, prove that the sequence $\{a^n\}$, $0 < a < 1$, is a Cauchy sequence.
3. Let $\{a_n\}$ and $\{b_n\}$ be Cauchy sequences. Without the aid of Theorem 3.4.3, prove that the sequences $\{a_n + b_n\}$ and $\{a_n b_n\}$ are Cauchy sequences.
4. Prove the "if" part of Theorem 3.4.1.
5. Prove Theorem 3.4.2.
6. If $\{a_n\} \to +\infty$ and $\{b_n\} \to +\infty$, prove that $\{a_n + b_n\} \to +\infty$ and $\{a_n b_n\} \to +\infty$.
7. If $\{a_n\} \to -\infty$ and $\{b_n\} \to -\infty$, prove that $\{a_n + b_n\} \to -\infty$ and $\{a_n b_n\} \to +\infty$.

For exercises 8–10, give an example of sequences $\{a_n\}$ and $\{b_n\}$ such that $\{a_n\} \to +\infty$ and $\{b_n\} \to -\infty$, and which satisfies the condition given in the exercise.

8. $\{a_n + b_n\} \to +\infty$.
9. $\{a_n + b_n\} \to A$, where A is any real number.
10. $\{a_n + b_n\} \to -\infty$.
11. Give an example of a bounded sequence with exactly three limit points.
12. Prove Theorem 3.4.6.

If the sequence $\{a_n\}$ is not bounded above (below), then $+\infty$ $(-\infty)$ is said to be a limit point of $\{a_n\}$. For Exercises 13–16, give examples of each of the following.

13. A sequence $\{a_n\}$ whose only limit point is $+\infty$.
14. A sequence $\{a_n\}$ whose only limit point is $-\infty$.
15. A sequence $\{a_n\}$ whose only limit points are $+\infty$ and $-\infty$.
16. A sequence $\{a_n\}$ whose only limit points are $+\infty$, $-\infty$, and 0.
17. Show that a sequence diverges to $+\infty$ $(-\infty)$ if and only if its only limit point is $+\infty$ $(-\infty)$.

For Exercises 18–21 find the limit superior and limit inferior of the given sequences.

18. $1, 3, -1, 1, 3, -1, 1, 3, -1, \ldots$.
19. $\{\cos(n\pi/2)\}$.

20. $\{(1 - 1/n)\}$.

21. $\{(1 - 1/n) \sin (n\pi/2)\}$.

22. Let $\{a_n\}$ be a bounded sequence. For each positive integer n, let $A_n = \{a_n, a_{n+1}, ...\}$, $\lambda_n = \sup A_n$, and $\gamma_n = \inf A_n$. Show that $\{\lambda_n\}$ is a decreasing sequence and that $\{\gamma_n\}$ is an increasing sequence. Show that both $\{\lambda_n\}$ and $\{\gamma_n\}$ are convergent sequences.

23. Let $\{a_n\}$ be a bounded sequence. If

$$\lim_n \sup a_n = a,$$

show that for each pair of positive numbers ε and N there is $k > N$ such that

$$a_k > a - \varepsilon.$$

24. Let $\{a_n\}$ be a bounded sequence. If

$$\lim_n \inf a_n = a,$$

show that for each pair of positive numbers ε and N there is a $k > N$ such that

$$a_k < a + \varepsilon.$$

25. Let $\{a_n\}$ be a bounded sequence. Prove that $\{a_n\}$ converges if and only if its limit superior and limit inferior are equal.

26. If $\{a_n\}$ is a bounded sequence, show that there is a subsequence of $\{a_n\}$ converging to $\lim_n \sup a_n$ and a subsequence of $\{a_n\}$ converging to $\lim_n \inf a_n$.

27. Prove the second part of Theorem 3.4.7.

28. Prove the second part of Theorem 3.4.8.

29. Let $\{a_n\}$ be a bounded sequence with $\lim_n \sup a_n = a$. If $\{a_{n_k}\}$ is any subsequence of $\{a_n\}$,show that

$$\lim_k \sup a_{n_k} \leqslant a.$$

30. Let $\{a_n\}$ be a bounded sequence with $\lim_n \inf a_n = a$. If $\{a_{n_k}\}$ is any subsequence of $\{a_n\}$, show that

$$\lim_k \inf a_{n_k} \geqslant a.$$

31. Give an example of bounded sequences $\{a_n\}$ and $\{b_n\}$ such that

$$\lim_n \sup(a_n + b_n) \neq \lim_n \sup a_n + \lim_n \sup b_n.$$

32. Give an example of bounded sequences $\{a_n\}$ and $\{b_n\}$ such that

$$\lim_n \inf(a_n + b_n) \neq \lim_n \inf a_n + \lim_n \inf b_n.$$

33. Give an example of bounded sequences $\{a_n\}$ and $\{b_n\}$ such that

$$\lim_n \sup(a_n b_n) \neq (\lim_n \sup a_n)(\lim_n \sup b_n).$$

34. Give an example of a bounded sequence $\{a_n\}$ and $\{b_n\}$ such that

$$\lim_n \inf(a_n b_n) \neq (\lim_n \inf a_n)(\lim_n \inf b_n).$$

35. Let $\{a_n\}$ and $\{b_n\}$ be bounded sequences. Prove that

$$\lim_n \sup(a_n + b_n) \leqslant \lim_n \sup a_n + \lim_n \sup b_n.$$

36. Let $\{a_n\}$ and $\{b_n\}$ be bounded sequences. Prove that

$$\lim_n \inf(a_n + b_n) \geqslant \lim_n \inf a_n + \lim_n \inf b_n.$$

37. Let $\{a_n\}$ and $\{b_n\}$ be nonnegative bounded sequences. Prove that

$$\lim_n \sup(a_n\, b_n) \leqslant (\lim_n \sup a_n)(\lim_n \sup b_n).$$

38. Let $\{a_n\}$ and $\{b_n\}$ be nonnegative bounded sequences. Prove that

$$\lim_n \inf(a_n\, b_n) \geqslant (\lim_n \inf a_n)(\lim_n \inf b_n).$$

39. Let $\{a_n\}$ be a sequence of real numbers. Explain what is meant by $\lim_n \sup a_n = +\infty\ (-\infty)$ and $\lim_n \inf a_n = -\infty\ (+\infty)$. Show how the word "bounded" can be omitted in Exercises 29–38.

3.5. Open and Closed Sets

In analysis we are interested in results that can be established for more general sets than open and closed intervals; these are *open* and *closed* sets.

> **Definition 3.5.1.** *Let S be a set of real numbers. Then S is said to be* **open** *if and only if for each $x \in S$, there is a neighborhood T of x such that $T \subseteq S$.*

Some examples of open sets are the following:

1. Any open interval.
2. Any set that is the union of open intervals.
3. Any set that is the union of open sets.
4. Any set that is the intersection of a finite family of open sets.

It is left as an exercise to show that the above sets are open (Exercises 1–5 of Problem Set 3.5).

> **Definition 3.5.2.** *Let S be a set of real numbers. Then S is said to be* **closed** *if and only if it contains all of its limit points.*

Some examples of closed sets are the following:

1. Any closed interval.
2. Any set that is the intersection of closed intervals.
3. Any set that is the intersection of closed sets.
4. Any set that is the union of a finite family of closed sets.

It is left as an exercise to show that the above sets are closed (Exercises 7–11 of Problem Set 3.5).

Observe that the infinite intersection of open sets need not be open, since

$$\bigcap_{n=1}^{\infty} (-1 - 1/n,\, 1 + 1/n) = [-1, 1],$$

which is not open. Also, infinite unions of closed sets need not be closed, since

$$\bigcup_{n=1}^{\infty} [-1 + 1/n, \, 1 - 1/n] = (-1, 1),$$

which is not closed.

The set of real numbers R and the empty set $\emptyset$ are the only sets of real numbers that are both open and closed. The set of rational numbers is a set that is neither open nor closed. Also, half-open intervals are neither open nor closed.

The following theorem establishes an important relationship between open and closed sets.

Theorem 3.5.1. *A set of real numbers is open if and only if its complement is closed.*

Equivalently, we can state this theorem as: *A set of real numbers is closed if and only if its complement is open.*

Proof. Assume that S is an open set. We need to show $\mathbf{C}(S)$ contains all its limit points. Let x be a limit point of $\mathbf{C}(S)$ that does not belong to $\mathbf{C}(S)$. Then $x \in S$, and since S is open there is a neighborhood T of x such that $T \subseteq S$. Thus T contains no points of $\mathbf{C}(S)$ and x cannot be a limit point of $\mathbf{C}(S)$. This is a contradiction, and so there are no limit points of $\mathbf{C}(S)$ that do not belong to $\mathbf{C}(S)$. Therefore, $\mathbf{C}(S)$ is closed.

Assume that $\mathbf{C}(S)$ is a closed set. We need to show that S is open. Let $x \in S$. Since $\mathbf{C}(S)$ is closed, x is not a limit point of $\mathbf{C}(S)$. Hence there exists a neighborhood T of x such that T contains no points of $\mathbf{C}(S)$. Thus $T \subseteq S$. Therefore, S is open.

Two sets S and T are said to be **disjoint** if and only if $S \cap T = \emptyset$. The family of sets $\{A_\lambda : \lambda \in \Lambda\}$ is said to be **pairwise disjoint** if and only if, for each pair, $\lambda, \mu \in \Lambda$ such that $\lambda \neq \mu$ implies that $A_\lambda \cap A_\mu = \emptyset$. The following theorem describes the structure of any open set.

Theorem 3.5.2. *Let $S \neq \emptyset$ be an open set of real numbers. Then S is the countable union of pairwise disjoint open intervals.*

Proof. Let x be any point of S. Let $U = \{y : (x, y) \subset S\}$. Since S is an open set, there is at least one open interval I containing x such that $I \subseteq S$; hence $U \neq \emptyset$. If U has an upper bound, then U has a least upper bound, say b_x. Then $(x, b_x) \subset S$. If U does not have an upper bound, then every real number greater than x belongs to S. In a similar manner, if $L = \{x : (z, x) \subset S\}$, then L has a greatest lower bound a_x and $(a_x, x) \subset S$ or L has no greatest lower bound and every real number less than x is in S. We then have an interval $I_x = (a_x, b_x)$, which may be unbounded, such that $x \in I_x$ and $I_x \subseteq S$. We see that $S \subseteq \bigcup_{x \in S} I_x$.

We must now demonstrate that $S = \bigcup_{x \in S} I_x$ and that there are only countably many such intervals. We first prove that for $x, y \in S$ and $x \neq y$, either $I_x = I_y$ or $I_x \cap I_y = \varnothing$. We assume, to the contrary, that $I_x \neq I_y$ and $I_x \cap I_y \neq \varnothing$. Since $I_x \cap I_y \neq \varnothing$, the set $I_x \cup I_y$ is an interval. Since $I_x \neq I_y$, either there is a $w \in I_x$ such that $w \notin I_y$ or there is $z \in I_y$ such that $z \notin I_x$. Without loss of generality, we may assume that there is a $z \in I_y$ such that $z \notin I_x$. Then I_x is a proper subset of $I_x \cup I_y$. Then either there is a $u \in U$ with $u > b_x$, or a $v \in L$ with $v < a_x$. In either case, we have a contradiction to the definition of U or L. Hence, either $I_x = I_y$ or $I_x \cap I_y = \varnothing$. Thus, S is the union of pairwise disjoint open intervals. Since each open interval contains a rational number, there is a 1–1 mapping from this set of open intervals into the set of rational numbers. Hence there are only countably many open intervals, which concludes the proof.

Utilizing the concepts of complements, Theorem 3.5.2 can be stated equivalently as follows.

Corollary 3.5.2. *Every closed set is the complement of a finite or denumerable number of disjoint open intervals.*

With each set of real numbers there is associated a closed set of real numbers, called its *closure* and defined as follows:

Definition 3.5.3. *Let S be a set of real numbers. The **closure** $\bar{S}$ of S is the intersection of all closed sets containing S.*

We list some of the properties of closures and leave their proofs as Exercises 22 and 23 of Problem Set 3.5.

Theorem 3.5.3. *The point $x \in \bar{S}$ if and only if $x \in S$ or x is a limit point of S.*

Theorem 3.5.4. *Let S and T be any sets of real numbers. Then*

(a) $S \subseteq \bar{S}$.

(b) $\overline{S \cup T} = \bar{S} \cup \bar{T}$.

(c) $\bar{\bar{S}} = \bar{S}$

We now define a special type of closed set. A set is said to be compact if and only if it is closed and bounded. We extend the Nested Interval Theorem and the Bolzano–Weirstrass Theorem to compact sets, and use these to prove the Heine–Borel Covering Theorem.

Definition 3.5.4. *Let $\{S_n\}$ be a sequence of sets. Then $\{S_n\}$ is said to be **decreasing** if and only if $S_{n+1} \subseteq S_n$ for each positive integer n.*

The following theorem, called the **Cantor Intersection Theorem**, is an extension of the nested interval theorem. The theorem is named in honor of the German mathematician G. Cantor (1845–1918), who first discovered it. Cantor is best known for his work in set theory, which he developed during the years 1874–1895.

Theorem 3.5.5. *If $\{S_n\}$ is a decreasing sequence of nonempty compact sets, then the set*

$$S = \bigcap_{n=1}^{\infty} S_n$$

is nonempty.

Proof. Since $\{S_n\}$ is decreasing, $S_{n+1} \subseteq S_n$ for each positive integer n. Since $S_n \neq \varnothing$, S_n contains at least one element; we label one of these elements by x_n. Then $x_k \in S_n$ for each $k \geq n$ and $x_n \in S_1$ for each positive integer n. Since S_1 is bounded, the sequence $\{x_n\}$ has a convergent subsequence, say, $\{x_{n_k}\}$ with limit x. We now show that $x \in S$. Consider any fixed positive integer n. If $x = x_m$ for $m \geq n$, then $x \in S_n$. If for every $m \geq n$, $x \neq x_m$, then x is a limit point of S_n since we have a subsequence of $\{x_n\}$ converging to x. The set S_n is closed, and so $x \in S_n$. Thus $x \in S_n$ for each positive integer n and $x \in S = \bigcap_{n=1}^{\infty} S_n$. Therefore, S is nonempty.

The following theorem is an extension of the Bolzano–Weierstrass Theorem. The proof is similar to that of Theorem 3.3.3 and is left as Exercise 26 of Problem Set 3.5.

Theorem 3.5.6. *Let T be a compact set. If S is an infinite subset of T, then S has a limit point in T.*

Let $\mathfrak{C}$ be a family of sets. Then $\mathfrak{C}$ is said to **cover** the set S if and only if, for each $x \in S$, there is an $F \in \mathfrak{C}$ such that $x \in F$, that is, $S \subseteq \bigcup_{F \in \mathfrak{C}} F$. If all of the sets of the cover family $\mathfrak{C}$ are open, then the family $\mathfrak{C}$ is called an **open cover**.

Example 3.5.1. Let N be the set of positive integers and

$$\mathfrak{C} = \{(n - 1/n, n + 1/n) : n \in N\}.$$

Then for each positive integer k, $k \in (k - 1/k, k + 1/k)$ and so $\mathfrak{C}$ is a cover of N.

Example 3.5.2. Let $I = (0, 1)$ and

$$\mathfrak{C} = \{(1/2n, 3/2n): n \in N\}.$$

Then for each $x \in I$; there is a positive integer k such that $1/2k < x < 3/2k$ and $x \in (1/2n, 3/2n)$. Thus $\mathfrak{C}$ is an open cover of I.

We now state and prove the important **Heine–Borel Covering Theorem**. This theorem is named to honor the German mathematician E. Heine (1821–1881) and the French mathematician E. Borel (1871–1956).

Theorem 3.5.7. *A nonempty set S is compact if and only if for each open cover $\mathfrak{D}$ of S there is a finite collection of sets in $\mathfrak{D}$ that cover S.*

Proof. Assume that S is a nonempty compact set and that $\mathfrak{D}$ is an open cover of S. Then S contains its greatest lower bound l and its least upper bound u. (Cf. Exercise 18 of Problem Set 3.5.) Define the set X as follows:

$X = \{x: x \in S$ and $S \cap [l, x]$ may be covered by a finite number of open sets of $\mathfrak{D}\}$.

The set $X \neq \varnothing$ since $l \in X$, and X is bounded since $X \subseteq S$. Thus X has a least upper bound, say w. We first show $w \in X$. Assume $w \notin X$. Then there is an open set $O \in \mathfrak{D}$ such that $w \in O$. But $\mathfrak{D}$ must contain points $y \in X$ such that $y < w$, for otherwise w would not be the least upper bound of X. By the definition of X, the set $S \cap [l, y]$ is covered by a finite number of open sets of $\mathfrak{D}$, say $O_1, O_2, ..., O_n$. Then $S \cap [l, w]$ is covered by the $n + 1$ sets $O_1, O_2, ..., O_n, O$. Thus $w \in X$. We next show that $w = u$. Assume that $w < u$. Then every open set of $\mathfrak{D}$ that contains w contains points larger than w. If each open set of $\mathfrak{D}$ that contains w does not contain points of S, then the set

$$Y = \{y: y \in S \text{ and } y > w\}$$

has a greatest lower bound $v > w$. Since S is closed, we get $v \in S$ and the set $X \cup \{v\}$ can be covered by a finite number of open sets of $\mathfrak{D}$. Hence $v \in X$, which is a contradiction. Thus, some open sets of $\mathfrak{D}$ containing w also contain a point z of S with $z > w$. But then $z \in X$ and w is not the least upper bound of X, which is a contradiction. In either case we are led to a contradiction. Therefore, $w = u$ and S can be covered by a finite number of open sets of $\mathfrak{D}$.

To prove the converse, we assume that for every open cover $\mathfrak{O}$ of S there is a finite collection of sets in $\mathfrak{O}$ that cover S and we assume that S is not compact. Since S is not compact, there is a sequence $\{x_n\}$ of distinct points in S such that no subsequence of $\{x_n\}$ converges to a point in S. For each positive integer n, let $X_n = \{x_{n+1}, x_{n+2}, \ldots\}$ and let $U_n = \mathbf{C}(\overline{X}_n)$. Then the collection of sets $\{U_n\}$ is an open cover for the set S, but there is no finite collection of sets in $\{U_n\}$ that cover S since $x_n \notin U_n$ for each positive integer n. Therefore the set S is compact.

Problem Set 3.5

1. Prove that any open interval is an open set.
2. Prove that the union of two open intervals is an open set.
3. Prove that the union of open sets is an open set.
4. Prove that the intersection of two open sets is an open set.
5. Prove that the intersection of a finite family of open sets is an open set.
6. Give an example, other than the one given in the text, to show that a countable intersection of open sets need not be an open set.
7. Prove that any closed interval is a closed set.
8. Prove that the intersection of two closed sets is a closed set.
9. Prove that the intersection of closed sets is a closed set.
10. Prove that the union of two closed sets is a closed set.
11. Prove that the union of a finite family of closed sets is a closed set.
12. Give an example, other than one given in the text, to show that a countable union of closed sets need not be a closed set.
13. Show that $\varnothing$ and R are both open and closed.
14. Show that the set of rational numbers is neither open nor closed.
15. For any set of real numbers S, let S' be the set of limit points of S. Show that S' is closed.
16. Give an example of a nonempty set S for which $S' = S$. (Cf. Exercise 15.)
17. Show that every uncountable set of real numbers has at least one limit point.
18. Let S be a nonempty compact set. Prove that sup S and inf S both belong to S.
19. Let $S_1, S_2, \ldots, S_k$ be compact sets. Prove that

$$\bigcup_{j=1}^{k} S_j \text{ and } \bigcap_{j=1}^{k} S_j$$

are both compact.
20. A set S is said to be **isolated** if and only if $S' \cap S = \varnothing$. Prove that every isolated set is countable.
21. Show that the set S is isolated if and only if every point of S is isolated. (Cf. Exercise 8 of Problem Set 3.3.)
22. Prove Theorem 3.5.3.

23. Prove Theorem 3.5.4.

24. Prove that a set S is compact if and only if every sequence of points of S contains a subsequence converging to a point S.

25. A set $S \subseteq R$ is said to be **totally bounded** if and only if, for each $\varepsilon > 0$, there is a finite set of real numbers $x_1, x_2, ..., x_n$ such that for each $x \in S$ at least one $x_k, k = 1, 2, ..., n$, satisfies $|x - x_k| < \varepsilon$. Prove that a set of real numbers is bounded if and only if it is totally bounded.

26. Prove Theorem 3.5.6.

27. Let S be a compact set, A a dense subset of S, and $\delta > 0$. Prove that there is a finite subset $\{x_1, x_2, ..., x_n\}$ of A such that the set $I = \{I_{x_i} : I_{x_i} = \{x : |x - x_i| < \delta\}\ i = 1, 2, ..., n\}$ covers S.

28. If S is a set of real numbers covered by a family $\mathcal{D}$ of open sets, prove that there is a countable subfamily of $\mathcal{D}$ that covers S.

Chapter Four

LIMITS AND CONTINUITY

4.1. Limits

Most students have had some experience with limits in their beginning calculus course. They usually learn some intuitive notions about limits, such as *the limit of the function f at $x = a$ is L if the values $f(x)$ are close to L as x approaches a,* together with some rules such as *the limit of the sum is the sum of the limits* or *the limit of the product is the product of the limits.* Notions and rules of this type are helpful in learning how to work with limits of many elementary functions, but they are not too valuable in really understanding the underlying theory of analysis. A thorough comprehension of the ideas and properties of limits is required to assimilate completely the concepts of continuity, differentiation, and integration.

As indicated above, an intuitive idea of "the limit of f at $x = a$ is L" is "$f(x)$ gets close to some number L as x approaches a." We now give a precise definition of the limit of a function at a point. Let f map a set D into the set of real numbers R. For the statement "x approaches a" or "x gets close to a" to have meaning, every neighborhood of a must contain points of D other than a itself, that is, a must be a limit point of D. Also, this statement should depend on values of x "near" a and not necessarily at a itself. Hence a need not belong to the domain D of the function. That is, the function need not be defined at a.

Definition 4.1.1. *Let $f: D \to R$, let a be a limit point of D, and let L be a real number. Then f is said to have **limit** L **at** a if and only if to each $\varepsilon > 0$ there is a $\delta = \delta(\varepsilon) > 0$ such that*

$$|f(x) - L| < \varepsilon$$

whenever $x \in D$ and $0 < |x - a| < \delta$. The limit of f at a is denoted b,

$$\lim_{x \to a} f(x) = L.$$

If there is a real number L such that

$$\lim_{x \to a} f(x) = L,$$

then we say that the limit of f at a exists.

The symbol "$\delta(\varepsilon)$" is standard notation for the value of the function δ at ε but in the above definition we also use the latter δ to represent a number. We shall continue to use the symbol δ to represent a number, but think o "δ as a function of ε" if it is used in the context of the above definition.

To aid in our discussion of limits, we introduce the notion of a *deleted neighborhood*. If T is a neighborhood of the point a, then $T \setminus \{a\}$, T with the point a removed, is called a **deleted neighborhood** of a. An open interval containing a is a neighborhood of a, while the interval I with the point a removed is a deleted neighborhood of a.

The concept of limit of a function is a generalization of the notion of limit of a sequence. Thus it is not surprising that many of the results on limits o functions are similar to results on limits of sequences. This is illustrated in the following theorem for constant functions.

Theorem 4.1.1. *Let $f: D \to R$, and let a be a limit point of D. If there is a deleted neighborhood T of a such that $f(x) = c$, where c is a real number, for each $x \in D \cap T$, then*

$$\lim_{x \to a} f(x) = c.$$

Proof. Let $f(x) = c$ for each x in $D \cap T$. Then $T^* = T \cup \{a\}$ is a neighborhood of a and there is a $\delta > 0$ such that

$$(a - \delta, a + \delta) \subseteq T^*.$$

(Cf. Exercise 5 of Problem Set 3.3.) Then for $x \in D$ with $0 < |x - a| < \delta$, we have

$$|f(x) - c| = 0.$$

Therefore

$$\lim_{x \to a} f(x) = c.$$

In working with limit problems, we must first decide whether or not the function $f: D \to R$ has a limit at the given point a, where a is a limit point of D. If we think the function has a limit at a, say L, we then check the definition to see if L is actually the limit of the function at a. If L satisfies the conditions of the definition, then we have found the limit, and the function has limit L at a. If L does not satisfy the conditions for the limit, we must show that some other real number is the limit or prove that the function does not have a limit at a. The following two examples illustrate this process.

Example 4.1.1. Consider the function $f(x) = x^2$. Then the domain of this function is the set of real numbers. We shall check to see if f has a limit at 3. Since the domain of f is the set of all real numbers 3 is a limit point of the domain of f. If we select values of x near 3, for example, $x = 3.01$ and $x = 3.001$, then $f(3.01) = 9.0601$ and $f(3.001) = 9.006001$. These two numbers are near 9 (the value of f at 3), and we select 9 as our candidate for the limit of $f(x) = x^2$ at $x = 3$. Then

$$|f(x) - L| = |x^2 - 9| = |(x + 3)(x - 3)| = |x + 3||x - 3|.$$

Since we are interested only in values of x near 3, we may assume that $2 < x < 4$, that is, $|x - 3| < 1$. Then

$$5 < |x + 3| < 7.$$

Thus,

$$|f(x) - L| = |x^2 - 9| \leqslant 7|x - 3|.$$

Let $\varepsilon > 0$ be given. If we choose $\delta = \min(1, \varepsilon/7)$, then, for $|x - 3| < \delta$, we have

$$|f(x) - L| = |x^2 - 9| \leqslant 7|x - 3| < 7 \cdot \varepsilon/7 = \varepsilon.$$

Therefore, $|x^2 - 9| < \varepsilon$ when $|x - 3| < \delta$ and

$$\lim_{x \to 3} x^2 = 9.$$

Example 4.1.2. As a second example, we consider the function $f(x) = (3x - 1)/(x + 2)$ for $x \in R \setminus \{-2\}$. We shall check if f has a limit at 2. Since the number 2 is a limit point of the domain of f, we look for a candidate for the limit of f at 2. If x is near 2, then $f(x)$ appears to be near 5/4 (the value of f at 2), and so our candidate might be 5/4. Then

$$|f(x) - L| = \left|\frac{3x - 1}{x + 2} - \frac{5}{4}\right| = \left|\frac{7x - 14}{4(x + 2)}\right| = \frac{7}{4}\left|\frac{x - 2}{x + 2}\right|.$$

Since we are interested only in values of x near 2, we may assume that $1 < x < 3$, that is, $|x - 2| < 1$. Then

$$3 < |x + 2| < 4.$$

Thus,

$$|f(x) - L| = \frac{7}{4}\left|\frac{x - 2}{x + 2}\right| \leqslant 7|x - 2|/12.$$

Let $\varepsilon > 0$ be given. If we choose $\delta = \min(1, 12\,\varepsilon/7)$, then, for $|x - 2| < \delta$, we have

$$|f(x) - L| \leqslant 7|x - 2|/12 < 7(12\,\varepsilon/7)/12 = \varepsilon.$$

Therefore,

$$\left|\frac{3x - 1}{x + 2}\right| < \varepsilon$$

when $|x - 2| < \delta$, and

$$\lim_{x \to 2} \frac{3x - 1}{x + 2} = \frac{5}{4}.$$

Examples 4.1.1 and 4.1.2 demonstrate the procedure for proving that a function has a limit at a given point. We can prove that the function $f: D \to R$ does not have a limit at the point a, where a is a limit point of D, by showing that for each real number L we can find a real number $\varepsilon_L > 0$ such that for each $\delta > 0$ there is a real number $x_\delta \in D$ such that

$$|x_\delta - a| < \delta$$

and

$$|f(x_\delta) - L| \geqslant \varepsilon_L.$$

This procedure is illustrated by the following example.

Example 4.1.3. Let

$$f(x) = \begin{cases} 1 & \text{if } 0 \leqslant x, \\ -1 & \text{if } x < 0. \end{cases}$$

We shall show that the function f does not have a limit at 0. Let L be any real number and let $\varepsilon_L = \varepsilon = 1$. We consider two cases, first where $L \geqslant 0$, and second where $L < 0$.

CASE 1. $(L \geqslant 0)$. Let δ be any positive real number. Let x_δ be any real number such that $-\delta < x_\delta < 0$. Since $L \geqslant 0$ and $f(x_\delta) = -1$, we have

$$|f(x_\delta) - L| = |-1 - L| = 1 + L \geqslant 1 = \varepsilon.$$

CASE 2. ($L < 0$). Let δ be any positive real number. Let x_δ be any real number such that $0 < x_\delta < \delta$. Since $L < 0$ and $f(x_\delta) = 1$, we have

$$|f(x_\delta) - L| = |1 - L| = 1 + |L| > 1 = \varepsilon.$$

Thus, in either case, for any $\delta > 0$ there is x_δ such that

$$0 < |x_\delta - 0| < \delta$$

and

$$|f(x_\delta) - L| \geqslant 1 = \varepsilon.$$

Therefore, the limit of the function f does not exist at 0.

If $a \neq 0$, then the function f is constant on a neighborhood of a, and by Theorem 4.1.1 the function f has limit 1 if $a > 0$ and limit -1 if $a < 0$.

The following theorem establishes that, *if a function f has a limit at a point, then that limit is unique.* The proof is similar to the proof of Theorem 3.1.1 and is left as Exercise 18 Problem Set 4.1.

Theorem 4.1.2. *If*

$$\lim_{x \to a} f(x)$$

exists, then it is unique; that is, if

$$\lim_{x \to a} f(x) = L \quad and \quad \lim_{x \to a} f(x) = M,$$

then $L = M$.

We now consider two special types of limits, namely, the *limit from the left* and the *limit from the right*, that is, *one-sided limits*, and exhibit the relation between these special limits and the concept of limit.

Definition 4.1.2. *Let $f : D \to R$ and let a be a limit point of $D \cap (a, +\infty)$. Then the real number L is said to be the **limit from the right** or the **right-hand limit** of f at a if and only if to each $\varepsilon > 0$ there is $\delta = \delta(\varepsilon) > 0$ such that*

$$|f(x) - L| < \varepsilon$$

whenever $x \in D$ and $a < x < a + \delta$. The limit from the right is denoted by

$$\lim_{x \to a+} f(x) = L.$$

Definition 4.1.3. *Let $f : D \to R$, and let a be a limit point of $D \cap (-\infty, a)$. Then the real number L is said to be the **limit from the left***

*or the **left-hand limit** of f at a if and only if to each $\varepsilon > 0$ there is $\delta = \delta(\varepsilon) > 0$ such that*

$$|f(x) - L| < \varepsilon$$

whenever $x \in D$ and $a - \delta < x < a$. The limit from the left is denoted by

$$\lim_{x \to a-} f(x) = L.$$

These limits are called **one-sided limits.**

Example 4.1.4. For the function defined in Example 4.1.3,

$$\lim_{x \to 0+} f(x) = 1,$$

$$\lim_{x \to 0-} f(x) = -1,$$

$$\lim_{x \to a+} f(x) = \lim_{x \to a-} f(x) = 1, \quad a > 0,$$

and

$$\lim_{x \to a+} f(x) = \lim_{x \to a-} f(x) = -1, \quad a < 0.$$

It is easily verified that, if $f(x) = c$ for $a < x < a + \delta$ $(a - \delta < x < a)$, then

$$\lim_{x \to a+} f(x) = c \left(\lim_{x \to a-} f(x) = c \right).$$

The proof that one-sided limits are unique is similar to the proof that limits are unique and is left as Exercise 21 of Problem Set 4.1.

Theorem 4.1.3. *If*

$$\lim_{x \to a+} f(x) \left(\lim_{x \to a-} f(x) \right)$$

exists, then it is unique.

The following theorem relates limits and one-sided limits. The proof is left as Exercise 22 of Problem Set 4.1.

Theorem 4.1.4. *Let $f: D \to R$, and let a be a limit point of both $D \cap (-\infty, a)$ and $D \cap (a, +\infty)$. Then*

$$\lim_{x \to a} f(x) = L$$

if and only if

$$\lim_{x \to a+} f(x) \text{ and } \lim_{x \to a-} f(x)$$

both exist and

$$\lim_{x \to a+} f(x) = \lim_{x \to a-} f(x) = L.$$

Observe that this theorem provides a strong tool for proving that limits do not exist. If either

$$\lim_{x \to a+} f(x) \quad \text{or} \quad \lim_{x \to a-} f(x)$$

does not exist, then

$$\lim_{x \to a} f(x)$$

does not exist, and, if the one-sided limits both exist but are not equal, then the limit does not exist.

Example 4.1.5. For the function defined in Example 4.1.3, it follows from Example 4.1.4 that

$$\lim_{x \to 0} f(x)$$

does not exist while

$$\lim_{x \to a} f(x) = 1, \quad a > 0,$$

and

$$\lim_{x \to a} f(x) = -1, \quad a < 0.$$

We conclude this section by presenting an example involving an important function in analysis.

Example 4.1.6. The **bracket function** or **greatest integer function**, denoted by $f(x) = [x]$, is defined for each real number x to be the largest integer less than or equal to x. For any real number x, Theorem 2.6.5 implies that there is an integer n such that $n - 1 \leqslant x < n$. Then $f(x) = [x] = n - 1$. For example, $[2] = 2$, $[-1] = -1$, $[1/2] = 0$, $[-1/2] = -1$, $[7/4] = 1$, $[-7/4] = -2$, and $[\pi] = 3$.
Let n be any integer, then

$$\lim_{x \to n+} [x] = n \quad \text{and} \quad \lim_{x \to n-} [x] = n - 1.$$

For example, if $n = 2$, then

$$\lim_{x \to 2+} [x] = 2 \quad \text{and} \quad \lim_{x \to 2-} [x] = 1.$$

Since

$$\lim_{x \to n+} [x] \neq \lim_{x \to n-} [x],$$

Theorem 4.1.4 implies that the bracket function does not have a limit at any integer.

Let a be any real number not an integer. Then there is an integer n such that $n - 1 < a < n$. Thus for any x, $n - 1 < x < n$, we have

$$[x] = [a] = n - 1,$$

and the bracket function is constant on a neighborhood of a. Thus

$$\lim_{x \to a^+} [x] = \lim_{x \to a^-} [x] = \lim_{x \to a} [x] = n - 1.$$

Problem Set 4.1

1. Evaluate $\lim_{x \to 3} \{x^2\}$.

2. Evaluate $\lim_{x \to a} \{x^2\}$, a any real number.

3. Evaluate $\lim_{x \to 3} \{x^2 + x\}$.

4. Evaluate $\lim_{x \to a} \{x^2 + x\}$.

5. Evaluate $\lim_{x \to 3} \left\{ \dfrac{x^2 + 2x}{x - 1} \right\}$.

6. Evaluate $\lim_{x \to 3} \left\{ \dfrac{x^2 - 9}{x + 3} \right\}$.

7. Evaluate $\lim_{x \to 3} \{x^5\}$.

8. Evaluate $\lim_{x \to 2} \{(x - 2)^n\}$, n a positive integer.

9. Evaluate $\lim_{x \to 2} \{(x - 2)^{1/3}\}$.

10. Evaluate $\lim_{x \to 0} \{\sqrt{1 + x}\}$.

11. Evaluate $\lim_{x \to 1} \left\{ \dfrac{x^3 - 1}{x - 1} \right\}$.

12. Evaluate $\lim_{x \to 0} \left\{ \dfrac{\sqrt{1 + x} - 1}{x} \right\}$. [*Hint*: Multiply numerator and denominator by $\sqrt{1 + x} + 1$.]

13. Evaluate $\lim_{x \to 10} \left\{ \dfrac{x^2 - 100}{x - 10} \right\}$.

14. Evaluate $\lim\limits_{x \to 0} \left\{ \dfrac{x^3 + 5x^2 + x}{x^2 - 2x} \right\}$.

15. Evaluate $\lim\limits_{x \to 0} \left\{ \dfrac{\sqrt{1 + ax} - 1}{x} \right\}$.

16. Let D be the set of nonzero real numbers. Let $f: D \to R$ such that

$$f(x) = [x]x.$$

Prove that f has a limit at 0.

17. If n is a positive integer, prove that $\lim\limits_{x \to a} x^n = a^n$.

18. Prove Theorem 4.1.2.

19. Evaluate $\lim\limits_{x \to a+} [3x^2]$ and $\lim\limits_{x \to a-} [3x^2]$, a any real number

20. Evaluate $\lim\limits_{x \to a+} [\sqrt{x^2 + 1}]$ and $\lim\limits_{x \to a-} [\sqrt{x^2 + 1}]$, a any real number.

21. Prove Theorem 4.1.3.

22. Prove Theorem 4.1.4.

23. Let $f(x) = x - [x]$, where $[x]$ denotes the bracket function. Find

$$\lim_{x \to a+} f(x) \quad \text{and} \quad \lim_{x \to a-} f(x)$$

for each real number a. For what real numbers a does $\lim\limits_{x \to a} f(x)$ exist?

24. Let $S \subseteq R$. The **characteristic function** of S, denoted by $\chi_S(x)$, is defined for each real number x by

$$\chi_S(x) = \begin{cases} 1 & \text{if } x \in S, \\ 0 & \text{if } x \notin S. \end{cases}$$

Show that $\chi_S(x) = 1 - \chi_{\mathbf{C}(S)}(x)$, where $\mathbf{C}(S)$ denotes the complement of S with respect to the set R of real numbers.

25. The **signum** or **sign function** is defined by

$$\text{sgn } x = \begin{cases} 1 & \text{if } x > 0, \\ 0 & \text{if } x = 0, \\ -1 & \text{if } x < 0. \end{cases}$$

For each real number a, find $\lim\limits_{x \to a+} \{\text{sgn } x\}$, $\lim\limits_{x \to a-} \{\text{sgn } x\}$, and $\lim\limits_{x \to a} \{\text{sgn } x\}$.

26. Let R^+ denote the set of positive real numbers, and let R^- denote the set of negative real numbers. Show that

$$\text{sgn } x = \chi_{R^+}(x) - \chi_{R^-}(x).$$

[Cf. Exercises 24 and 25.]

27. Let $S = (0, 1)$. For each real number a, find

$$\lim_{x \to a+} \chi_S(x) \quad \text{and} \quad \lim_{x \to a-} \chi_S(x).$$

For what real numbers a does $\lim_{x \to a} \chi_S(x)$ exist? [Cf. Exercise 24.]

28. Let Z be the set of integers. For each real number a, find

$$\lim_{x \to a+} \chi_Z(x) \quad \text{and} \quad \lim_{x \to a-} \chi_Z(x).$$

For what real numbers a does $\lim_{x \to a} \chi_Z(x)$ exist? [Cf. Exercise 24.]

29. Let Q be the set of rational numbers. For what real numbers a does

$$\lim_{x \to a+} \chi_Q(x), \quad \lim_{x \to a-} \chi_Q(x), \quad \text{and} \quad \lim_{x \to a} \chi_Q(x)$$

exist? [Cf. Exercise 24.] [*Hint*: Use the density of the set of rational numbers and the density of the set of irrational numbers.]

30. Let $f: D \to R$, and let a be a limit point of D. The function f is said to have **limit** $+\infty$ $(-\infty)$ at a if and only if to each real number M there is a $\delta = \delta(M) > 0$ such that

$$f(x) > M \ \big(f(x) < M\big)$$

whenever $x \in D$ and $0 < |x - a| < \delta$. This limit is denoted by

$$\lim_{x \to a} f(x) = +\infty \ \left(\lim_{x \to a} f(x) = -\infty\right).$$

Give examples of functions f and g such that

$$\lim_{x \to 2} f(x) = +\infty \quad \text{and} \quad \lim_{x \to -2} g(x) = -\infty.$$

31. For each real number a, evaluate

$$\lim_{x \to a} \left|\frac{x + 1}{x - 1}\right|.$$

32. A set A is said to be a **neighborhood** of $+\infty$ $(-\infty)$ if and only if there is a real number b such that the interval $(b, +\infty) \subseteq A \big((-\infty, b) \subseteq A\big)$. We say $+\infty$ $(-\infty)$ is a **limit point** of the set S if and only if each neighborhood of $+\infty$ $(-\infty)$ contains infinitely many points of S. Show that $+\infty$ $(-\infty)$ is a limit point of the set S if and only if every neighborhood of $+\infty$ $(-\infty)$ contains at least one point of S.

33. Show that every infinite set has at least one limit point (see Theorem 3.3.3).

34. Let $f: D \to R$ and let $+\infty$ $(-\infty)$ be a limit point of D. Then f is said to have **limit** L at $+\infty$ $(-\infty)$ if and only if to each real number $\varepsilon > 0$ there is a real number $M = M(\varepsilon)$ such that

$$|f(x) - L| < \varepsilon$$

whenever $x \in D$ and $x > M$ ($x < M$). This limit is denoted by

$$\lim_{x \to +\infty} f(x) = L \left(\lim_{x \to -\infty} f(x) = L \right).$$

Give examples of functions f and g such that

$$\lim_{x \to +\infty} f(x) = 2 \quad \text{and} \quad \lim_{x \to -\infty} g(x) = -2.$$

35. Prove Theorems analogous to Theorems 4.1.1 and 4.1.2 for the limits defined in Exercise 34.

36. Explain what is meant by the symbols $\lim\limits_{x \to +\infty} f(x) = +\infty$, $\lim\limits_{x \to +\infty} f(x) = -\infty$, $\lim\limits_{x \to -\infty} f(x) = +\infty$, and $\lim\limits_{x \to -\infty} f(x) = -\infty$. Illustrate with examples.

37. Explain what is meant by the symbols $\lim\limits_{x \to a+} f(x) = +\infty$, $\lim\limits_{x \to a-} f(x) = +\infty$, $\lim\limits_{x \to a+} f(x) = -\infty$, and $\lim\limits_{x \to a-} f(x) = -\infty$. Illustrate with examples.

The remaining exercises of this section deal with the *trigonometric functions*. In order to define the trigonometric functions, we consider the unit circle $\xi^2 + \eta^2 = 1$ in the $\xi\eta$ plane and denote the centre of this circle by O. Let $P(\xi, \eta)$ be a point on this circle, and let x be the angle measured counterclockwise in radians, $0 \leqslant x < 2\pi$, from the positive ξ-axis to the ray OP. Then the **sine function** and **cosine function** are defined for $0 \leqslant x < 2\pi$ by the equations:

$$s(x) = \sin x = \eta, \qquad c(x) = \cos x = \xi.$$

We extend the domain of the trigonometric functions to the set of real numbers by the equations:

$$\sin(x + 2\pi) = \sin x, \qquad \cos(x + 2\pi) = \cos x.$$

38. Evaluate $\sin 0$, $\cos 0$, $\sin(\pi/6)$, $\cos(\pi/6)$, $\sin(\pi/4)$, $\cos(\pi/4)$, $\sin(\pi/3)$, $\cos(\pi/3)$, $\sin(\pi/2)$, $\cos(\pi/2)$, $\sin(\pi)$, and $\cos(\pi)$.

In Exercises 39–58, show that the given equation is true for all real x and y.

39. $\sin^2 x + \cos^2 x = 1$.
40. $\sin(-x) = -\sin x$.
41. $\cos(-x) = \cos x$.
42. $\sin(\pi - x) = \sin x$.
43. $\cos(\pi - x) = -\cos x$.
44. $\sin(\pi/2 - x) = \cos x$.
45. $\cos(\pi/2 - x) = \sin x$.
46. $\sin(x + y) = \sin x \cos y + \sin y \cos x$.
47. $\sin(x - y) = \sin x \cos y - \sin y \cos x$.
48. $\sin(2x) = 2 \sin x \cos x$.
49. $\cos(x + y) = \cos x \cos y - \sin x \sin y$.
50. $\cos(x - y) = \cos x \cos y + \sin x \sin y$.

51. $\cos(2x) = \cos^2 x - \sin^2 x$.

52. $\sin x + \sin y = 2 \sin\left[\frac{1}{2}(x + y)\right] \cos\left[\frac{1}{2}(x - y)\right]$.

53. $\sin x - \sin y = 2 \sin\left[\frac{1}{2}(x - y)\right] \cos\left[\frac{1}{2}(x + y)\right]$.

54. $\cos x + \cos y = 2 \cos\left[\frac{1}{2}(x + y)\right] \cos\left[\frac{1}{2}(x - y)\right]$.

55. $\cos x - \cos y = - 2 \sin\left[\frac{1}{2}(x + y)\right] \sin\left[\frac{1}{2}(x - y)\right]$.

56. $\sin x \sin y = \frac{1}{2} \cos(x - y) - \frac{1}{2} \cos(x + y)$.

57. $\cos x \cos y = \frac{1}{2} \cos(x - y) + \frac{1}{2} \cos(x + y)$.

58. $\sin x \cos y = \frac{1}{2} \sin(x - y) + \frac{1}{2} \sin(x + y)$.

59. Show that $\lim_{x \to 0} \{\sin x\} = 0$. [*Hint*: $|\sin x| \leqslant |x|$.]

60. Show that $\lim_{x \to 0} \{\cos x\} = 1$. [*Hint*: $1 - |x| \leqslant |\cos x| \leqslant 1$.].

61. Show that $\lim_{x \to 0} \{\cos(x + y)\} = \cos y$.

4.2. Limit Theorems

Many of the theorems on limits of sequences are special cases of more general limit theorems for functions. Thus a number of the theorems presented here are similar to some of those of Chapter 3.

> **Definition 4.2.1.** *A function f is said to be **bounded above** (**below**) on a set S if and only if there is a real number M such that $f(x) \leqslant M$ $(f(x) \geqslant M)$ for each $x \in S$. A function f is said to be **bounded** on a set S if and only if there is a real number M such that $|f(x)| \leqslant M$ for each $x \in S$.*

If the function f is bounded on the set S, then there is a real number M such that $|f(x)| \leqslant M$ for each $x \in S$. Thus for each $x \in S$, we have $-M \leqslant f(x)$ and $f(x) \leqslant M$, and f is bound above and f is bounded below on S. Conversely, if f is bounded above and bounded below on S, then there are real numbers A and B such that $f(x) \leqslant A$ and $f(x) \geqslant B$ for each $x \in S$. If we let $M = \max(|A|, |B|)$, then it follows that $|f(x)| \leqslant M$ for each $x \in S$ and f is bounded on S.

The following theorem gives a necessary condition for a function to have a limit.

> **Theorem 4.2.1.** *If $\lim_{x \to a} f(x)$ exists and is finite, then there is a deleted neighborhood T of a such that f is bounded on the set $T \cap D$, where D is the domain of f.*

> *Proof.* Let $\varepsilon = 1$. Since $\lim_{x \to a} f(x)$ exists, assume that $\lim_{x \to a} f(x) = L$.

Then there is a δ such that $x \in D$ and $0 < |x - a| < \delta$ imply that $|f(x) - L| < \varepsilon = 1$. Define $T = \{x : 0 < |x - a| < \delta\}$. Then, for $x \in T \cap D$,

$$-1 = -\varepsilon < f(x) - L < \varepsilon = 1,$$

or

$$-(1 + |L|) \leqslant -1 + L < f(x) < 1 + L \leqslant (1 + |L|).$$

Thus

$$|f(x)| < (1 + |L|)$$

for each $x \in T \cap D$. Therefore, f is bounded on $T \cap D$.

The following theorems show that *the limit of the sum is the sum of the limits* and that *the limit of the product is the product of the limits.*

Theorem 4.2.2. *Let $f: D \to R$, let $g: D \to R$, and let a be a limit point of D. If $\lim_{x \to a} f(x)$ and $\lim_{x \to a} g(x)$ both exist and are finite, then $\lim_{x \to a} [f(x) + g(x)]$ exists and*

$$\lim_{x \to a} [f(x) + g(x)] = \lim_{x \to a} f(x) + \lim_{x \to a} g(x).$$

This theorem may be paraphrased as follows: *If the limits each exist, then the limit of the sum exists and is equal to the sum of the limits.*

Proof. Assume that $\lim_{x \to a} f(x) = L$ and that $\lim_{x \to a} g(x) = M$. We need to show that

$$\lim_{x \to a} [f(x) + g(x)] = L + M,$$

that is, to each $\varepsilon > 0$ there is $\delta > 0$ such that $x \in D$ and $0 < |x - a| < \delta$ imply that

$$|(f(x) + g(x)) - (L + M)| < \varepsilon.$$

Let $\varepsilon > 0$. Since $\lim_{x \to a} f(x) = L$, there is $\delta_1 > 0$ such that $x \in D$ and $0 < |x - a| < \delta_1$ imply that

$$|f(x) - L| < \varepsilon/2.$$

Since $\lim_{x \to a} g(x) = M$, there is $\delta_2 > 0$ such that $x \in D$ and $0 < |x - a| < \delta_2$ imply that

$$|g(x) - M| < \varepsilon/2.$$

Let $\delta = \min(\delta_1, \delta_2)$. Then for $x \in D$ and $0 < |x - a| < \delta$, we have that

$$\begin{aligned}
|(f(x) + g(x)) - (L + M)| &= |(f(x) - L) + (g(x) - M)| \\
&\leqslant |f(x) - L| + |g(x) - M| \\
&< \varepsilon/2 + \varepsilon/2 = \varepsilon.
\end{aligned}$$

Therefore,

$$\lim_{x \to a} [f(x) + g(x)] = L + M = \lim_{x \to a} f(x) + \lim_{x \to a} g(x).$$

Example 4.2.1. Evaluate

$$\lim_{x \to 3} \left\{ \frac{x^3 + 2x}{x - 1} \right\}.$$

We may write

$$\frac{x^3 + 2x}{x - 1} = x^2 + \frac{x^2 + 2x}{x - 1}.$$

Then

$$\lim_{x \to 3} x^2 = 9 \quad \text{and} \quad \lim_{x \to 3} \left\{ \frac{x^2 + 2x}{x - 1} \right\} = \frac{15}{2}$$

(cf. Exercises 1 and 5 of Problem Set 4.1), and so

$$\lim_{x \to 3} \left\{ \frac{x^3 + 2x}{x - 1} \right\} = \lim_{x \to 3} x^2 + \lim_{x \to 3} \left\{ \frac{x^2 + 2x}{x - 1} \right\}$$

$$= 9 + 15/2 = 33/2.$$

Theorem 4.2.3. *Let $f: D \to R$, let $g: D \to R$, and let a be a limit point of D. If $\lim_{x \to a} f(x)$ and $\lim_{x \to a} g(x)$ both exist and are finite, then $\lim_{x \to a} [f(x) \cdot g(x)]$ exists and*

$$\lim_{x \to a} [f(x) \cdot g(x)] = \left[\lim_{x \to a} f(x) \right] \left[\lim_{x \to a} g(x) \right].$$

We may paraphrase this theorem as follows: *If the limits each exist, then the limit of the product exists and is equal to the product of the limits.*

Proof. Assume that $\lim_{x \to a} f(x) = L$ and that $\lim_{x \to a} g(x) = M$. We need to show that $\lim_{x \to a} [f(x) \cdot g(x)] = LM$, that is, to each $\varepsilon > 0$ there is $\delta > 0$ such that $x \in D$ and $0 < |x - a| < \delta$ imply that

$$|f(x) g(x) - LM| < \varepsilon.$$

Let $\varepsilon > 0$. Since $\lim_{x \to a} f(x) = L$, Theorem 4.2.1 implies that there is $\delta_1 > 0$ and a real number $B > 0$ such that for $x \in D$ and $0 < |x - a| < \delta_1$, we have $|f(x)| \leqslant B$. Also, since $\lim_{x \to a} f(x) = L$, there is $\delta_2 > 0$ such that $x \in D$ and $0 < |x - a| < \delta_2$ imply that

$$|f(x) - L| < \varepsilon/2(|M| + 1).$$

Since $\lim_{x \to a} g(x) = M$, there is $\delta_3 > 0$ such that $x \in D$ and

$$0 < |x - a| < \delta_3$$

imply that

$$|g(x) - M| < \varepsilon/2B.$$

Let $\delta = \min(\delta_1, \delta_2, \delta_3)$. Then for $x \in D$ and $0 < |x - a| < \delta$, we have that

$$\begin{aligned}
|f(x)g(x) - LM| &= |f(x)g(x) - f(x)M + f(x)M - LM| \\
&\leqslant |f(x)||g(x) - M| + |M||f(x) - L| \\
&\leqslant B|g(x) - M| + |M||f(x) - L| \\
&< B\varepsilon/2B + |M|\varepsilon/2(|M| + 1) \\
&< \varepsilon/2 + \varepsilon/2 = \varepsilon.
\end{aligned}$$

Thus

$$\lim_{x \to a} [f(x)g(x)] = LM = \left[\lim_{x \to a} f(x)\right]\left[\lim_{x \to a} g(x)\right].$$

Example 4.2.2. Evaluate

$$\lim_{x \to 0} \{x^n \sqrt{1 + x}\},$$

n a positive integer. Since

$$\lim_{x \to 0} x^n = 0 \quad \text{and} \quad \lim_{x \to 0} \sqrt{1 + x} = 1$$

(cf. Exercises 10 and 17 of Problem Set 4.1), Theorem 4.2.3 implies that

$$\lim_{x \to 0} \{x^n \sqrt{1 + x}\} = \left(\lim_{x \to 0} \{x^n\}\right)\left(\lim_{x \to 0} \sqrt{1 + x}\right) = 0 \cdot 1 = 0.$$

Let $\lim_{x \to a} g(x) = M$. Then Theorem 4.2.3 implies that

$$\lim_{x \to a} [-g(x)] = \left[\lim_{x \to a} (-1)\right]\left[\lim_{x \to a} g(x)\right] = -M.$$

Furthermore, if $\lim_{x \to a} f(x) = L$, then using the above fact and Theorem 4.2.2, we have

$$\begin{aligned}
\lim_{x \to a} [f(x) - g(x)] &= \lim_{x \to a} [f(x) + (-g(x))] \\
&= \lim_{x \to a} f(x) + \lim_{x \to a} [-g(x)] \\
&= \lim_{x \to a} f(x) - \lim_{x \to a} g(x).
\end{aligned}$$

Corollary 4.2.3. *Let* $f: D \to R$, *let* $g: D \to R$, *and let* a *be a limit point of* D. *If* $\lim\limits_{x \to a} f(x)$ *and* $\lim\limits_{x \to a} g(x)$ *both exist and are finite, then*

$$\lim_{x \to a} [f(x) - g(x)]$$

exists and

$$\lim_{x \to a} [f(x) - g(x)] = \lim_{x \to a} f(x) - \lim_{x \to a} g(x).$$

In order to prove a theorem about quotients similar to the one for products, we need the following lemma.

Lemma 4.2.1. *If* $\lim\limits_{x \to a} f(x) = L \neq 0$, *then there is a deleted neighborhood* T *of* a *such that* f *is bounded away from* 0 *on the set* $T \cap D$, *where* D *is the domain of* f. *That is, there is a deleted neighborhood* T *of* a *and a real number* $m > 0$ *such that* $|f(x)| \geqslant m$ *for each* $x \in T \cap D$.

Proof. Let $\varepsilon = |L|/2 \neq 0$. Then there is a δ such that $x \in D$ and $0 < |x - a| < \delta$ imply that $|f(x) - L| < \varepsilon = |L|/2$. Define

$$T = \{x: 0 < |x - a| < \delta\}.$$

Then for $x \in T \cap D$,

$$-|L|/2 = -\varepsilon < f(x) - L < \varepsilon = |L|/2$$

or

$$L - |L|/2 < f(x) < L + |L|/2.$$

If $L > 0$, then $L = |L|$ and the left-hand side of the above inequality implies that $f(x) > |L|/2$. If $L < 0$, then $L = -|L|$ and the right-hand side of the above inequality implies that $f(x) < -|L|/2$ or $-f(x) > |L|/2$. In either case, $|f(x)| > |L|/2$ for $x \in T \cap D$.

The proof of the following theorem on quotients of limits follows in a fashion analogous to the proof of Theorem 4.2.3 by using Lemma 4.2.1 and is left as Exercise 8 of Problem Set 4.2.

Theorem 4.2.4. *Let* $f: D \to R$, *let* $g: D \to R$, *and let* a *be a limit point of* D. *If* $\lim\limits_{x \to a} f(x)$ *and* $\lim\limits_{x \to a} g(x)$ *both exist and are finite, and* $\lim\limits_{x \to a} g(x) \neq 0$, *then* $\lim\limits_{x \to a} [f(x)/g(x)]$ *exists and*

$$\lim_{x \to a} [f(x)/g(x)] = \left[\lim_{x \to a} f(x) \right] \bigg/ \left[\lim_{x \to a} g(x) \right].$$

Example 4.2.3. Evaluate

$$\lim_{x \to 0} \left\{ \frac{x^n}{\sqrt{1 + x}} \right\},$$

n a positive integer. Since

$$\lim_{x \to 0} \{x^n\} = 0 \quad \text{and} \quad \lim_{x \to 0} \sqrt{1 + x} = 1 \neq 0,$$

(cf. Exercises 10 and 17 of Problem Set 4.1), Theorem 4.2.4 implies that

$$\lim_{x \to 0} \left\{ \frac{x^n}{\sqrt{1 + x}} \right\} = \frac{\lim\limits_{x \to 0} \{x^n\}}{\lim\limits_{x \to 0} \sqrt{1 + x}} = \frac{0}{1} = 0.$$

We now prove a theorem relating limits of sequences and limits of functions.

Theorem 4.2.5. Let $f: D \to R$, and let a be a limit point of D. Then $\lim\limits_{x \to a} f(x)$ exists and is finite if and only if for each sequence $\{x\} \to a$, with $x_n \in D$ and $x_n \neq a$ for each positive integer n, the sequence $\{f(x_n)\}$ converges. Furthermore, if $\lim\limits_{x \to a} f(x) = L$, then $\lim\limits_{n} f(x_n) = L$ for every sequence $\{x_n\} \to a$ with $x_n \in D$ and $x_n \neq a$.

Proof. Assume that $\lim\limits_{x \to a} f(x) = L$. Let $\{x_n\}$ be any sequence such that $x_n \in D$, $x_n \neq a$ for each positive integer n, and $\{x_n\} \to a$. Assume that $\varepsilon > 0$ is given. Since $\lim\limits_{x \to a} f(x) = L$, there is $\delta > 0$ such that $|f(x) - L| < \varepsilon$ whenever $x \in D$ and $0 < |x - a| < \delta$. Since $\{x_n\} \to a$, there is N such that $|x_n - a| < \delta$ when $n > N$. Thus for $n > N$, $|f(x_n) - L| < \varepsilon$, and $\{f(x_n)\} \to L$. That is, for any sequence $\{x_n\}$ satisfying the conditions of this theorem, the sequence $\{f(x_n)\} \to L$.

Now suppose that for each sequence $\{x_n\}$ satisfying the conditions of this theorem, the sequence $\{f(x_n)\}$ converges. We first show that if $\{x_n\}$ and $\{y_n\}$ are any two sequences satisfying the conditions of this theorem, then $\{f(x_n)\}$ and $\{f(y_n)\}$ have the same limit. Assume that $\{f(x_n)\} \to L_1$ and $\{f(y_n)\} \to L_2$. Define a new sequence $\{z_n\}$ as follows:

$$z_n = \begin{cases} x_n & \text{if } n \text{ is even,} \\ y_n & \text{if } n \text{ is odd} \end{cases}$$

Then the sequence $\{z_n\}$ satisfies the conditions of the theorem, and so $\{f(z_n)\}$ converges.

Since $\{f(x_{2n})\} = \{f(z_{2n})\}$, the sequence $\{f(z_{2n})\}$ is a subsequence of $\{f(x_n)\}$ and so $\{f(z_{2n})\} \to L_1$. Since $\{f(y_{2n-1})\} = \{f(z_{2n-1})\}$, the sequence $\{f(z_{2n-1})\}$ is a subsequence of $\{f(y_n)\}$ and so $\{f(z_{2n-1})\} \to L_2$. Theorem 3.2.1 implies that $L_1 = L_2$. From this it follows that for all sequences $\{x_n\}$ satisfying the conditions of the theorem, the sequences $\{f(x_n)\}$ have the same limit, say L. Assume that L is not the limit of f at a. Then there is an $\varepsilon > 0$ such that for every $\delta > 0$ there is $x_\delta \in D$ such that $0 < |x_\delta - a| < \delta$ and $|f(x_\delta) - L| \geqslant \varepsilon$. In particular, if we let $\delta_n = 1/n$ for each positive integer n, then there is a sequence $\{x_n\}$ such that $0 < |x_n - a| < 1/n$ and $|f(x_n) - L| \geqslant \varepsilon$ for each positive integer n. Thus $\{x_n\}$ satisfies the conditions of the theorem, but $\{f(x_n)\}$ does not converge to L. This is a contradiction. Therefore, $\lim\limits_{x \to a} f(x) = L$.

This theorem is of particular value for showing that a function does not have a limit at a given point. If $f : D \to R$ and a is a limit point of D, the question arises as to whether f has a limit at a. If we can find two sequences $\{x_n\}$ and $\{y_n\}$ with $x_n, y_n \in D$, $x_n, y_n \neq a$, for each positive integer n, and if $\{x_n\} \to a$ and $\{y_n\} \to a$ such that $\lim\limits_{n} f(x_n) \neq \lim\limits_{n} f(y_n)$, then Theorem 4.2.5 implies that $\lim\limits_{x \to a} f(x)$ does not exist.

Example 4.2.4. Using the above criteria, it is now quite easy to prove that the function $f(x) = \sin(1/x)$ for $x \neq 0$ does not have a limit at $x = 0$. Let $x_n = 2/\pi(4n - 1)$ and $y_n = 2/\pi(4n + 1)$ for each positive integer n. Then $\{x_n\} \to 0$, and $\{y_n\} \to 0$. Also,

$$\sin(1/x_n) = \sin[\pi(4n - 1)/2] = -1$$

and

$$\sin(1/y_n) = \sin[\pi(4n + 1)/2] = +1$$

for each positive integer n. Thus, $\{\sin(1/x_n)\} \to -1$, and $\{\sin(1/y_n)\} \to +1$, and the limit of $\sin(1/x)$ does not exist at $x = 0$.

We now define a property for functions that is similar to the monotone property for sequences.

Definition 4.2.2. *Let* $f : D \to R$. *Then* f *is said to be* **increasing** *(nondecreasing) on* D *if and only if* $x, y \in D$ *and* $x < y$ *imply that* $f(x) \leqslant f(y)$. *The function* f *is said to be* **decreasing** *(nonincreasing) on* D *if and only if* $x, y \in D$ *and* $x < y$ *imply that* $f(y) \leqslant f(x)$. *The function* f *is said to be* **monotone** *if and only if* f *is either increasing or decreasing.*

Theorem 4.2.6. *Let* $f : [a, b] \to R$ *be a monotone function. Then* f *has one-sided limits at each point of* (a, b) *and* $\lim\limits_{x \to a^+} f(x)$ *and* $\lim\limits_{x \to b^-} f(x)$ *both exist.*

The proof is left as Exercise 33 of Problem Set 4.2.

We conclude this section by giving a *Cauchy condition* for limits of functions.

Theorem 4.2.7. *(Cauchy Criterion for Functions) Let* $f: D \to R$, *and let* a *be a limit point of* D. *Then*

$$\lim_{x \to a} f(x)$$

exists and is finite if and only if to each $\varepsilon > 0$ *there is* $\delta = \delta(\varepsilon) > 0$ *such that* $x', x'' \in D$, $0 < |x' - a| < \delta$, *and* $0 < |x'' - a| < \delta$ *imply that*

$$|f(x') - f(x'')| < \varepsilon.$$

This theorem is similar to the Cauchy criterion for sequences in the sense that it guarantees that the function has a limit, but it does not give the value of the limit. The proof of this theorem is left as Exercise 35 of Problem Set 4.2.

Problem Set 4.2

1. Let $f: D \to R$, let $g: D \to R$, and let a be a limit point of D. If $f(x) \geqslant g(x)$ for each $x \in D$, and $\lim_{x \to a} f(x)$ and $\lim_{x \to a} g(x)$ both exist, then prove that $\lim_{x \to a} f(x) \geqslant \lim_{x \to a} g(x)$. Show it is possible to have $f(x) > g(x)$ for each $x \in D$, but $\lim_{x \to a} f(x) = \lim_{x \to a} g(x)$.

2. Let $f: D \to R$, let $g: D \to R$, and let a be a limit point of D. If $f(x) > 0$, $|g(x)| \leqslant f(x)$ for each $x \in D$, and $\lim_{x \to a} f(x) = 0$, show that $\lim_{x \to a} g(x) = 0$.

3. Let $f: D \to R$, let $g: D \to R$, let $h: D \to R$, and let a be a limit point of D. If $f(x) \leqslant g(x) \leqslant h(x)$ for each $x \in D$, and $\lim_{x \to a} f(x) = \lim_{x \to a} h(x) = L$, show that

$$\lim_{x \to a} g(x) = L.$$

4. Let $f: D \to R$, and let a be a limit point of D. Show that $\lim_{x \to a} f(x) = 0$ if and only if $\lim_{x \to a} |f(x)| = 0$.

5. Let $f: D \to R$, and let a be a limit point of D. If $\lim_{x \to a} f(x) = L$, show that $\lim_{x \to a} |f(x)| = |L|$. Is the converse true? That is, does $\lim_{x \to a} |f(x)| = |L|$ imply that $\lim_{x \to a} f(x) = L$? Is it possible for $\lim_{x \to a} |f(x)|$ to exist and $\lim_{x \to a} f(x)$ not to exist?

6. Let $f: D \to R$, let $g: D \to R$, and let a be a limit point of D. If f is bounded in a neighborhood of a and $\lim_{x \to a} g(x) = 0$, show that $\lim_{x \to a} [f(x) g(x)] = 0$.

7. Let $f: D \to R$, and let a be a limit point of D. Prove that $\lim_{x \to a} f(x) = L$ if and only if $\lim_{x \to a} [f(x) - L] = 0$.

8. Prove Theorem 4.2.4.

9. Let $f: D \to R$ with $f(x) \geqslant 0$ for each $x \in D$, and let a be a limit point of D. if $\lim_{x \to a} f(x) = L$, show that $\lim_{x \to a} \sqrt{f(x)} = \sqrt{L}$. [*Hint*: If $L \geqslant 0$, then

$$\sqrt{f(x)} - \sqrt{L} = (f(x) - L)(\sqrt{f(x)} + \sqrt{L}).]$$

10. Let $f: D \to R$, let $g: D \to R$, and let a be a limit point of D. Give an example of functions f and g such that f and g do not have limits at a, but $f + g$ has a limit at a.

11. Let $f: D \to R$, $g: D \to R$, and let a be a limit point of D. Give an example of functions f and g such that f and g do not have limits at a but fg has a limit at a.

12. Let $f: D \to R$, let $g: D \to R$, and let a be a limit point of D. If $\lim_{x \to a} f(x)$ and $\lim_{x \to a} [f(x) + g(x)]$ both exist, show that $\lim_{x \to a} g(x)$ exists.

13. Let $f: D \to R$, $g: D \to R$, and let a be a limit point of D. If $\lim_{x \to a} f(x)$ and $\lim_{x \to a} [f(x) g(x)]$ both exist and $\lim_{x \to a} f(x) \neq 0$, show that $\lim_{x \to a} g(x)$ exists.

14. Let $f: R \to R$ such that $f(x + y) = f(x) f(y)$ for all $x, y \in R$. If $\lim_{x \to 0} f(x)$ exists, show that $\lim_{x \to a} f(x)$ exists for each $a \in R$ and either $\lim_{x \to 0} f(x) = 1$ or $f(x) = 0$ for each $x \in R$.

15. Let P be a polynomial, and let a be a real number. Prove that

$$\lim_{x \to a} P(x) = P(a).$$

16. If $\lim_{x \to a+} f(x)$ and $\lim_{x \to a+} g(x)$ both exist, prove that $\lim_{x \to a+} [f(x) + g(x)]$ exists and that

$$\lim_{x \to a+} [f(x) + g(x)] = \lim_{x \to a+} f(x) + \lim_{x \to a+} g(x).$$

17. If $\lim_{x \to a+} f(x)$ and $\lim_{x \to a+} g(x)$ both exist, prove that $\lim_{x \to a+} [f(x) g(x)]$ exists and that

$$\lim_{x \to a+} [f(x) g(x)] = \left[\lim_{x \to a+} f(x) \right]\left[\lim_{x \to a+} g(x) \right].$$

18. Prove results similar to Problems 16 and 17 for the limit from the left.

19. If $\lim_{x \to a} f(x) = +\infty$ and $\lim_{x \to a} g(x) = +\infty$, prove that $\lim_{x \to a} [f(x) + g(x)] = +\infty$.

20. If $\lim_{x \to a} f(x) = +\infty$ and $\lim_{x \to a} g(x) = +\infty$, prove that $\lim_{x \to a} [f(x) g(x)] = +\infty$.

21. If $\lim\limits_{x \to a} f(x) = -\infty$ and $\lim\limits_{x \to a} g(x) = -\infty$, prove that $\lim\limits_{x \to a} [f(x) + g(x)] = -\infty$.

22. If $\lim\limits_{x \to a} f(x) = -\infty$ and $\lim\limits_{x \to a} g(x) = -\infty$, prove that $\lim\limits_{x \to a} [f(x)\, g(x)] = +\infty$.

23. Prove results similar to Exercise 19, where a is replaced by $+\infty$ $(-\infty)$.

24. Prove results similar to Exercise 20, where a is replaced by $+\infty$ $(-\infty)$.

25. Prove results similar to Exercise 21, where a is replaced by $+\infty$ $(-\infty)$.

26. Prove results similar to Exercise 22, where a is replaced by $+\infty$ $(-\infty)$.

For Exercises 27–30, give examples of functions f and g with $\lim\limits_{x \to a} f(x) = +\infty$ and $\lim\limits_{x \to a} g(x) = -\infty$ and such that the given condition is also valid.

27. $\lim\limits_{x \to a} [f(x) + g(x)] = -\infty$.

28. $\lim\limits_{x \to a} [f(x) + g(x)] = L$, where L is any real number.

29. $\lim\limits_{x \to a} [f(x) + g(x)] = +\infty$.

30. $\lim\limits_{x \to a} [f(x) + g(x)]$ does not exist in any sense, finite or infinite.

31. Let $f: D \to R$, $g: E \to R$, and $\operatorname{Ran} f \subseteq E$. Also, assume that a is a limit point of D and that b is a limit point of E. If $\lim\limits_{x \to a} f(x) = b$ where $f(x) \neq b$ in some deleted neighborhood of a, and $\lim\limits_{y \to b} g(y) = L$, show that $\lim\limits_{x \to a} (g \circ f)(x) = L$.

32. Give an example to show that the result in Exercise 31 is false if the condition "$f(x) \neq b$ in a deleted neighborhood of a" is dropped.

33. Prove Theorem 4.2.6.

34. Prove that $\lim\limits_{x \to 0} \left\{ \dfrac{\sin x}{x} \right\} = 1$.

35. Prove Theorem 4.2.7.

36. Prove results similar to Theorem 4.2.7 for $\lim\limits_{x \to +\infty} f(x)$ and $\lim\limits_{x \to -\infty} f(x)$.

37. Show that the sine and cosine functions are bounded for all real numbers x.

38. Evaluate $\lim\limits_{x \to 0} [x \sin(1/x)]$. [Hint: See Exercise 6.]

4.3. Continuity

One of the most important classes of functions in analysis is the class of *continuous functions*. This section presents many of the basic properties of continuous functions. There are three methods of introducing the concept of *continuity*: using an $\varepsilon - \delta$ definition, using the concept of limit, or using sequences. We shall use the $\delta - \varepsilon$ definition of continuity and show that the other two definitions are equivalent to this.

In defining the limit of a function f at the point a, no assumption was made regarding the value of the function at a. In fact, it was emphasized that a need not be in the domain of f, but it must be a limit point of the domain of f. The concept of continuity at a point a now brings the value of f at a back into the picture.

> **Definition 4.3.1.** *Let $f : D \to R$, and let $a \in D$. The function f is **continuous** at a if and only if to each $\varepsilon > 0$ there is a $\delta = \delta(\varepsilon) > 0$ such that $x \in D$ and $|x - a| < \delta$ imply that*
>
> $$|f(x) - f(a)| < \varepsilon.$$
>
> *The function f is **continuous on** D, or **continuous**, if and only if f is continuous at each point $a \in D$.*

In looking at this definition, we see that a must be an element of D, that is, the function f must be defined at a. Then either a is a limit point of D, or a is an isolated point of D. If a is an isolated point of D, then there is a $\delta > 0$ such that a is the only point common to both the sets $T = \{x : |x - a| < \delta\}$ and D. Thus $x \in D \cap T$ implies that

$$|f(x) - f(a)| = 0 < \varepsilon$$

for every $\varepsilon > 0$. Hence any isolated point a of D is a point of *continuity* of f, that is, f is continuous at a. Therefore, relative to the definition of continuity, the only interesting points are limit points of D.

> **Example 4.3.1.** Let $f : R \to R$ such that $f(x) = c$, c a real number, for each real number x. Let $\varepsilon > 0$ be given. Since
>
> $$|f(x) - f(a)| = |c - c| = 0 < \varepsilon$$
>
> for all real numbers x and a, we may let δ be any positive real number. We then have that
>
> $$|f(x) - f(a)| = |c - c| = 0 < \varepsilon$$
>
> whenever $|x - a| < \delta$. Thus the constant function is continuous on

R, that is, the constant function is continuous for each real number x.

Example 4.3.2. Let $f: R \to R$ such that $f(x) = x^2$ for each real number x. We shall show that f is continuous on R.

Let a be any real number. For any given real number $\varepsilon > 0$, we must find a $\delta = \delta(\varepsilon) > 0$ such that

$$|f(x) - f(a)| = |x^2 - a^2| = |x - a||x + a| < \varepsilon$$

whenever $|x - a| < \delta$. Since we are interested only in values of x near a, we may assume that $-1 + a < x < 1 + a$. Then

$$|x + a| < 1 + 2|a|$$

and

$$|f(x) - f(a)| = |x^2 - a^2| < (1 + 2|a|)|x - a|.$$

If $|x - a| < \min [1, \varepsilon/(1 + 2|a|)]$, then we have the desired result. Thus we let $\delta = \min [1, \varepsilon/(1 + 2|a|)]$. Then

$$\begin{aligned}
|f(x) - f(a)| &= |x^2 - a^2| \\
&< (1 + 2|a|)|x - a| \\
&< (1 + 2|a|)\delta \\
&< (1 + 2|a|)\, \varepsilon/(1 + 2|a|) \\
&= \varepsilon
\end{aligned}$$

whenever $|x - a| < \delta$. Thus f is continuous at a. Since a was an arbitrary real number, f is continuous on R.

With the ideas presented above, we are now ready to present a condition equivalent to continuity.

Theorem 4.3.1. *Let $f: D \to R$, let $a \in D$, and let a be a limit point of D. The function f is continuous at a if and only if*

$$\lim_{x \to a} f(x) = f(a).$$

Proof. Assume that f is continuous at a, where a is a limit point of D. Then $f(a)$ exists, and, for each $\varepsilon > 0$, there is a $\delta > 0$ such that $x \in D$ and $|x - a| < \delta$ imply that

$$|f(x) - f(a)| < \varepsilon.$$

Thus, for $x \in D$ and $|x - a| < \delta$, we have

$$|f(x) - f(a)| < \varepsilon,$$

and, by the definition of limit,

$$\lim_{x \to a} f(x) = f(a).$$

Therefore, if f is continuous at a, where a is a limit point of D, then $\lim_{x \to a} f(x) = f(a)$.

Assume that $\lim_{x \to a} f(x) = f(a)$. Then $f(a)$ exists, and, by the definition of limit, to each $\varepsilon > 0$ there is a $\delta > 0$ such that

$$|f(x) - f(a)| < \varepsilon$$

whenever $x \in D$ and $0 < |x - a| < \delta$. Since $|x - a| = 0$ implies that $x = a$, we have that

$$|f(x) - f(a)| = 0 < \varepsilon$$

whenever $|x - a| = 0$. Therefore, to each $\varepsilon > 0$ there is a $\delta > 0$ such that $x \in D$ and $|x - a| < \delta$ imply that

$$|f(x) - f(a)| < \varepsilon.$$

Hence f is continuous at a.

Example 4.3.3. Let

$$f(x) = \begin{cases} x \sin (1/x) & \text{if } x \neq 0, \\ 0 & \text{if } x = 0. \end{cases}$$

Then f has domain R. Is f continuous at 0? We know that

$$\lim_{x \to 0} [x \sin (1/x)] = 0$$

(cf. Exercise 38 of Problem Set 4.2), and therefore Theorem 4.3.1 implies that f is continuous at 0.

One application of Theorem 4.3.1 is to show that a function f is not continuous at a given point. For example, if a is a limit point of the domain D of f and either $\lim_{x \to a} f(x)$ does not exist or $\lim_{x \to a} f(x) \neq f(a)$, then f is not continuous at a.

Example 4.3.4. Let f be the bracket function defined in Example 4.1.6, that is, $f(x) = [x]$ for each $x \in R$. If n is an integer, then

$$\lim_{x \to n+} [x] = n \quad \text{and} \quad \lim_{x \to n-} [x] = n - 1.$$

Thus $\lim_{x \to n} [x]$ does not exist and the bracket function f is *not* continuous at n. Let a be any real number other than an integer and let

$\delta = \min\{(a - [a], ([a + 1] - a)\}$. Then for any $\varepsilon > 0$ and $|x - a| < \delta$, we have

$$|f(x) - f(a)| = |[x] - [a]| = 0 < \varepsilon.$$

Thus the bracket function f is continuous at each non-integral real number a.

From a slightly different point of view, this theorem may be used to show that certain limits exist. If we are interested in the limit of a function f at a point a which is a limit point of the domain of f, then we know that the limit exists if the function is continuous at a (actually we also know the value of the limit, namely, $f(a)$).

We now present a condition in terms of limits of sequences which is equivalent to continuity.

Theorem 4.3.2. *Let $f: D \to R$ and let $a \in D$. The function f is continuous at a if and only if for each sequence $\{x_n\} \to a$, $x_n \in D$, the sequence $\{f(x_n)\} \to f(a)$.*

The proof of this result follows from Theorems 4.2.5 and 4.3.1, and is left as Exercise 6 of Problem Set 4.3.

An application of this theorem could be to show that certain sequences converge and to find their limits. If f is continuous at a and $\{x_n\} \to a$ with x_n in the domain of f, then we know that the sequence $\{f(x_n)\}$ converges and has limit $f(a)$, that is, $\{f(x_n)\} \to f(a)$. This theorem can also be used to show that a function is not continuous at a given point. For example, if a is a limit point of the domain D of f and there exists a sequence $\{x_n\} \to a$ with $x_n \in D$, then f is not continuous at a if either $\{f(x_n)\}$ does not converge or $\{f(x_n)\}$ converges to a limit different from $f(a)$.

Example 4.3.5. Let

$$f(x) = \begin{cases} \sin(1/x) & \text{if } x \neq 0, \\ 0 & \text{if } x = 0. \end{cases}$$

Then f has domain R. We show that f is not continuous at 0. To do this we construct a sequence $\{x_n\} \to 0$ such that $\lim_n f(x_n) \neq 0$. Let $x_n = 2/(4n + 1)\pi$ for each positive integer n. Then

$$\lim_n x_n = \lim_n \{2/(4n + 1)\pi\} = 0.$$

Also, $f(x_n) = \sin[(4n + 1)\pi/2] = 1$ for each positive integer n, and

$$\lim_n f(x_n) = \lim_n \{1\} = 1 \neq 0 = f(0).$$

Thus we see that f is not continuous at 0.

If $f: D \to R$, $a \in D$, and f is not continuous at a, then f is said to be **discontinuous** at a. The bracket function is discontinuous at each integer (cf. Example 4.3.4) while the function f defined in Example 4.3.5 is discontinuous at 0.

We conclude this section by showing that the composite of two continuous functions is a continuous function.

Theorem 4.3.3. *Let* $f : D \to R$, *let* $g : E \to R$, *and let* $\operatorname{Ran} f \subseteq E$. *If* f *is continuous at* a *and* g *is continuous at* $f(a)$, *then* $g \circ f$ *is continuous at* a.

Proof. By Theorem 4.3.1, we need to show that

$$\lim_{x \to a} (g \circ f)(x) = (g \circ f)(a).$$

Since f is continuous at a,

$$\lim_{x \to a} f(x) = f(a) \in E.$$

Since g is continuous at $f(a)$,

$$\lim_{x \to f(a)} g(x) = g(f(a)) = (g \circ f)(a).$$

Then

$$\lim_{x \to a} (g \circ f)(x) = (g \circ f)(a).$$

(Cf. Exercise 31 of Problem Set 4.2.) Thus $g \circ f$ is continuous at a.

Problem Set 4.3

1. Let $f(x) = x + [x]$. Where is f continuous? (See Example 4.1.6.)
2. Let $f(x) = 1/x^2$ for $x \in (0, +\infty)$. Show that f is continuous on $(0, +\infty)$.
3. Let n be a positive integer, and let $f(x) = x^n$ for each real number x. Prove that f is continuous on R.
4. Let $f(x) = \sqrt{x}$ for each nonnegative real number x. Prove that f is continuous for each real number $a \geqslant 0$.
5. Let n be a positive integer and let $f(x) = \sqrt[n]{x}$ for all nonnegative real numbers x. Prove that f is continuous for each real number $a \geqslant 0$.
6. Prove Theorem 4.3.2.
7. Let $f: D \to R$. If S is any set, the **inverse image**, $f^{-1}[S]$, is the set

$$f^{-1}[S] = \{x : f(x) \in S\}.$$

Prove the following statement: A function $f: R \to R$ is continuous if and only if, for every open set G, the inverse image $f^{-1}[G]$ is open.

8. Let $f: R \to R$. The value $f(a)$ is said to be a **relative minimum** of f if there exists a neighborhood T of a such that $f(x) \geqslant f(a)$ for every $x \in T$. If f is continuous, and for every $x \in R$, $f(x)$ is a relative minimum of f, show that f is a constant.

9. Show that $f(x) = 1/x$ is continuous for all $x \neq 0$.

10. Show that $f(x) = \sin x$ is continuous for all x.

11. Show that the function $f(x) = \sin(1/x)$ is continuous for all $x \neq 0$.

12. Let $f: D \to R$, and let a be a limit point of D. The function f is said to have a **removable discontinuity** at a if and only if $\lim_{x \to a} f(x)$ exists and either $a \notin D$ or $a \in D$ and $\lim_{x \to a} f(x) \neq f(a)$.

Give three examples of functions with removable discontinuities.

13. Show how a function with a removable discontinuity at a can be refined to make it continuous at a.

14. Let $f: D \to R$, and let a be a limit point of D. The function f is said to have a **jump discontinuity** at a if and only if both $\lim_{x \to a+} f(x)$ and $\lim_{x \to a-} f(x)$ exist, but $\lim_{x \to a+} f(x) \neq \lim_{x \to a-} f(x)$.

Give three examples of functions with jump discontinuities.

15. Prove that the only discontinuities a monotone function may possess in its domain are jump discontinuities.

16. Prove that the set of points of discontinuity of a monotone function is countable.

4.4. The Algebra of Continuous Functions

In this section we define new functions in terms of given functions and establish where they are continuous by checking the continuity of the original functions. We first define the *sum, difference, product,* and *quotient functions* of two given functions. Let $f: D \to R$ and let $g: D \to R$. The **sum function** $(f + g): D \to R$ (**difference function** $(f - g): D \to R$) of the functions f and g is defined for each $x \in D$ by the equation $(f + g)(x) = f(x) + g(x)$ $((f - g)(x) = f(x) - g(x))$. The equation $(fg)(x) = f(x) g(x)$ for each $x \in D$ defines the **product function** $fg: D \to R$ of the functions f and g. If $g(x) \neq 0$ for each $x \in D$, then we define the **quotient function** $f/g: D \to R$ of the functions f and g by the equation $(f/g)(x) = f(x)/g(x)$ for each $x \in D$. We now show that continuity is preserved by these standard algebraic operations on functions.

Theorem 4.4.1. *Let* $f: D \to R$, *and let* $g: D \to R$. *If* f *and* g *are continuous at* $a \in D$ *and* c *is a constant, then the functions* $f + g, f - g$, cf, *and* fg *are continuous at* a.

We shall prove the result for $f + g$ and fg and leave the proof of the remaining parts as Exercise 2 of Problem Set 4.4.

Proof. We use the definition of continuity to prove that $f + g$ is continuous at a. Assume that $\varepsilon > 0$ is given. Since f is continuous at $a \in D$, there is $\delta_1 > 0$ such that $x \in D$ and $|x - a| < \delta_1$ imply that

$$|f(x) - f(a)| < \varepsilon/2.$$

Similarly, there is a $\delta_2 > 0$ such that $x \in D$ and $|x - a| < \delta_2$ imply that

$$|g(x) - g(a)| < \varepsilon/2.$$

Define $\delta = \min(\delta_1, \delta_2)$. Then, for $x \in D$ and $|x - a| < \delta$, we have that

$$
\begin{aligned}
|(f + g)(x) - (f + g)(a)| &= |[f(x) + g(x)] - [f(a) + g(a)]| \\
&= |[f(x) - f(a)] + [g(x) - g(a)]| \\
&\leqslant |f(x) - f(a)| + |g(x) - g(a)| \\
&< \varepsilon/2 + \varepsilon/2 = \varepsilon.
\end{aligned}
$$

Therefore, $f + g$ is continuous at a.

We use Theorem 4.3.2 to show that fg is continuous at a. If a is an isolated point of D, then fg is defined at a and fg is automatically continuous at a. Thus assume that a is a limit point of D.

Let $\{x_n\} \to a$ be any sequence such that $x_n \in D$ for each n. Then, by Theorem 4.3.2, $\{f(x_n)\} \to f(a)$ and $\{g(x_n)\} \to g(a)$. Using Theorem 3.2.2, we have

$$\{(fg)(x_n)\} = \{f(x_n)\,g(x_n)\} \to f(a)\,g(a) = (fg)(a).$$

Since $\{(fg)(x_n)\} \to (fg)(a)$, Theorem 4.3.2 implies that fg is continuous at a.

In the proof of Theorem 4.4.1, two methods of proof were used to prove that a function is continuous at a point, namely, the definition of continuity and Theorem 4.3.2. These two methods of proof were given only for illustrative purposes, but when actually proving a function is continuous at a point, one can use either the definition of continuity, Theorem 4.3.1, or Theorem 4.3.2. In the proof of Theorem 4.4.2, we demonstrate the use of Theorem 4.3.1.

We see that Theorem 4.4.1 assures us that, if two functions are continuous at a, then their sum, difference, and product are continuous at a. Also, Theorem 4.4.1 shows that, if a function is continuous at a, then a constant times that function is also continuous at a.

The following result concerns quotients of continuous functions; we employ the concept of limits to prove it.

Theorem 4.4.2. *Let* $f: D \to R$, $g: D \to R$, *and* $g(a) \neq 0$. *If* f *and* g *are continuous at* $a \in D$, *then the function* f/g *is continuous at* a.

Proof. If a is an isolated point of D, then f/g is defined at a and is automatically continuous at a. Hence we assume that a is a limit point of D.

Since f is continuous at a, Theorem 4.3.1 implies that

$$\lim_{x \to a} f(x) = f(a).$$

Similarly,

$$\lim_{x \to a} g(x) = g(a) \neq 0.$$

Theorem 4.2.4 now implies that

$$\lim_{x \to a} [(f/g)(x)] = \lim_{x \to a} [f(x)/g(x)]$$

$$= f(a)/g(a) = (f/g)(a).$$

Therefore, f/g is continuous at a.

The next theorem assures us that, if f is continuous at a, then the function $|f|$, defined by $|f(x)| = |f|(x)$, is continuous at a.

Theorem 4.4.3. *Let* $f: D \to R$, *and let* $a \in D$. *If* f *is continuous at* a, *then* $|f|$ *is continuous at* a.

The proof is left as Exercise 3 of Problem Set 4.4.

Example 4.4.1. We note that the converse of Theorem 4.4.3 is not true. For example, let

$$f(x) = \begin{cases} 1 & \text{if } x \text{ is rational,} \\ -1 & \text{if } x \text{ is irrational.} \end{cases}$$

Then $|f|(x) = |f(x)| = 1$ for each real x. Thus, $|f|$ is continuous on R, while f is discontinuous at each $x \in R$.

If $f: D \to R$ and $g: D \to R$, the functions max (f, g) and min (f, g) are defined by

$$\max (f, g)(x) = \max [f(x), g(x)]$$
$$\min (f, g)(x) = \min [f(x), g(x)]$$

for each $x \in D$. Then max $(f, g): D \to R$ and min $(f, g): D \to R$. We can now establish the following theorem concerning the continuity of these functions.

Theorem 4.4.4. *Let $f: D \to R$, and let $g: D \to R$. If f and g are both continuous at a, then the functions max (f, g) and min (f, g) are both continuous at a.*

Proof. We can write

$$\max (f, g) = (f + g)/2 + |f - g|/2,$$

and

$$\min (f, g) = (f + g)/2 - |f - g|/2.$$

Theorems 4.4.1 and 4.4.3 then give the desired results.

Problem Set 4.4

1. Show that any constant function is continuous.
2. Prove the remaining parts of Theorem 4.4.1.
3. Prove Theorem 4.4.3. [*Hint*: Use Theorem 4.3.3.]
4. If $f: D \to R$ and $g: D \to R$, verify that

$$\max(f, g) = (f + g)/2 + |f - g|/2,$$

and

$$\min(f, g) = (f + g)/2 - |f - g|/2.$$

5. Let P be a polynomial. Prove that P is continuous at each real number.
6. Let P and Q be polynomials. Define the **rational function** R by

$$R(x) = P(x)/Q(x)$$

for each real number x such that $Q(x) \neq 0$. Show that R is continuous at each point of its domain.

7. Let $f: D \to R$. If f is continuous at a, and if $\varepsilon > 0$ is any positive real number, show that there exists a neighborhood T of a such that for any two points $x, y \in T \cap D, |f(x) - f(y)| < \varepsilon$.

8. Show that the characteristic function χ_Q of the set of rational numbers is discontinuous at each real number.

9. Let $f: R \rightarrow R$ be the function defined by

$$f(x) = \begin{cases} x & \text{if } x \text{ is rational,} \\ 1-x & \text{if } x \text{ is irrational.} \end{cases}$$

Show that f is discontinuous at each $x \in R$ except $x = 1/2$.

10. Let $f: R \rightarrow R$. If f is continuous, show that the set $\{x: f(x) \leqslant k\}$, where k is a constant, is a closed set.

11. Let $f: D \rightarrow R$ and $a \in D$. The function f is said to be **continuous from the right** at a if and only if to each $\varepsilon > 0$ there is a $\delta = \delta(\varepsilon) > 0$ such that $x \in D$ and $a \leqslant x < x + \delta$ imply that

$$|f(x) - f(a)| < \varepsilon.$$

Prove that f is continuous from the right at a if and only if

$$\lim_{x \to a+} f(x) = f(a)$$

or there exists $\sigma > 0$ such that for each $x \in D$, with $x > a$, $x - a \geqslant \sigma$.

12. Let $f: D \rightarrow R$ and $a \in D$. Show that f is continuous from the right at a if and only if for each sequence $\{x_n\} \rightarrow a$, $x_n \in D$, and $a \leqslant x_n$, the sequence $\{f(x_n)\} \rightarrow f(a)$.

13. Show that the bracket function is continuous from the right for each real number. (See Example 4.1.6.)

14. Let $f: D \rightarrow R$ and $a \in D$. The function f is said to be **continuous from the left** at a if and only if to each $\varepsilon > 0$ there is a $\delta = \delta(\varepsilon) > 0$ such that $x \in D$ and $a - \delta < x \leqslant a$ imply that

$$|f(x) - f(a)| < \varepsilon.$$

Prove that f is continuous from the left at a if and only if

$$\lim_{x \to a-} f(x) = f(a)$$

or there exists $\sigma > 0$ such that for each $x \in D$, with $x < a$, $a - x \geqslant \sigma$.

15. Let $f: D \rightarrow R$ and $a \in D$. Show that f is continuous from the left at a if and only if for each sequence $\{x_n\} \rightarrow a$, $x_n \in D$, and $x_n \leqslant a$, the sequence $\{f(x_n)\} \rightarrow f(a)$.

16. For what real numbers is the bracket function continuous from the left?

17. Let $f: D \rightarrow R$ and $a \in D$. Prove that f is continuous at a if and only if f is continuous from the left at a and f is continuous from the right at a.

18. Let $f: D \rightarrow R$ and $a \in D$. Then the function f is said to be **lower-semicontinuous** at a if and only if for each $\varepsilon > 0$, there is a $\delta = \delta(\varepsilon) > 0$ such that $x \in D$ and $|x - a| < \delta$ imply that

$$f(a) - f(x) < \varepsilon.$$

Prove: The function $f: D \rightarrow R$ is lower-semicontinuous at $a \in D$ if and only if for each sequence $\{x_n\} \rightarrow a$ with $x_n \in D$, we have that $\lim_n \inf f(x_n) \geqslant f(a)$.

19. Let $f: R \to R$. Show that f is lower-semicontinuous at each point $a \in R$ if and only if the set $\{x: f(x) \leqslant k\}$, where k is a constant, is closed.

20. Let $f: [a, b] \to R$. If f is lower-semicontinuous, show that f is bounded below, that is, there is an m such that $f(x) \geqslant m$ for each $x \in [a, b]$.

21. Let $f: D \to R$ and $g: D \to R$. If f and g are lower-semicontinuous at a, show that $f + g$ is lower-semicontinuous at a.

22. Let $f: D \to R$ and $a \in D$. Then the function f is said to be **upper-semicontinuous** at a if and only if for each $\varepsilon > 0$, there is a $\delta = \delta(\varepsilon) > 0$ such that $x \in D$ and $|x - a| < \delta$ imply that

$$f(x) - f(a) < \varepsilon.$$

Prove: The function $f: D \to R$ is upper-semicontinuous at $a \in D$ if and only if for each sequence $\{x_n\} \to a$, with $x_n \in D$, we have that $\lim_n \sup f(x_n) \leqslant f(a)$.

23–25. Prove the statements analogous to Exercises 19, 20, and 21 for upper-semicontinuity.

26. Let $f: D \to R$. Show that f is continuous at $a \in D$ if and only if f is lower-semicontinuous and f is upper-semicontinuous at a.

27. For what real numbers is the characteristic function χ_Q of the set of rational numbers upper-semicontinuous?

28. For what real number is the characteristic function χ_Q of the set of rational numbers lower-semicontinuous?

29. Let $f: R \to R$ be defined by

$$f(x) = \begin{cases} 0 & \text{if } x \text{ is an irrational number,} \\ 1/q & \text{if } x = p/q, \text{ where } p \text{ and } q \text{ are} \\ & \text{relative prime positive integers.} \end{cases}$$

For what real numbers is f upper-semicontinuous?

30. For what real numbers is the function f defined in Exercise 29 lower-semicontinuous?

31. For what real numbers is the function f defined in Exercise 29 continuous?

32. Let $f: R \to R$. If f is continuous on R and $f(x) = 0$ for each x in some dense subset of R, show that f is the constant function 0.

33. Let $f: D \to R$ and $C \subset D$. Define $g: C \to R$ by $g(x) = f(x)$ for each $x \in C$. If f is continuous at $a \in C$, show that g is continuous at a. Give an example to show that continuity of g at $a \in C$ need not imply continuity of f at a.

34. Prove that any monotone function can be refined at the points of discontinuity of its domain so that it becomes everywhere continuous from the right (from the left).

4.5. Properties of Continuous Functions

The seemingly simple concept of continuity becomes, on careful examination, exceedingly more interesting. We now study some of the subtle properties of continuous functions.

We begin this study by showing that the continuous image of a compact set is compact.

Theorem 4.5.1. *Let D be a compact set, and let $f: D \to R$. If f is continuous on D, then the image of D, $f[D]$, is compact.*

Proof. Let $\{y_n\}$ be any sequence of points in $f[D]$. Then for each positive integer n, there is x_n in D such that $f(x_n) = y_n$. The sequence $\{x_n\}$ has a subsequence $\{x_{n_k}\}$ which converges to a point x in D. (Cf. Exercise 24 of Problem Set 3.5.) Since f is continuous on D, it follows that

$$\lim_k f(x_{n_k}) = f(x).$$

That is, the subsequence $\{y_{n_k}\}$ of $\{y_n\}$ converges to $f(x)$ in $f[D]$. Thus, any sequence $\{y_n\}$ of points in $f[D]$ has a subsequence which converges to a point in $f[D]$ and so $f[D]$ is compact. (Cf. Exercise 24 of Problem Set 3.5.)

Since compact sets are bounded, the following corollary is immediate.

Corollary 4.5.1. *Let D be a compact set, and let $f: D \to R$. If f is continuous on D, then f is bounded on D.*

Let $f: D \to R$. Then the function f is said to have a **maximum** (or a **maximum value**), $f(y)$, on D if and only if $f(y) \geqslant f(x)$ for each $x \in D$. Analogously, the function f is said to have a **minimum** (or a **minimum value**), $f(y)$, on D if and only if $f(y) \leqslant f(x)$ for each $x \in D$.

Example 4.5.1. Let $D = [0, 1]$, and let $f(x) = x^2$ for each $x \in D$. Then $f(1) = 1$ is a maximum value for f on D, and $f(0) = 0$ is a minimum value for f on D.

Example 4.5.2. Let $D = (1, +\infty)$, and let $f(x) = 1 - 1/x$ for each $x \in D$. Then f does not have a maximum value or a minimum value on D.

Example 4.5.2 illustrates a bounded and continuous function which has neither a maximum nor minimum value. In this case the domain D is an open set. The situation is different for a continuous function with a compact domain.

Theorem 4.5.2. *Let D be a compact set, and let $f: D \to R$. If f is continuous on D, then f has a maximum value and a minimum value on D.*

Proof. We prove that f has a maximum value on D and leave the proof that f has a minimum value on D as Exercise 7 of Problem Set 4.5.

Corollary 4.5.1 implies that f is bounded on D. Let

$$L = \sup f[D].$$

By the definition of least upper bound, the number L is a limit point of the set $f[D]$. Theorem 4.5.1 implies that $f[D]$ is compact and, hence, closed. But closed sets contain all their limit points, and so $L \in f[D]$. That is, there is $y \in D$ such that $L = f(y)$. Hence $f(y)$ is a maximum value for f on D.

The following theorem is called the *Intermediate Value Property for continuous function*. It says that if f is continuous on the interval $[a, b]$, then f takes on every value between $f(a)$ and $f(b)$.

Theorem 4.5.3. (*Intermediate Value Theorem*) *Let $f: [a, b] \to R$. If f is continuous on $[a, b]$, $f(a) \neq f(b)$, and c is any real number between $f(a)$ and $f(b)$, then there is a real number $y \in (a, b)$ such that $f(y) = c$.*

Proof. Suppose that $f(a) < c < f(b)$. Let $A = \{x : x \in [a, b]$ and $f(x) \leqslant c\}$ and $B = \{x : x \in [a, b]$ and $f(x) \geqslant c\}$. Since $f(a) < c$, $a \in A$; and since $f(b) > c$, $b \in B$. The set A is a nonempty bounded set, and the Completeness Property ensures us that the set A has a least upper bound, say $y = \sup A$. If $f(y) = c$, then the proof is complete, so suppose that $f(y) \neq c$. If $f(y) < c$, let $\varepsilon = (c - f(y))/2 > 0$. Since $y \in [a, b]$ implies that f is continuous at y, there is a $\delta > 0$ such that $x \in [a, b]$ and $|x - y| < \delta$ imply that $|f(x) - f(y)| < \varepsilon$. This last inequality may be written as

$$-\varepsilon < f(x) - f(y) < \varepsilon$$

or

$$f(y) - \varepsilon < f(x) < f(y) + \varepsilon.$$

Since $\varepsilon = (c - f(y))/2$, we have

$$f(x) < f(y) + (c - f(y))/2 = c/2 + f(y)/2 < c.$$

Since $f(y) < c$, y cannot belong to B and so $y < b$. Thus, there is an $x \in [a, b]$ such that $y < x < y + \delta$ and $f(x) < c$; this and $x \in A$ contradict the choice of y as the least upper bound of A. Therefore $f(y)$ cannot be less than c, that is, $f(y) \geqslant c$. In a similar manner, it is possible to show that $f(y) \leqslant c$. These two inequalities imply that $f(y) = c$.

Corollary 4.5.3. *Let $f: [a, b] \to R$. If f is continuous on $[a, b]$ and $f(a) \neq f(b)$, then $f[[a, b]]$ is an interval.*

Let $f: D \to R$, and let $f[D]$ denote the image of D. If f is 1–1, then Theorem 1.3.5 implies that f has an inverse f^{-1}, and $f^{-1}: f[D] \to D$. If f is continuous on D, can anything be said about the continuity of f^{-1}? The following theorem tells us that, if D is compact, then f^{-1} is continuous.

Theorem 4.5.4. *Let D be a compact set, and let $f: D \to R$. If f is 1–1 and continuous, then the function $f^{-1}: f[D] \to D$ is continuous.*

Proof. We use sequences to show that f^{-1} is continuous. Let y be any point of $f[D]$. If y is an isolated point of $f[D]$, then f^{-1} is continuous at y. Suppose that y is a limit point of $f[D]$. We need to show that, for any sequence $\{y_n\} \to y$, where $y_n \in f[D]$ for each n, that $\{f^{-1}(y_n)\} \to f^{-1}(y)$. For convenience, let $x = f^{-1}(y)$ and $x_n = f^{-1}(y_n)$ for each n. Assume that $\{f^{-1}(y_n)\}$ does not converge to $f^{-1}(y)$, that is, the sequence $\{x_n\}$ does not converge to x. Since $\{x_n\}$ does not converge to x, there is a neighborhood T of x and infinitely many terms of $\{x_n\}$ that belong to $D \setminus T$. Since D is compact, there is a subsequence $\{x_{n_k}\}$ of $\{x_n\}$ that converges to some value z different from x. That is, $\lim_k x_{n_k} = z \neq x$. Since f is continuous, the sequence $\{f(x_{n_k})\} = \{y_{n_k}\} \to f(z)$. On the other hand, $\{y_n\} \to y$ implies that $\{y_{n_k}\} \to y = f(x)$. Since the limit of a convergent sequence is unique, $f(z) = f(x)$. The function f was assumed to be 1–1, and so $z = x$. Thus, the sequence $\{f^{-1}(y_n)\} \to f^{-1}(y)$, and f^{-1} is continuous.

Problem Set 4.5

1. Show that if P is a polynomial of odd degree with real coefficients, then $P(x) = 0$ has at least one real number x.
2. Show that if $f: [0, 1] \to [0, 1]$, $f[[0, 1]] = [0, 1]$, and f is continuous, then there is $x_0 \in [0, 1]$ such that $f(x_0) = x_0$ (x_0 is called a **fixed point** of f). [*Hint*: Consider the function $g(x) = f(x) - x$.]
3. Show that the equation $x = \cos x$ has a solution in the closed interval $[0, \pi/2]$.
4. Prove the converse of Theorem 4.5.1. That is, prove that if D is a compact set, $f: D \to R$, and $f[D]$ is a compact set, then f is continuous on D.
5. Let $f: D \to R$. If f is continuous and $f(a) > 0$, show that there exists a neighborhood T of a and a positive number k such that

$$f(x) \geqslant k$$

for every $x \in D \cap T$.
6. Let D be a compact set and let $f: D \to R$. If f is continuous and $f(x) > 0$ for each $x \in D$, show that there exists a positive number k such that

$$f(x) \geqslant k$$

for every $x \in D$.

7. Prove the remaining part of Theorem 4.5.2.

8. Let D be nonempty closed set. Show that there exists a sequence $\{a_n\}$, where $a_n \in D$, $n = 1, 2, \ldots$, such that the set $A = \{a_n : n = 1, 2, \ldots\}$ is dense in D, that is, $\bar{A} = D$.

4.6. Uniform Continuity

An important concept related to continuity is *uniform continuity*. Continuity is a local property, that is, continuity is a property at the individual points of a set. Uniform continuity is a global property, that is, it is a property on a set as a whole.

> **Definition 4.6.1.** *Let $f: D \to R$. The function f is **uniformly continuous** on D if and only if to each $\varepsilon > 0$ there is a $\delta = \delta(\varepsilon) > 0$ such that $x, y \in D$ and $|x - y| < \delta$ imply that $|f(x) - f(y)| < \varepsilon$.*

We see from the above definition that, if f is uniformly continuous on D and x_0 is any point of D, then for any given $\varepsilon > 0$ there is a $\delta > 0$ such that $y \in D$ and $|y - x_0| < \delta$ imply that $|f(y) - f(x_0)| < \varepsilon$, and so the function f is continuous at x_0. Thus, *if f is uniformly continuous on its domain D, then f is continuous on D*. If the converse were true, then uniform continuity and continuity would be equivalent. Examples will be given of functions that are continuous but not uniformly continuous.

Before presenting such examples, however, let us illustrate two uniformly continuous functions.

> **Example 4.6.1.** Let $f(x) = x^2$ with domain $[0, 1]$. Let $\varepsilon > 0$ be given, and let $\delta = \varepsilon/2$. Then for $x, y \in [0, 1]$ with $|x - y| < \delta$, we have
> $$|f(x) - f(y)| = |x^2 - y^2|$$
> $$= |x + y||x - y|$$
> $$\leqslant 2|x - y| < 2\delta = \varepsilon.$$

Therefore, f is uniformly continuous on $[0, 1]$.

> **Example 4.6.2.** Let $f: R \to R$ be defined by $f(x) = \sin x$ for each $x \in R$. We shall make use of the following relations:
>
> (1) for any two real numbers x and y;
> $$\sin x - \sin y = 2 \cos [(x + y)/2] \sin [(x - y)/2];$$
>
> (2) for any real number x, $|\cos x| \leqslant 1$; and
>
> (3) for any real number x, $|\sin x| \leqslant |x|$.

Let $\varepsilon > 0$, and let $\delta = \min(\varepsilon, \pi/4)$. Then, for $|x - y| < \delta$,

$$|\sin x - \sin y| \leqslant 2|\cos[(x + y)/2]||\sin[(x - y)/2]|,$$

and, using (2) and (3), we have

$$|\sin x - \sin y| \leqslant 2|x - y|/2 = |x - y| < \delta \leqslant \varepsilon.$$

Thus, f is uniformly continuous on R.

To show that the function f is not uniformly continuous on its domain D, we must find an $\varepsilon > 0$ such that for each $\delta > 0$ there are points $x_\delta, y_\delta \in D$ with

$$|x_\delta - y_\delta| < \delta \quad \text{and} \quad |f(x_\delta) - f(y_\delta)| \geqslant \varepsilon.$$

The following two examples illustrate this and show the difference between continuity and uniform continuity.

Example 4.6.3. Let $f: (0, 1) \to R$ such that $f(x) = 1/x$ for each $x \in (0, 1)$. Exercise 6 of Problem Set 4.4 implies that f is continuous on $(0, 1)$, since it is a rational function.

In order to show that the function $f(x) = 1/x$ is not uniformly continuous, let $\varepsilon = 1$ and let δ be any positive real number. Let h be any real number such that $h < \delta$ and $0 < h < 1/2$. Let $x_\delta = h/2$ and $y_\delta = x_\delta + h = 3h/2$. Then

$$\begin{aligned}
|f(x_\delta) - f(y_\delta)| &= |1/x_\delta - 1/y_\delta| \\
&= |1/(h/2) - 1/(3h/2)| \\
&= 4/(3h) > 1.
\end{aligned}$$

Thus, for each $\delta > 0$, we have found points x_δ and y_δ in $(0, 1)$ such that $|x_\delta - y_\delta| = h < \delta$ and $|f(x_\delta) - f(y_\delta)| > 1$. Therefore, the function f is *not* uniformly continuous on $(0, 1)$.

This is an example where the domain of the function is bounded, but the function is unbounded. The next example presents a function with an unbounded domain which is not uniformly continuous.

Example 4.6.4. Let $f: R \to R$ be defined by $f(x) = x^2$ for each $x \in R$. Since f is a polynomial, it is continuous on R. We show that f is not uniformly continuous on R. (However, f is uniformly continuous on any bounded subset of real numbers.) Let $\varepsilon = 2$, and let δ be any positive real number. Let h be any real number such that $0 < h < \delta$, and let $x = 1/h$ and $y = h + 1/h$. Then

$$\begin{aligned}
|f(x) - f(y)| &= |(1/h)^2 - (h + 1/h)^2| \\
&= 2 + h^2 > 2 = \varepsilon.
\end{aligned}$$

Thus, for each $\delta > 0$, we have found real numbers x and y such that $|x - y| = h < \delta$ and $|f(x) - f(y)| > 2$. Therefore, f is not uniformly continuous on R.

For functions defined on bounded sets the following theorem gives necessary condition for uniform continuity.

Theorem 4.6.1. *Let D be a bounded set, and let $f: D \to R$. If f is uniformly continuous on D, then f is bounded on D.*

The proof is left as Exercise 10 of Problem Set 4.6.

Theorem 4.6.1 simplifies the proof that the function $f: (0, 1) \to R$ such that $f(x) = 1/x$ for each $x \in (0, 1)$ is not uniformly continuous on $(0, 1)$ (Cf. Example 4.6.3.)

The following theorem gives a property common to all uniformly continuous functions, namely, *if f is uniformly continuous on D and a is a limit point of D, then the function f has a limit at a.*

Theorem 4.6.2. *Let $f: D \to R$, and let a be a limit point of D. If f is uniformly continuous on D, then $\lim\limits_{x \to a} f(x)$ exists.*

Proof.   Let a be a limit point of D, and let $\{x_n\} \to a$, where $x_n \in D$ and $x_n \neq a$ for each positive integer n. We show that the sequence $\{f(x_n)\}$ is a Cauchy sequence and thus is convergent. Choose $\varepsilon > 0$. By the uniform continuity of f on D, there is a $\delta > 0$ such that

$$|f(x) - f(y)| < \varepsilon$$

when $x, y \in D$ and $|x - y| < \delta$. Since the sequence $\{x_n\}$ is convergent, it is a Cauchy sequence and there is an N such that $n > m > N$ implies that $|x_n - x_m| < \delta$. Hence for $n > m > N$, we have

$$|f(x_n) - f(x_m)| < \varepsilon$$

and the sequence $\{f(x_n)\}$ is a Cauchy sequence. Therefore, the sequence $\{f(x_n)\}$ is convergent and Theorem 4.3.2 implies that f has a limit at a.

Employing Theorem 4.6.2 we construct a bounded continuous function with a bounded domain that is not uniformly continuous.

Example 4.6.5. Let $f: (0, 1) \to R$ be defined by $f(x) = \sin(1/x)$ for each $x \in (0, 1)$. Exercise 11 of Problem Set 4.3 proves that f is continuous on $(0, 1)$. Since $\sin x$ is bounded for all real numbers x, f is bounded on $(0, 1)$. The point 0 is a limit point of $(0, 1)$, but $\lim\limits_{x \to 0} f(x)$

does not exist (see Example 4.2.4). Theorem 4.6.2 shows that f is *not* uniformly continuous on $(0, 1)$.

We conclude this section by presenting a theorem that gives conditions that ensure that a continuous function is uniformly continuous.

Theorem 4.6.3. *Let $f : D \to R$, and let D be a compact set. If f is continuous on D, then f is uniformly continuous on D.*

Proof. We assume, to the contrary, that f is *not* uniformly continuous on D and show that this leads to a contradiction. Since f is not uniformly continuous on D, there is an $\varepsilon > 0$ such that for each positive integer n, there exist $x_n, y_n \in D$ with

$$|x_n - y_n| < 1/n \quad \text{and} \quad |f(x_n) - f(y_n)| \geq \varepsilon.$$

Since the set D is compact, Theorem 3.4.4 implies that the sequence $\{x_n\}$ has a convergent subsequence $\{x_{n_k}\}$ whose limit, say a, belongs to D. The inequality $|x_{n_k} - y_{n_k}| < 1/n_k$ for each positive integer k assures us that the sequence $\{y_{n_k}\}$ also converges to a. The continuity of f on D implies that the sequences $\{f(x_{n_k})\}$ and $\{f(y_{n_k})\}$ both converge to $f(a)$. However, the condition $|f(x_{n_k}) - f(y_{n_k})| \geq \varepsilon$ for each positive integer k shows that $\lim_k f(x_{n_k}) \neq \lim_k f(y_{n_k})$. This is a contradiction. Therefore, f is uniformly continuous on D.

Problem Set 4.6

1. Show that the function $f(x) = x$ for real each number x is uniformly continuous on R.

2. Show that the function $f(x) = \cos x$ for each real number x is uniformly continuous on R.

3. Show that the function $f(x) = x^3$ for each $x \in [0, 1]$ is uniformly continuous on $[0, 1]$.

4. Show that the function $f(x) = 1/(1 + x^2)$ for each $x \in [0, +\infty)$ is uniformly continuous on $[0, +\infty)$.

5. Show that the function $f(x) = \sqrt{x}$ for $x \in (0, +\infty)$ is uniformly continuous on $(0, +\infty)$.

6. Show that the function $f(x) = x^{3/2}$ for each $x \in (0, +\infty)$ is not uniformly continuous on $(0, +\infty)$.

7. If f is uniformly continuous on the set D, show that f is uniformly continuous on each subset of D.

8. If f is uniformly continuous on an open interval (a, b), prove that $\lim_{x \to a+} f(x)$ and $\lim_{x \to b-} f(x)$ both exist.

9. If f is continuous on an open interval (a, b), and if $\lim\limits_{x \to a+} f(x)$ and $\lim\limits_{x \to b-} f(x)$ both exist, prove that f is uniformly continuous on (a, b).

10. Prove Theorem 4.6.1.

11. Let $f: R \to R$ have the property that $f(x + y) = f(x) + f(y)$ for all $x, y \in R$. Prove that if f is continuous at zero, then f is uniformly continuous.

12. Let $f: D \to R$, and let $g: D \to R$. If f and g are both uniformly continuous on D, show that $f + g$ is uniformly continuous on D.

13. Give an example of two functions f and g with domain R that are uniformly continuous on R, but whose product fg is not uniformly continuous on R.

14. Let $f: D \to R$, let $g: D \to R$, and let D be a bounded set. If f and g are both uniformly continuous on D, show that fg is uniformly continuous on D.

15. Let $f: R \to R$. The function f is said to be **periodic** if and only if there is a real number $p \neq 0$ such that $f(x + p) = f(x)$ for each $x \in R$. If f is continuous and periodic, prove that f is uniformly continuous.

16. Let P be a polynomial. Show that P is uniformly continuous on any bounded set of real numbers.

17. Let f be continuous on R, and let

$$\lim_{x \to +\infty} f(x) = 0 = \lim_{x \to -\infty} f(x).$$

Prove that f is uniformly continuous on R.

Chapter Five

DIFFERENTIATION

5.1. Differentiable Functions

In beginning courses in calculus, there are many theorems that are stated and used but are not proved. This chapter is devoted to defining and proving the standard theorems of differential calculus. We begin by defining the concept of a *derivative*.

Definition 5.1.1. *Let* $f: D \to R$. *The function* f *is said to be* ***differentiable*** *at a point* a *if and only if* $a \in D$, a *is a limit point of* D, *and the limit*

$$\lim_{x \to a} \left[\frac{f(x) - f(a)}{x - a} \right]$$

exists and is finite. If this limit exists and is finite, then its value is called the ***derivative*** *of* f *at* a *and is denoted by* $f'(a)$. *If the function* f *is differentiable at each point of the set* $S \subseteq D$, *then* f *is said to be* ***differentiable on*** S *and the function* $f': S \to R$ *is called the* ***derivative*** *of* f *on* S.

We present three examples to illustrate the concept of differentiable and nondifferentiable functions.

Example 5.1.1. Let $f(x) = x^2$ for each $x \in R$. Then, for any $a \in R$, we have

$$\lim_{x \to a} \left[\frac{f(x) - f(a)}{x - a} \right] = \lim_{x \to a} \left[\frac{x^2 - a^2}{x - a} \right]$$

$$= \lim_{x \to a} \left[\frac{(x - a)(x + a)}{(x - a)} \right]$$

$$= \lim_{x \to a} (x + a) = 2a.$$

Thus the function f is differentiable on R, and $f'(a) = 2a$.

Example 5.1.2. Let f be defined by

$$f(x) = \begin{cases} 1/x & \text{if } x \neq 0, \\ 0 & \text{if } x = 0. \end{cases}$$

Then f has domain R. If $a \neq 0$, then

$$\lim_{x \to a} \left[\frac{f(x) - f(a)}{x - a} \right] = \lim_{x \to a} \left[\frac{1/x - 1/a}{x - a} \right]$$

$$= \lim_{x \to a} \frac{1}{ax} \left[\frac{a - x}{x - a} \right]$$

$$= \lim_{x \to a} (-1/ax) = -1/a^2.$$

Hence, the function f is differentiable on $R \backslash \{0\}$, and if $a \neq 0$ then $f'(a) = -1/a^2$. If $a = 0$, then we have

$$\lim_{x \to 0} \left[\frac{f(x) - f(0)}{x - 0} \right] = \lim_{x \to 0} \left[\frac{1/x}{x} \right] = \lim_{x \to 0} (1/x^2) = +\infty$$

and this limit is not finite. Thus f is not differentiable at 0. This is an example of a function that is differentiable at each point of its domain except one.

Example 5.1.3. Let f be defined by

$$f(x) = \begin{cases} x^2 & \text{if } x \text{ is a rational number,} \\ 0 & \text{if } x \text{ is an irrational number.} \end{cases}$$

Then f has domain R.

Let a be any nonzero rational number. Let $\{x_n\}$ be any sequence of irrational numbers such that $\lim_n x_n = a$. Then

$$\frac{f(x_n) - f(a)}{x_n - a} = \frac{0 - a^2}{x_n - a} = \frac{-a^2}{x_n - a}$$

and

$$\lim_{n} \left[\frac{f(x_n) - f(a)}{x_n - a} \right]$$

does not exist, thus theorem 4.2.5 implies that

$$\lim_{x \to a} \left[\frac{f(x) - f(a)}{x - a} \right]$$

does not exist. Hence, if a is any nonzero rational number, then f is not differentiable at a.

Let a be any irrational number. Let $\{x_n\}$ be any sequence of rational numbers such that $\lim_{n} x_n = a$. Then

$$\frac{f(x_n) - f(a)}{x_n - a} = \frac{x_n^2 - 0}{x_n - a} = \frac{x_n^2}{x_n - a},$$

and so

$$\lim_{n} \left[\frac{f(x_n) - f(a)}{x_n - a} \right]$$

does not exist; thus Theorem 4.2.5 implies that

$$\lim_{x \to a} \left[\frac{f(x) - f(a)}{x - a} \right]$$

does not exist. Hence, if a is any irrational number, then f is not differentiable at a.

Finally, let $a = 0$. Then, for any real number x,

$$\left| \frac{f(x) - f(0)}{x - 0} \right| = \left| \frac{f(x)}{x} \right| \leqslant \left| \frac{x^2}{x} \right| \leqslant |x|.$$

Since $\lim_{x \to 0} |x| = 0$, we have

$$\lim_{x \to 0} \frac{f(x) - f(0)}{x - 0} = 0.$$

(Cf. Exercise 2 of Problem Set 4.2.) Thus, $f'(0) = 0$. This is an example of a function that is differentiable at exactly one point of its domain.

In the above example we used sequences to show that the derivative did not exist at a given point. As in the case of limits and continuity, there is a condition for differentiability in terms of sequences. The proof of the next theorem, which gives this condition, follows directly from Theorem 4.2.5.

Theorem 5.1.1. *Let* $f: D \to R$, *let* $a \in D$, *and let* a *be a limit point of* D. *Then the function* f *is differentiable at* a *if and only if for each sequence* $\{x_n\} \to a$, *with* $x_n \in D$ *and* $x_n \neq a$ *for each positive integer* n, *the sequence*

$$\left\{ \frac{f(x_n) - f(a)}{x_n - a} \right\}$$

converges.

If f is differentiable at a, then for any sequence $\{x_n\} \to a$, with $x_n \in D$ and $x_n \neq a$ for each positive integer n,

$$\left\{ \frac{f(x_n) - f(a)}{x_n - a} \right\} \to f'(a).$$

It is easily observed in each of the above examples that at each point where the function is differentiable, the function is also continuous. This is more than a mere coincidence, as the following theorem points out. This theorem also gives a necessary condition for differentiability.

Theorem 5.1.2. *Let* $f: D \to R$, *and let* f *be differentiable at* $a \in D$. *Then* f *is continuous at* a.

Proof. Since f is differentiable at $a \in D$, a is a limit point of D. We shall show that $\lim_{x \to a} f(x) = f(a)$. Write

$$f(x) = \left[\frac{f(x) - f(a)}{x - a} \right] (x - a) + f(a)$$

for each $x \in D$, $x \neq a$. Since

$$\lim_{x \to a} (x - a) = 0, \qquad \lim_{x \to a} f(a) = f(a),$$

and

$$\lim_{x \to a} \left[\frac{f(x) - f(a)}{x - a} \right] = f'(a),$$

we have

$$\lim_{x \to a} f(x) = \lim_{x \to a} \left\{ \left[\frac{f(x) - f(a)}{x - a} \right] (x - a) + f(a) \right\}$$

$$= \lim_{x \to a} \left\{ \left[\frac{f(x) - f(a)}{x - a} \right] (x - a) \right\} + \lim_{x \to a} f(a)$$

$$= \lim_{x \to a} \left[\frac{f(x) - f(a)}{x - a} \right] \cdot \lim_{x \to a} (x - a) + f(a)$$

$$= f'(a) \cdot 0 + f(a) = f(a).$$

Thus, f is continuous at a.

Problem Set 5.1

1. By using the definition of the derivative, show that any constant function is differentiable on R and that its derivative is the zero function.

2. A function $f: R \to R$ such that $f(x) = cx + d$, where c and d are real numbers, is called a **linear function**. By using the definition of the derivative, show that any linear function is differentiable on R and that its derivative is a constant function

3. By using the definition of the derivative, show that the function $f: R \to R$ such that $f(x) = 2x^2 + 4$ is differentiable on R and that $f'(x) = 4x$ for each $x \in R$.

4. By using the definition of the derivative, show that the function $f: R^+ \to R^+$ such that $f(x) = \sqrt{x}$ is differentiable on R^+ and that $f'(x) = 1/2\sqrt{x}$ for each $x \in R^+$

5. Let $f: R \to R$ be defined by

$$f(x) = \begin{cases} x^3 \sin (1/x) & \text{if } x \neq 0, \\ 0 & \text{if } x = 0. \end{cases}$$

Prove that f is differentiable on R, and find f'. Where is f' continuous? Where is f' differentiable?

6. Let r_1 and r_2 be distinct real numbers. Give an example of a function $f: R \to R$ such that f is differentiable on $R \backslash \{r_1, r_2\}$ but is *not* differentiable at r_1 and r_2. Give an example of a function $f: R \to R$ such that f is differentiable only at r_1 and r_2.

7. Let S be a finite set of real numbers. Give an example of a function $f: R \to R$ such that f is differentiable on $R \backslash S$ but is *not* differentiable at any point of S.

8. Where is the signum function $f(x) = \text{sgn } x$ differentiable? (Cf. Exercise 25 of Problem Set 4.1.)

9. Where is the bracket function $f(x) = [x]$ differentiable?

10. Let $f: R \to R$ be differentiable on R. If f is increasing on R, show that $f'(x) \geqslant 0$ for each $x \in R$.

11. Let $f: [a, b] \to R$. If $c \in (a, b)$ and $f'(c) > 0$, show that there exists a neighborhood T of c such that $x \in [a, c] \cap T$ implies that $f(x) < f(c)$ and $x \in (c, b] \cap T$ implies that $f(c) < f(x)$.

12. Let $f: R \to R$ be differentiable on R. Show that, for each real number a,

$$\lim_{h \to 0} \frac{f(a + h) - f(a - h)}{2h}$$

exists and equals $f'(a)$. Give an example of a function $f: R \to R$ for which the above limit exists for some real number a, but for which the function is *not* differentiable at a.

13. Let $f: D \to R$, let $a \in D$, and let a be a limit point of $D \cap (-\infty, a)$. The function f is said to be **differentiable from the left** or to have a **left-hand derivative** at a if and only if the limit

$$\lim_{x \to a-} \left[\frac{f(x) - f(a)}{x - a} \right]$$

exists and is finite. If f has a left-hand derivative at a it is denoted by $f'_-(a)$. Prove that if f has a left-hand derivative at a, then f is continuous from the left at a.

14. Let $f: D \to R$, let $a \in D$, and let a be a limit point of $D \cap (a, +\infty)$. The function f is said to be **differentiable from the right** or to have a **right-hand derivative** at a if and only if the limit

$$\lim_{x \to a+} \left[\frac{f(x) - f(a)}{x - a} \right]$$

exists and is finite. If f has a right-hand derivative at a it is denoted by $f'_+(a)$. Prove that if f has a right-hand derivative at a, then f is continuous from the right at a.

15. Give an example of a function $f: R \to R$ such that for some real number a, both $f'_-(a)$ and $f'_+(a)$ exist and are finite but are unequal.

16. Let $f: R \to R$, and let a be a limit point of $D \cap (-\infty, a)$ and $D \cap (a, +\infty)$. Show that f is differentiable at a if and only if $f'_-(a)$ and $f'_+(a)$ both exist and $f'_-(a) = f'_+(a)$.

17. Show that the function $f(x) = \sin x$ is differentiable for each $x \in R$, and find f'. (Cf. Exercise 53 of Problem Set 4.1.)

18. Show that the function $f(x) = \cos x$ is differentiable for each $x \in R$, and find f'. (Cf. Exercise 55 of Problem Set 4.1.)

5.2. The Class of Differentiable Functions

The purpose of this section is to show that differentiability is preserved by the standard algebraic operations of functions.

Theorem 5.2.1. *Let $f: D \to R$, and let $g: D \to R$. If f and g are differentiable at $a \in D$ and c is a constant, then the functions $f + g$, $f - g$, cf, and fg are differentiable at a and*

$$(f + g)'(a) = f'(a) + g'(a),$$
$$(f - g)'(a) = f'(a) - g'(a),$$
$$(cf)'(a) = cf'(a),$$

and

$$(fg)'(a) = f(a)g'(a) + f'(a)g(a).$$

We shall prove the result for $f + g$ and leave the remaining parts as Exercise 7 of Problem Set 5.2.

Proof. We need to show that the limit

$$\lim_{x \to a} \left[\frac{(f + g)(x) - (f + g)(a)}{x - a} \right]$$

exists and equals $f'(a) + g'(a)$. Since

$$\lim_{x \to a} \left[\frac{f(x) - f(a)}{x - a} \right] = f'(a)$$

and

$$\lim_{x \to a} \left[\frac{g(x) - g(a)}{x - a} \right] = g'(a),$$

Theorem 4.2.2 implies that

$$\lim_{x \to a} \left[\frac{(f + g)(x) - (f + g)(a)}{x - a} \right] = \lim_{x \to a} \left[\frac{f(x) + g(x) - f(a) - g(a)}{x - a} \right]$$

$$= \lim_{x \to a} \left[\frac{f(x) - f(a)}{x - a} + \frac{g(x) - g(a)}{x - a} \right]$$

$$= \lim_{x \to a} \left[\frac{f(x) - f(a)}{x - a} \right] + \lim_{x \to a} \left[\frac{g(x) - g(a)}{x - a} \right]$$

$$= f'(a) + g'(a).$$

Thus, the above limit exists and

$$(f + g)'(a) = f'(a) + g'(a).$$

The following theorem provides a similar result for quotients.

Theorem 5.2.2. *Let $f: D \to R$, let $g : D \to R$, and let $g(x) \neq 0$ for each $x \in D$. If f and g are differentiable at $a \in D$, then f/g is differentiable at a, and*

$$(f/g)'(a) = [f'(a)g(a) - f(a)g'(a)]/[g(a)]^2.$$

Proof. We need to show that the limit

$$\lim_{x \to a} \left[\frac{(f/g)(x) - (f/g)(a)}{x - a} \right]$$

exists and equals $[f'(a)g(a) - f(a)g'(a)]/[g(a)]^2$.

We may rewrite the difference quotient as follows:

$$\frac{(f/g)(x) - (f/g)(a)}{x - a} = \frac{f(x)/g(x) - f(a)/g(a)}{x - a}$$

$$= \frac{f(x)g(a) - f(a)g(a) + f(a)g(a) - f(a)g(x)}{g(a)g(x)(x - a)}$$

$$= \left\{ g(a)\left[\frac{f(x) - f(a)}{x - a}\right] - f(a)\left[\frac{g(x) - g(a)}{x - a}\right]\right\} \frac{1}{g(a)g(x)}.$$

Since f and g are differentiable at a,

$$f'(a) = \lim_{x \to a}\left[\frac{f(x) - f(a)}{x - a}\right], \qquad g'(a) = \lim_{x \to a}\left[\frac{g(x) - g(a)}{x - a}\right],$$

and

$$\lim_{x \to a} g(x) = g(a) \neq 0,$$

and we have

$$\lim_{x \to a}\left[\frac{(f/g)(x) - (f/g)(a)}{x - a}\right]$$

$$= \lim_{x \to a}\left\{ g(a)\left[\frac{f(x) - f(a)}{x - a}\right] - f(a)\left[\frac{g(x) - g(x)}{x - a}\right]\right\} \frac{1}{g(a)g(x)}$$

$$= [f'(a)g(a) - f(a)g'(a)]/[g(a)]^2.$$

Thus, the above limit exists and

$$(f/g)'(a) = [f'(a)g(a) - f(a)g'(a)]/[g(a)]^2,$$

which concludes the proof.

The following theorem presents conditions which determine when the composite of two functions is differentiable.

Theorem 5.2.3. *Let $f: D \to E$, and let $g : E \to F$. If f is differentiable at $a \in D$ and g is differentiable at $f(a) \in E$, then $g \circ f$ is differentiable at a and*

$$(g \circ f)'(a) = g'(f(a)) \cdot f'(a).$$

Proof.   We need to show that the limit

$$\lim_{x \to a} \left[\frac{(g \circ f)(x) - (g \circ f)(a)}{x - a} \right]$$

exists and equals $g'(f(a))f'(a)$.

Let $b = f(a)$, and define the function $\phi: E \to F$ such that

$$\phi(y) = \begin{cases} \dfrac{g(y) - g(b)}{y - b} - g'(b) & \text{if } y \in E, y \neq b \\ \\ 0 & \text{if } y = b. \end{cases}$$

Since g is differentiable at b, we see that $\lim_{y \to b} \phi(y) = 0$, that is, ϕ is continuous at b. The function f being differentiable at a implies that f is continuous at a. Since f is continuous at a and ϕ is continuous at $b = f(a)$, Theorem 4.3.3 implies that $\phi \circ f$ is continuous at a. Thus,

$$\lim_{x \to a} \phi(f(x)) = \lim_{x \to a} (\phi \circ f)(x)$$

$$= (\phi \circ f)(a)$$

$$= \phi(f(a))$$

$$= \phi(b) = 0.$$

For each $x \in D$ such that $f(x) \neq b = f(a)$, we have

$$\phi(f(x)) = \frac{g(f(x)) - g(f(a))}{f(x) - f(a)} - g'(f(a)),$$

and so

$$g(f(x)) - g(f(a)) = [\phi(f(x)) + g'(f(a))][f(x) - f(a)].$$

This equation is also true for $f(x) = f(a)$, and so it holds for all $x \in D$. Therefore,

$$\lim_{x \to a} \left[\frac{(g \circ f)(x) - (g \circ f)(a)}{x - a} \right] = \lim_{x \to a} \left[\frac{g(f(x)) - g(f(a))}{x - a} \right]$$

$$= \lim_{x \to a} \left[\{\phi(f(x)) + g'(f(a))\} \left\{ \frac{f(x) - f(a)}{x - a} \right\} \right]$$

$$= \lim_{x \to a} \left[\phi(f(x)) + g'(f(a)) \right] \lim_{x \to a} \left[\frac{f(x) - f(a)}{x - a} \right]$$

$$= g'(f(a)) \cdot f'(a),$$

which completes the proof.

Example 5.2.1. As an application of Theorem 5.2.3, we find the derivative of the function

$$h(x) = \sqrt{2x^2 + 4}.$$

To apply Theorem 5.2.3, we write $h = g \circ f$, where

$$g(x) = \sqrt{x} \quad \text{and} \quad f(x) = 2x^2 + 4.$$

Then

$$g'(x) = 1/2\sqrt{x} \quad \text{and} \quad f'(x) = 4x.$$

(Cf. Exercises 3 and 4 of Problem Set 5.1.) Theorem 5.2.3 now implies that

$$h'(x) = g'(f(x))f'(x) = [1/2\sqrt{f(x)}][f'(x)]$$

$$= \frac{4x}{2\sqrt{2x^2 + 4}} = \frac{2x}{\sqrt{2x^2 + 4}}.$$

Problem Set 5.2

1. For each integer $n > 0$, show that the function $f(x) = x^n$ is differentiable and that $f'(x) = nx^{n-1}$.
2. For each integer $n < 0$, show that the function $f(x) = x^n$, with domain $D = R\backslash\{0\}$, is differentiable on D and that $f'(x) = nx^{n-1}$.
3. Show that every polynomial is differentiable.
4. Let $R(x) = P(x)/Q(x)$ for each x such that $Q(x) \neq 0$, where $P(x)$ and $Q(x)$ are polynomials. Show that $R(x)$ is differentiable at each x such that $Q(x) \neq 0$, and find R'.
5. Let $f: R \to R$ be the function defined by

$$f(x) = \begin{cases} x & \text{if } x \text{ is rational,} \\ x^2 + x & \text{if } x \text{ is irrational.} \end{cases}$$

   Show that f is differentiable only at the point 0, and find $f'(0)$.
6. Let $f: D \to R$, and let $C \subset D$. Define $g: C \to R$ by $g(x) = f(x)$ for each $x \in C$. Let $a \in C$, and let a be a limit point of C. If f is differentiable at a, show that g is differentiable at a. Give an example to show that differentiability of g at $a \in C$ need not imply differentiability of f at a.
7. Prove the remaining parts of Theorem 5.2.1.
8. Let $f: D \to R$, and let $a \in D$. Use the definition of derivative to show that if $f(a) \neq 0$ and f is differentiable at a, then

$$(1/f)'(a) = -f'(a)/[f(a)]^2.$$

9. Let $f: D \to R$. Show that the set of points where the left-hand derivative and the right-hand derivative both exist, but are not equal, is countable.

10. Let $f: R \to R$ be defined by

$$f(x) = \begin{cases} 0 & \text{if } x \leqslant 0, \\ x^n & \text{if } x > 0, \end{cases}$$

where n is an integer. For what values of n is the function f differentiable on R? For what values of n is the function f' continuous on R? For what values on n is the function f' differentiable on R?

11. Let $f: R \to R$ be defined by

$$f(x) = \begin{cases} x^2 \sin (1/x) & \text{if } x \neq 0, \\ 0 & \text{if } x = 0. \end{cases}$$

Show that f is differentiable and find f'. Where is f' continuous? Where is f' differentiable?

12. Let $f: R \to R$ be defined by

$$f(x) = \begin{cases} x + 2x^2 \sin (1/x) & \text{if } x \neq 0, \\ 0 & \text{if } x = 0. \end{cases}$$

Show that f is differentiable on R. Show that there exists a number a such that $f'(a) > 0$, and show that there does not exist a neighborhood of a where f is increasing.

13. Let $f: D \to R$, let $a \in D$, and let a be a limit point of D. For each natural number n define:

$$d_n^+(f; a) = \sup \left\{ \frac{f(a) - f(x)}{a - x} : x \in D \text{ and } 0 < x - a < 1/n \right\},$$

$$d_{n+}(f; a) = \inf \left\{ \frac{f(a) - f(x)}{a - x} : x \in D \text{ and } 0 < x - a < 1/n \right\},$$

$$d_n^-(f; a) = \sup \left\{ \frac{f(a) - f(x)}{a - x} : x \in D \text{ and } 0 < a - x < 1/n \right\},$$

$$d_{n-}(f; a) = \inf \left\{ \frac{f(a) - f(x)}{a - x} : x \in D \text{ and } 0 < a - x < 1/n \right\}.$$

If the sup or inf does not exist, then we assign the value $+\infty$ or $-\infty$, respectively. Show that the sequences $\{d_n^+(f; a)\}$ and $\{d_n^-(f; a)\}$ are nonincreasing and that the sequences $\{d_{n+}(f; a)\}$ and $\{d_{n-}(f; a)\}$ are nondecreasing.

14. Let $d^+(f; a) = \lim_{n \to +\infty} d_n^+(f; a)$, and let $d_+(f; a) = \lim_{n \to +\infty} d_{n+}(f; a)$. The number $d^+(f; a)$ is called the **upper-right Dini derivate**, while $d_+(f; a)$ is called the **lower-right Dini derivate**. Show that $d_+(f; a) \leqslant d^+(f; a)$.

15. Let $d^-(f; a) = \lim\limits_{n \to +\infty} d_n{}^-(f; a)$, and let $d_-(f; a) = \lim\limits_{n \to +\infty} d_{n-}(f; a)$. The number $d^-(f; a)$ is called the **upper-left Dini derivate**, while $d_-(f; a)$ is called the **lower-left Dini derivate**. Show that $d^-(f; a) \leqslant d_-(f; a)$.

16. Let $f: D \to R$, and let $a \in D$. Show that f has a right-hand derivative at a if and only if the upper-right and lower-right Dini derivates at a both exist and are finite and equal.

17. Let $f: D \to R$, and let $a \in D$. Show that f has a left-hand derivative at a if and only if the upper-left and lower-left Dini derivates at a both exist and are finite and equal.

18. Let $f: D \to R$, and let $a \in D$. Show that f is differentiable at a if and only if all four Dini derivates exist and are finite and equal.

19. Let $f: R \to R$ be differentiable at a. Let $\{x_n\}$ and $\{y_n\}$ be two sequences in $R \backslash \{a\}$ which converge to a such that the sequence

$$\left\{ \frac{y_n - a}{y_n - x_n} \right\}$$

is bounded. Prove that the sequence

$$\left\{ \frac{f(y_n) - f(x_n)}{y_n - x_n} \right\}$$

converges to $f'(a)$.

20. If n is a rational number with $n \geqslant 1$, show that the function $f(x) = x^n$ is differentiable and that $f'(x) = n\,x^{n-1}$.

21. If n is a rational number with $n < 1$, show that the function $f(x) = x^n$, with domain $D = R \backslash \{0\}$, is differentiable on D and that $f'(x) = n\,x^{n-1}$.

5.3. Properties of Differentiable Functions

We now study some of the other properties of differentiable functions. We begin with the standard theorem on maxima and minima.

Definition 5.3.1. *Let* $f: D \to R$. *The point* $a \in D$ *is called a **relative (local) maximum point** of f if and only if there is a neighborhood T of a such that* $x \in T \cap D$ *implies that*

$$f(x) \leqslant f(a).$$

The value $f(a)$ *is called a **relative maximum** of f. The point $a \in D$ is called a **relative (local) minimum point** of f if and only if there is a neighborhood T of a such that* $x \in T \cap D$ *implies that*

$$f(x) \geqslant f(a).$$

The value $f(a)$ *is called a **relative minimum** of f.*

In elementary calculus courses, students usually determine relative maxima and relative minima by finding where the derivative is zero. If we consider the function $f(x) = |x|$ on the closed interval $[-1, 1]$, we see that f has a relative maximum at 1 and a relative minimum at 0, but $f'(x) = 1$ for every $x \in (0, 1]$, $f'(x) = -1$ for every $x \in [-1, 0)$, and $f'(0)$ does not exist. Thus, we see that it is not necessary for a function to have zero derivative to have a relative maximum or a relative minimum. The following theorem points out one relation between relative maxima or minima and zero derivatives.

Theorem 5.3.1. *Let $f: [a, b] \rightarrow R$. If $c \in (a, b)$ is a relative maximum point or a relative minimum point of f, and if f is differentiable at c, then $f'(c) = 0$.*

Proof. We consider the case where c is a relative maximum point of f. The case where c is a relative minimum point can be handled in a similar manner. Assume that f is differentiable at c. Then

$$\lim_{x \to c} \left[\frac{f(x) - f(c)}{x - c} \right]$$

exists, so

$$\lim_{x \to c+} \left[\frac{f(x) - f(c)}{x - c} \right]$$

and

$$\lim_{x \to c-} \left[\frac{f(x) - f(c)}{x - c} \right]$$

both exist and are equal to $f'(c)$. We shall show that

$$\lim_{x \to c+} \left[\frac{f(x) - f(c)}{x - c} \right] \leqslant 0$$

and

$$\lim_{x \to c-} \left[\frac{f(x) - f(c)}{x - c} \right] \geqslant 0,$$

which together imply that $f'(c) = 0$.

Since c is a relative maximum point, there is a neighborhood T of c such that $x \in T \cap [a, b]$ implies that

$$f(x) \leqslant f(c) \quad \text{or} \quad f(x) - f(c) \leqslant 0.$$

For $x \in T \cap [a, c)$, $x - c < 0$, and so

$$\frac{f(x) - f(c)}{x - c} \geqslant 0.$$

Thus,

$$\lim_{x \to c-} \left[\frac{f(x) - f(c)}{x - c} \right] = f'(c) \geqslant 0.$$

Likewise, for $x \in T \cap (c, b], x - c > 0$, and so

$$\frac{f(x) - f(c)}{x - c} \leqslant 0.$$

Hence,

$$\lim_{x \to c+} \left[\frac{f(x) - f(c)}{x - c} \right] = f'(c) \leqslant 0.$$

Therefore $f'(c) = 0$.

We use Theorem 5.3.1 to prove the following result which is called *Rolle's Theorem* and is named for the French mathematician Michell Rolle (1652–1719).

Theorem 5.3.2. *(Rolle's Theorem) Let f: $[a, b] \to R$. If f is continuous on $[a, b]$ and differentiable on (a, b), and if $f(a) = f(b)$, then there is at least one point $c \in (a, b)$ such that $f'(c) = 0$.*

Proof. If f is constant, then $f'(c) = 0$ for each point $c \in (a, b)$. (Cf. Exercise 1 of Problem Set 5.1.) Thus, we assume that f is not a constant function. Since f is continuous on $[a, b]$, Theorem 4.5.2 implies that there is a point $\zeta \in [a, b]$ such that $f(\zeta)$ is the maximum of f on $[a, b]$ and hence is a relative maximum of f on $[a, b]$. If $\zeta \in (a, b)$, then Theorem 5.3.1 implies that $f'(\zeta) = 0$. If $\zeta \notin (a, b)$, then either $\zeta = a$ or $\zeta = b$. In this case, the maximum value of f on $[a, b]$ is $f(a) = f(b)$, and, since f is not a constant function, f takes on values less than $f(a)$. Theorem 4.5.2 implies that there is a point $\xi \in [a, b]$ such that $f(\xi)$ is the minimum of f on $[a, b]$. Then $f(\xi) < f(a)$ and $\xi \in (a, b)$. Since $f(\xi)$ is the minimum of f on $[a, b]$, $f(\xi)$ is also a relative minimum of f on $[a, b]$. Then Theorem 5.3.1 implies that $f'(\xi) = 0$. In any case, there exists a point $c \in (a, b)$ such that $f'(c) = 0$.

We use Rolle's Theorem to prove the *Generalized Mean Value Theorem* from which the *Mean Value Theorem* follows as a corollary.

Theorem 5.3.3. *(Generalized Mean Value Theorem) Let f: $[a, b] \to R$, and let g : $[a, b] \to R$. If f and g are continuous on $[a, b]$, if f and g are* *differentiable on* (a, b), *and if* $g'(x) \neq 0$ *for each* $x \in (a, b)$, *then there is at least one point* $c \in (a, b)$ *such that*

$$\frac{f'(c)}{g'(c)} = \frac{f(b) - f(a)}{g(b) - g(a)}.$$

Proof. If $g(b) - g(a) = 0$, then the function G defined by $G(x) = g(x) - g(a)$ for $x \in [a, b]$ satisfies the conditions of Rolle's Theorem, and so there is a $\zeta \in (a, b)$ such that $G'(\zeta) = g'(\zeta) = 0$. But this contradicts the fact that $g'(x) \neq 0$ for each $x \in (a, b)$. Thus $g(b) - g(a) \neq 0$.

We now define an auxiliary function ϕ by the equation

$$\phi(x) = f(x) - f(a) - \left[\frac{f(b) - f(a)}{g(b) - g(a)}\right][g(x) - g(a)]$$

for each $x \in [a, b]$. Then ϕ is continuous on $[a, b]$ and differentiable on (a, b). It is easy to verify that $\phi(a) = 0 = \phi(b)$. Thus ϕ satisfies the conditions of Rolle's Theorem, and so there exists a point $c \in (a, b)$ such that $\phi'(c) = 0$. Calculating ϕ', we have

$$\phi'(x) = f'(x) - \left[\frac{f(b) - f(a)}{g(b) - g(a)}\right] g'(x).$$

Substituting c for x and solving for $f'(c)/g'(c)$ in the equation $\phi'(c) = 0$, we have

$$\frac{f'(c)}{g'(c)} = \frac{f(b) - f(a)}{g(b) - g(a)}.$$

As a corollary to Theorem 5.3.3 we have the *Mean Value Theorem*.

Theorem 5.3.4. (*Mean Value Theorem*). *Let* $f: [a, b] \to R$. *If* f *is continuous on* $[a, b]$ *and differentiable on* (a, b), *then there is at least one point* $c \in (a, b)$ *such that*

$$f'(c) = \frac{f(b) - f(a)}{b - a}.$$

Proof. This corollary follows from the Generalized Mean Value Theorem by taking the function g to be the identity function on $[a, b]$.

From Exercise 1, Problem Set 5.1, we saw that the derivative of a constant is zero. As an application of the Mean Value Theorem, we show that the converse of this is also true.

Theorem 5.3.5. *Let* $f: [a, b] \to R$. *If* f *is continuous on* $[a, b]$ *and* $f'(x) = 0$ *for each* $x \in (a, b)$, *then* f *is constant on* $[a, b]$.

Proof. We assume, to the contrary, that f is not constant on $[a, b]$. Then there are points $x, y \in [a, b]$ such that $x < y$ and $f(x) \neq f(y)$. The function f is continuous on $[x, y]$ and differentiable on (x, y) and hence satisfies the hypothesis of the Mean Value Theorem. Thus, there is a point $c \in (x, y)$ such that

$$f'(c) = \frac{f(y) - f(x)}{y - x} \neq 0.$$

This is a contradiction, and so f must be constant on $[a, b]$.

As a corollary to Theorem 5.3.5 we have the following result.

Theorem 5.3.6. *Let* $f: [a, b] \to R$, *and let* $g: [a, b] \to R$. *If* f *and* g *are continuous on* $[a, b]$ *and* $f'(x) = g'(x)$ *for each* $x \in (a, b)$, *then there exists a constant* C *such that*

$$f(x) = g(x) + C.$$

Proof. Consider the function $F: [a, b] \to R$ such that

$$F(x) = f(x) - g(x)$$

for each $x \in [a, b]$. It follows from the hypotheses that F is continuous on $[a, b]$ and differentiable on (a, b), and that $F'(x) = 0$ for each $x \in (a, b)$. Thus, Theorem 5.3.5 implies that there exists a constant C such that

$$F(x) = f(x) - g(x) = C.$$

Hence

$$f(x) = g(x) + C.$$

In Chapter six, we shall prove that every continuous function is a derivative of some function, that is, *if* f *is continuous on* $[a, b]$, *then there is a function* $g: [a, b] \to R$ *such that* $f = g'$. The following example shows that not all derivatives are continuous.

Example 5.3.1. Let $f: R \to R$ be defined by

$$f(x) = \begin{cases} x^2 \sin (1/x) & \text{if } x \neq 0, \\ 0 & \text{if } x = 0. \end{cases}$$

Then

$$f'(x) = \begin{cases} 2x \sin (1/x) - \cos (1/x) & \text{if } x \neq 0, \\ 0 & \text{if } x = 0. \end{cases}$$

(Cf. Exercise 11 of Problem Set 5.2.) Thus, f' is not continuous at 0.

Even though derivatives need not be continuous, they do have one property in common with continuous function—the *Intermediate Value Property*—which we now define.

Definition 5.3.2. *Let I be an open interval, and let $f: I \to R$. The function f is said to have the **Intermediate Value Property** (or the **Darboux property**) if and only if for each $a,b \in I$, such that $f(a) \neq f(b)$, and C between $f(a)$ and $f(b)$, there is a c between a and b such that*

$$f(c) = C.$$

The Darboux property is named for the French mathematician Gaston Darboux (1842–1917) who was the first to prove the Intermediate Value Property for derivatives.

Theorem 4.5.3 shows that if f is continuous on the open interval I, then f has the Intermediate Value Property. We now show that all derivatives have the Intermediate Value Property.

Theorem 5.3.7. (*Intermediate Value Theorem for Derivatives*) *Let I be an open interval and $f: I \to R$. If f is a derivative, then f has the Intermediate Value Property.*

Proof. We first show that if $a,b \in I$, $a < b$, and $f(a) < 0 < f(b)$, then there is $c \in (a, b)$ such that $f(c) = 0$.

Since f is a derivative, there is a function $g: I \to R$ such that $g'(x) = f(x)$ for each $x \in I$. By Theorem 5.1.2 we know that the function g is continuous on I, and Theorem 4.5.2 implies that g has a minimum on $[a, b]$. Since $g'(b) = f(b) > 0$, there exists $\delta_1 > 0$ such that $a < b - \delta_1$ and $g(b - \delta_1) < g(b)$. (Cf. Exercise 11 of Problem Set 5.1.) Similarly, since $g'(a) = f(a) < 0$, there is $\delta_2 > 0$ such that $a + \delta_2 < b$ and $g(a) > g(a + \delta_2)$. Thus, the minimum of g is at some point $c \in (a, b)$. Theorem 5.3.1 implies that

$$g'(c) = f(c) = 0.$$

In general, let $a,b \in I$, $a < b$, and let C be any real number between $f(a)$ and $f(b)$. We shall use the first part of this proof to prove the general case. Either $f(a) < C < f(b)$ or $f(b) < C < f(a)$.

CASE 1. $(f(a) < C < f(b))$. Since f is a derivative on I, there is $g: I \to R$ such that $g'(x) = f(x)$ for each $x \in I$. Let $h: I \to R$ such that

$$h(x) = g(x) - Cx.$$

Then

$$h'(x) = g'(x) - C = f(x) - C,$$

$$h'(a) = g'(a) - C = f(a) - C < 0,$$

and

$$h'(b) = g'(b) - C = f(b) - C > 0.$$

The first part of this proof now implies that there is $c \in (a, b)$ such that $h'(c) = 0$. Thus,

$$h'(c) = 0 = g'(c) - C = f(c) - C$$

and

$$f(c) = C.$$

CASE 2. $(f(b) < C < f(a))$. In this case, let $h: I \to R$ such that

$$h(x) = Cx - g(x).$$

The remainder of the proof is similar to Case 1.

Thus, the derivative f has the Intermediate Value Property on I, and this concludes the proof.

Problem Set 5.3

1. Let $f: R \to R$ be defined by

$$f(x) = \begin{cases} x^2 \cos(1/x) & \text{if } x \neq 0, \\ 0 & \text{if } x = 0. \end{cases}$$

Show that f is differentiable and find f'. Where is f' continuous? Where is f' differentiable?

2. Let $f: R \setminus \{-d/c\} \to R$ be defined by $f(x) = (ax + b)/(cx + d)$ for each $x \in R \setminus \{-d/c\}$. Show that if f has a relative maximum or a relative minimum, then f is constant.

3. Let $a_1, a_2, ..., a_n$ be real numbers. Let $f: R \to R$ be defined by

$$f(x) = \sum_{i=1}^{n} (a_i - x)^2$$

for each $x \in R$. Show that the only relative minimum point of f is

$$x = \frac{1}{n} \sum_{i=1}^{n} a_i.$$

4. Let a and b be two real numbers with $a \neq b$. Let $f: R \to R$ be defined by

$$f(x) = a|x| + b|x - 1|$$

for each $x \in R$. Show that f has a minimum value if and only if $a + b \geqslant 0$. If $a + b \geqslant 0$, show that the minimum value of f is min (a, b).

5. Let $h > 0$. Use the Mean Value Theorem to show that $\sqrt{1 + h} < 1 + h/2$.

6. Let $h > 0$. If p is a rational number with $0 < p < 1$, use the Mean Value Theorem to show that $(1 + h)^p < 1 + ph$.

7. Let $h > 0$. If p is a rational number with $p > 1$, use the Mean Value Theorem to show that $(1 + h)^p > 1 + ph$.

8. Let $f: [a, b] \to R$. If f is continuous on $[a, b]$ and differentiable on (a, b), and if f' is nonzero on (a, b), use the Mean Value Theorem to show that f is 1–1.

9. Let $f: [a, b] \to R$. If f is continuous on $[a, b]$ and differentiable on (a, b), and if $f'(x) > 0$ for each $x \in (a, b)$, use the Mean Value Theorem to show that f is **strictly increasing** on $[a, b]$, that is, if $x > y$ then $f(x) > f(y)$. If $f'(x) \geqslant 0$ for each $x \in (a, b)$, show that f is increasing on $[a, b]$.

10. Let $f: [a, b] \to R$. If f is continuous on $[a, b]$ and differentiable on $[a, b]$, and if $f'(x) \neq 0$ for each $x \in (a, b)$, use the Intermediate Value Theorem to show that $f'(x) > 0$ for all $x \in [a, b]$ or $f'(x) < 0$ for all $x \in [a, b]$.

11. Let $f: [a, b] \to R$. If f is continuous on $[a, b]$ and differentiable on $[a, b]$, and if $f'(x) \neq 0$ for each $x \in (a, b)$, show that f is 1–1, f^{-1} is continuous and differentiable on $f[[a, b]]$, and

$$(f^{-1})'(f(x)) = 1/f'(x).$$

12. Let $f: (a, b) \to R$, and let $g: (a, b) \to R$. If f and g are differentiable on (a, b),

$$\lim_{x \to b^-} g'(x) \neq 0,$$

and

$$\lim_{x \to b^-} f(x) = \lim_{x \to b^-} g(x) = 0,$$

use the Generalized Mean Value Theorem to show that, if

$$\lim_{x \to b^-} \frac{f'(x)}{g'(x)}$$

exists (finite, $+\infty$, or $-\infty$), then

$$\lim_{x \to b^-} \frac{f(x)}{g(x)} \text{ exists and equals } \lim_{x \to b^-} \frac{f'(x)}{g'(x)}.$$

13. Let $f: (a, b) \to R$, and let $g: (a, b) \to R$. If f and g are differentiable on (a, b),

$$\lim_{x \to b^-} g'(x) \neq 0,$$

and

$$\lim_{x \to b^-} f(x) = \lim_{x \to b^-} g(x) = +\infty \, (-\infty),$$

use the Generalized Mean Value Theorem to show that, if

$$\lim_{x \to b^-} \frac{f'(x)}{g'(x)}$$

exists (finite, $+\infty$, or $-\infty$), then

$$\lim_{x \to b^-} \frac{f(x)}{g(x)} \text{ exists and equals } \lim_{x \to b^-} \frac{f'(x)}{g'(x)}.$$

14. Let $f:(a, +\infty) \to R$, and let $g:(a, +\infty) \to R$. If f and g are differentiable on $(a, +\infty)$,

$$\lim_{x \to +\infty} g'(x) \neq 0,$$

and

$$\lim_{x \to +\infty} f(x) = \lim_{x \to +\infty} g(x) = 0,$$

use the Generalized Mean Value Theorem to show that, if

$$\lim_{x \to +\infty} \frac{f'(x)}{g'(x)}$$

exists (finite, $+\infty$, or $-\infty$), then

$$\lim_{x \to +\infty} \frac{f(x)}{g(x)} \text{ exists and equals } \lim_{x \to +\infty} \frac{f'(x)}{g'(x)}.$$

15. Let $f:(a, +\infty) \to R$, and let $g:(a, +\infty) \to R$. If f and g are differentiable on $(a, +\infty)$,

$$\lim_{x \to +\infty} g'(x) \neq 0,$$

and

$$\lim_{x \to +\infty} f(x) = \lim_{x \to +\infty} g(x) = +\infty \, (-\infty),$$

use the Generalized Mean Value Theorem to show that, if

$$\lim_{x \to +\infty} \frac{f'(x)}{g'(x)}$$

exists (finite, $+\infty$, or $-\infty$), then

$$\lim_{x \to +\infty} \frac{f(x)}{g(x)} \text{ exists and equals } \lim_{x \to +\infty} \frac{f'(x)}{g'(x)}.$$

[The results in Exercises 12, 13, 14, and 15 are forms of **L'Hopital's rules**].

16. Let $f:[a, b] \to R$, and let $c \in (a, b)$. Show that if f is continuous on $[a, b]$ and differentiable on (a, c) and (c, b), and if $\lim_{x \to c} f'(x)$ exists, then $f'(c)$ exists and equals the value of the above limit.

17. Let $f:[a, b] \to R$, let $g:[a, b] \to R$, and let $[c, d] \subset (a, b)$. If f and g are differentiable on (a, b), $x \in [c, d]$, and $f'(x) \leqslant g'(x)$ for **each** $x \in [c, d]$, show that $f(x) - f(c) \leqslant g(x) - g(c)$.

18. Let $f:(a, +\infty) \to R$, and let f be differentiable on $(a, +\infty)$. If $\lim_{x \to +\infty} f(x)$ and $\lim_{x \to +\infty} f'(x)$ both exist and are finite, prove that $\lim_{x \to +\infty} f'(x) = 0$.

19. Let $f:(a, +\infty) \to R$, and let f be differentiable on $(a, +\infty)$. If $\lim_{x \to +\infty} f'(x) = 0$, prove that

$$\lim_{x \to +\infty} \frac{f(x)}{x} = 0.$$

20. Let $f: (a, +\infty) \to R$, and let f be differentiable on $(a, +\infty)$. If $\lim\limits_{x \to +\infty} f'(x) = b \neq 0$, prove that

$$\lim_{x \to +\infty} \frac{f(x)}{bx} = 1.$$

21. Let $f: (a, +\infty) \to R$, and let f be differentiable on $(a, +\infty)$. If $\lim\limits_{x \to +\infty} f'(x)$ exists, prove that for each nonzero number h,

$$\lim_{x \to +\infty} \frac{f(x + h) - f(x)}{h}$$

exists.

22. Let $f: (a, b) \to R$. If f is differentiable on (a, b) and f' is bounded on (a, b), show that f is uniformly continuous on (a, b).

23. Give an example of a function $f: (0, 1) \to R$ such that f is differentiable on $(0, 1)$, and f is uniformly continuous on $(0, 1)$, but f' is not bounded on $(0, 1)$.

24. Let $f: D \to R$. The function f is said to satisfy a **Lipschitz condition** at $a \in D$ if and only if there are constants A and δ such that $x \in D$ and $|x - a| < \delta$ imply that $|f(x) - f(a)| < A|x - a|$. Give an example of a function $f: D \to R$ such that at some point $a \in D, f$ is continuous but f does not satisfy a Lipschitz condition at a.

25. Let $f: D \to R$. If f is differentiable at $a \in D$, show that f satisfies a Lipschitz condition at a.

26. Let $f: D \to R$. The function f is said to satisfy a **Lipschitz condition** on D if and only if there is a constant A such that $x, y \in D$ implies that

$$|f(x) - f(y)| < A|x - y|.$$

If $f: D \to R$ satisfies a Lipschitz condition on D, show that f is uniformly continuous on D. Give an example to show that the converse is not true.

27. Let $f: D \to R$. The function f is said to be **uniformly differentiable** on D if and only if f is differentiable on D and for each $\varepsilon > 0$ there exists $\delta > 0$ such that $x, y \in D$ and $0 < |x - y| < \delta$ imply that

$$\left| \frac{f(x) - f(y)}{x - y} - f'(x) \right| < \varepsilon.$$

Show that if f is uniformly differentiable on D then f' is continuous on D.

28. Give an example of a function that has the Intermediate Value Property but is continuous at no more than one point.

29. Let $f: R \to R$ and $g: R \to R$ where

$$f(x) = \begin{cases} 1/x & \text{if } x \neq 0, \\ 0 & \text{if } x = 0, \end{cases}$$

and

$$g(x) = \begin{cases} 1/x + \text{sgn } x & \text{if } x \neq 0, \\ 0 & \text{if } x = 0. \end{cases}$$

The functions f and g have the same derivative on $R \setminus \{0\}$, but do not differ by a constant. Explain how this is possible in the presence of Theorem 5.3.6.

5.4. Higher Derivatives

If $f: D \to R$ is differentiable on D, then $f': D \to R$. If f' is differentiable at $a \in D$, then the derivative of f' at a is called the **second derivative** of f at a, and f is said to be **twice differentiable** at a. The second derivative of f at a is denoted by $f''(a)$ or $f^{(2)}(a)$. If f' is differentiable on D then f is said to be **twice differentiable** on D, and the derivative of f', denoted by f'' or $f^{(2)}$, is called the **second derivative** of f on D. It is easy to see that the idea of taking the derivative of a derivative can be extended. If, for some positive integer n, the function $f: D \to R$ can be differentiated n times, then we say that f is **n-times differentiable** or that f has an **nth derivative**; the nth derivative of f is denoted by $f^{(n)}$.

Example 5.4.1. As an example of a function that is 5-times differentiable, but not 6-times differentiable, consider the function $f: R \to R$ defined by

$$f(x) = \begin{cases} x^6 & \text{if } x \geqslant 0, \\ 0 & \text{if } x < 0. \end{cases}$$

The fifth derivative of f is given by

$$f^{(5)}(x) = \begin{cases} 6!x & \text{if } x \geqslant 0, \\ 0 & \text{if } x < 0. \end{cases}$$

The sixth derivative of f does not exist at 0. Hence f is 5-times differentiable on R, but not 6-times differentiable on R.

If a function f is $(n + 1)$-times differentiable on an interval, then there is an extension of the Mean Value Theorem to higher derivatives. This result is called the *Extended Mean Value Theorem*.

Theorem 5.4.1. (*Extended Mean Value Theorem*). *Let* $f: [a, b] \to R$. *If* $f, f', f'', \ldots, f^{(n)}$ *are continuous on* $[a, b]$ *and* f *is* $(n + 1)$-*times differentiable on* (a, b), *then there is* $c \in (a, b)$ *such that*

$$f(b) = f(a) + (b - a)f'(a) + \frac{(b - a)^2}{2!} f^{(2)}(a) + \ldots$$

$$+ \frac{(b - a)^n}{n!} f^{(n)}(a) + \frac{(b - a)^{n+1}}{(n + 1)} f^{(n+1)}(c).$$

Proof. The method used to establish the Generalized Mean Value Theorem (Theorem 5.3.3) will be extended to this theorem.

Define the auxiliary function $F: [a, b] \to R$ by the formula

$$F(x) = f(b) - \left\{ f(x) + (b - x)f'(x) + \ldots + \frac{(b - x)^n}{n!}f^{(n)}(x) \right\}$$
$$+ \frac{A(b - x)^{n+1}}{(n + 1)!},$$

where A is a constant to be specified. Since $f, f', f'', \ldots, f^{(n)}$ are continuous on $[a, b]$ and f is $(n + 1)$-times differentiable on (a, b), we have that F is continuous on $[a, b]$ and differentiable on (a, b). Also, $F(b) = 0$. If we let

$$A = \frac{(n + 1)!}{(b - a)^{n+1}} \left\{ -f(b) + f(a) + (b - a)f'(a) + \ldots + \frac{(b - a)^n}{n!}f^{(n)}(a) \right\},$$

then $F(a) = 0$. Thus, we see that F satisfies the hypotheses of Rolle's Theorem (Theorem 5.3.2) on $[a, b]$. Hence there is $c \in (a, b)$ such that $F'(c) = 0$.

From the definition of F, we see that

$$F'(x) = - \left\{ f'(x) + (b - x)f^{(2)}(x) + \ldots + \frac{(b - x)^n}{n!}f^{(n+1)}(x) \right\}$$
$$+ \left\{ f'(x) + (b - x)f^{(2)}(x) + \ldots + \frac{(b - x)^{n-1}}{(n - 1)!}f^{(n)}(x) \right\}$$
$$- \frac{A(b - x)^n}{n!}$$
$$= - \frac{(b - x)^n}{n!}f^{(n+1)}(x) - \frac{A(b - x)^n}{n!} .$$

Thus,

$$- \frac{(b - c)^n}{n!}f^{(n+1)}(c) = \frac{A(b - c)^n}{n!}$$

or

$$A = -f^{(n+1)}(c).$$

Substituting back in for A, we have

$$-f^{(n+1)}(c) = \frac{(n+1)!}{(b-a)^{n+1}} \left\{ -f(b) + f(a) + (b-a)f'(a) + \right.$$

$$\left. \dots + \frac{(b-a)^n}{n!} f^{(n)}(a) \right\},$$

or

$$f(b) = f(a) + (b-a)f'(a) + \dots + \frac{(b-a)^n}{n!} f^{(n)}(a)$$

$$+ \frac{(b-a)^{n+1}}{(n+1)!} f^{(n+1)}(c),$$

which concludes the proof.

This result establishes the fact that if f is $(n+1)$-times differentiable at a point, then in a neighborhood of that point the function f can be approximated by a polynomial of degree n, with the error given in terms of the $(n+1)$st derivative.

If we let $b = x$ and $a = 0$, or we let $a = x$ and $b = 0$, then Theorem 5.4.1 yields the following formula, called **Taylor's formula with a remainder**,

$$f(x) = f(0) + xf'(0) + \dots + \frac{x^n}{n!} f^{(n)}(0) + \frac{x^{n+1}}{(n+1)!} f^{(n+1)}(c),$$

where c is between 0 and x.

Example 5.4.2. Let $f: R \to R$ such that $f(x) = \sin x$. Then, for each positive integer n,

$$f^{(n)}(x) = \begin{cases} (-1)^{n/2} \sin x & \text{if } n \text{ is even,} \\ (-1)^{(n-1)/2} \cos x & \text{if } n \text{ is odd.} \end{cases}$$

Thus,

$$\sin x = x - \frac{x^3}{5!} + \frac{x^5}{5!} - \cos(c)\frac{x^7}{7!},$$

where c is between 0 and x. We see that $\sin(.1) = .0998333416$, with an error less than .0000000001.

Let $f: D \to R$, and let $a \in D$. If for each natural number n, f is n-times differentiable at a, then f is said to be **infinitely differentiable** at a. If at each $a \in D$, f is infinitely differentiable, then f is said to be **infinitely differentiable** on D.

Example 5.4.3. The function f defined in Example 5.4.2 is infinitely differentiable on R.

Problem Set 5.4

1. Let $f: R \to R$ be defined by $f(x) = \cos x$. Show that f is infinitely differentiable, and for each natural number n find $f^{(n)}$.
2. Let P be a polynomial. Show that P is infinitely differentiable, and for each natural number n find $P^{(n)}$.
3. Let $f: D \to R$, let $a \in D$, and let D be a neighborhood of a. If the limit

$$\lim_{h \to 0} \left[\frac{f(a + h) + f(a - h) - 2f(a)}{h^2} \right]$$

exist and is finite, then f is said to have a **generalized second derivative** at a and its value is given by the above limit. If f has a second derivative at a, show that f has a generalized second derivative at a and that the two are equal.
4. Give an example of a function which has a generalized second derivative at a but which does *not* have a second derivative at a.
5. Use the Extended Mean Value Theorem to show that

$$\cos x \geqslant 1 - x^2/2$$

for all $x \in R$ and that

$$\cos x > 1 - x^2/2$$

if $x \neq 0$.
6. Let $f: I \to R$, where I is an open interval, and let $a \in I$. If f is $(n + 1)$-times differentiable on I, show that for each $x \in I$ there exists a point ζ between a and x ($\zeta = a$ if $x = a$) such that

$$f(x) = f(a) + f'(a)(x - a) + \frac{f^{(2)}(a)}{2!} (x - a)^2 + \ldots$$

$$+ \frac{f^{(n)}(a)}{n!} (x - a)^n + \frac{f^{(n+1)}(\zeta)}{(n + 1)!} (x - a)^{n+1}.$$

7. Let $f: I \to R$, where I is an open interval, and let $a \in I$. If f is $(n + 1)$-times differentiable on I, show that for any h such that $a + h \in I$ there exists θ, where $0 \leqslant \theta < 1$, such that

$$f(a + h) = f(a) + f'(a)h + \frac{f^{(2)}(a)h^2}{2!} + \ldots$$

$$+ \frac{f^{(n)}(a)}{n!} h^n + \frac{f^{(n+1)}(a + \theta h)}{(n + 1)!} h^{n+1}.$$

8. Give an example of a function $f: R \to R$ such that f is positive on the open interval (a, b), zero elsewhere, and infinitely differentiable on R.

9. Let f and g have n continuous derivatives on the closed interval $[a, b]$ and be $(n + 1)$-times differentiable on (a, b). Prove that to each $x \in (a, b]$ there exists $c \in (a, x)$ such that

$$\left[f(x) - \sum_{k=0}^{n} \frac{f^{(k)}(a)}{k!} (x - a)^k \right] g^{(n+1)}(c)$$

$$= \left[g(x) - \sum_{k=0}^{n} \frac{g^{(k)}(a)}{k!} (x - a)^k \right] f^{(n+1)}(c).$$

By choosing g properly, obtain Theorem 5.4.1.

10. Let

$$f(x) = \begin{cases} e^{-1/x^2} & \text{if } x \neq 0 \\ 0 & \text{if } x = 0 \end{cases}$$

Show that f is infinitely differentiable on R and that $f^{(n)}(0) = 0$ for each positive integer n.

Chapter Six

INTEGRATION

6.1. The Integral

This chapter develops the concept of the operation of integration, which roughly speaking is the inverse of the operation of differentiation. It is assumed that the student is already familiar with some of the properties and applications of integrals. The purpose here is not to explain these applications but to give a precise definition of the integral and develop some of its simpler properties with analytical proofs that do not depend on the persuasion of a picture. If, however, the reader's understanding is facilitated by drawing suitable pictures, then he is urged to do so. We begin with several definitions.

Definition 6.1.1. *Let* $I = [a, b]$ *be a compact interval. A **net** $\mathfrak{n}$ of I is a finite ordered set*

$$\mathfrak{n} = \{a = x_0 < x_1 < \dots < x_{n-1} < x_n = b\}.$$

The closed intervals

$$I_k = [x_{k-1}, x_k], \qquad k = 1, 2, \dots, n,$$

*are called **subintervals** of the net $\mathfrak{n}$. The symbol $|I_k|$ will be used to denote the length of the interval I_k. The **norm** of the net $\mathfrak{n}$, denoted by $\|\mathfrak{n}\|$, is defined by*

$$\|\mathfrak{n}\| = \max \{|I_k| : k = 1, 2, \dots, n\}.$$

Definition 6.1.2. *Let* $\mathfrak{n}$ *and* $\mathfrak{n}^*$ *be nets of the compact interval* I. *Then* $\mathfrak{n}^*$ *is called a* **refinement** *of* $\mathfrak{n}$ *if and only if* $\mathfrak{n} \subseteq \mathfrak{n}^*$. *If* $\mathfrak{n}_1$ *and* $\mathfrak{n}_2$ *are any nets of* I, *then a net* $\mathfrak{n}_3$ *of* I *is called a* **common refinement** *of* $\mathfrak{n}_1$ *and* $\mathfrak{n}_2$ *if and only if* $\mathfrak{n}_1 \cup \mathfrak{n}_2 \subseteq \mathfrak{n}_3$. *The common refinement* $\mathfrak{n}_3$ *of* $\mathfrak{n}_1$ *and* $\mathfrak{n}_2$ *is called the* **smallest common refinement** *of* $\mathfrak{n}_1$ *and* $\mathfrak{n}_2$ *if and only if* $\mathfrak{n}_3 = \mathfrak{n}_1 \cup \mathfrak{n}_2$.

It is clear that a common refinement of two nets is a refinement of each of those nets. Also, the norm of a common refinement is less than or equal to the minimum of the norms of the given nets. It is left as an exercise to show that if $\mathfrak{n}^*$ is a refinement of $\mathfrak{n}$, then each subinterval of $\mathfrak{n}$ is the union of one or more subintervals of $\mathfrak{n}^*$ (cf. Exercise 2 of Problem Set 6.1). We now present an example to illustrate the above concepts.

Example 6.1.1. Let $I = [0, 1]$. Three nets of I are

$$\mathfrak{n}_1 = \{0, 1/4, 1/2, 3/4, 1\},$$
$$\mathfrak{n}_2 = \{0, 1/3, 2/3, 1\},$$

and

$$\mathfrak{n}_3 = \{0, 1/4, 1/3, 1/2, 2/3, 3/4, 1\}.$$

For the net $\mathfrak{n}_1$, the subintervals are $I_1 = [0, 1/4]$, $I_2 = [1/4, 1/2]$, $I_3 = [1/2, 3/4]$, and $I_4 = [3/4, 1]$. Thus

$$|I_1| = |I_2| = |I_3| = |I_4| = 1/4,$$

and the norm of $\mathfrak{n}_1$ is $\|\mathfrak{n}_1\| = 1/4$. The norm of $\mathfrak{n}_2$ is $\|\mathfrak{n}_2\| = 1/3$. For the net $\mathfrak{n}_3$, the subintervals are $J_1 = [0, 1/4]$, $J_2 = [1/4, 1/3]$, $J_3 = [1/3, 1/2]$, $J_4 = [1/2, 2/3]$, $J_5 = [2/3, 3/4]$, and $J_6 = [3/4, 1]$. Thus $|J_1| = |J_6| = 1/4$, $|J_2| = |J_5| = 1/12$, and $|J_3| = |J_4| = 1/6$, and the norm of $\mathfrak{n}_3$ is $\|\mathfrak{n}_3\| = 1/4$.

The net $\mathfrak{n}_3$ is a common refinement of the nets $\mathfrak{n}_1$ and $\mathfrak{n}_2$; in fact, $\mathfrak{n}_3$ is the smallest common refinement of $\mathfrak{n}_1$ and $\mathfrak{n}_2$.

Finally we note that $I_1 = J_1$, $I_2 = J_2 \cup J_3$, $I_3 = J_4 \cup J_5$, and $I_4 = J_6$. That is, the intervals of $\mathfrak{n}_1$ are the unions of intervals of $\mathfrak{n}_3$, a refinement of $\mathfrak{n}_1$.

We now turn to the problem of defining *upper and lower sums* of a bounded function on a compact interval. Let I be a compact interval, and let f be a bounded function on I. For the remainder of this chapter we shall use the following notation:

$$M(f, I) = \sup \{f(x): x \in I\},$$

$$m(f, I) = \inf \{f(x): x \in I\}.$$

For the function $f: [2, 3] \to R$ such that $f(x) = x^2$ for each $x \in [2, 3]$, we have $M(f, [2, 3]) = 9$ and $m(f, [2, 3]) = 4$.

Definition 6.1.3. *Let* $\mathfrak{n}$ *be a net of the compact interval* $I = [a, b]$ *with subintervals* I_k, *where* $k = 1, 2, ..., n$, *and let* $f: I \to R$ *such that* f *is bounded on* I. *Let*

$$u(f, \mathfrak{n}) = \sum_{k=1}^{n} M(f, I_k)|I_k|$$

and

$$l(f, \mathfrak{n}) = \sum_{k=1}^{n} m(f, I_k)|I_k|.$$

The numbers $u(f, \mathfrak{n})$ *and* $l(f, \mathfrak{n})$ *are respectively called the* **upper** *and* **lower sums** *of* f *for the net* $\mathfrak{n}$.

Since the least upper bound of a nonempty set is always greater than or equal to the greatest lower bound of the set, it follows, for any bounded function f with domain the compact interval I, and for any net $\mathfrak{n}$ of I, that $l(f, \mathfrak{n}) \leqslant u(f, \mathfrak{n})$. We now present an example to demonstrate the concepts of upper and lower sums.

Example 6.1.2. Let $I = [0, 1]$, and let $f: I \to R$ such that $f(x) = x^2$. For each natural number n, define the net $\mathfrak{n}_n$ of I by

$$\mathfrak{n}_n = \{0 = 0/n, 1/n, 2/n, ..., n/n = 1\}.$$

The subintervals of $\mathfrak{n}_n$ are

$$I_{n,k} = [(k-1)/n, k/n], \qquad k = 1, 2, ..., n.$$

Thus $|I_{n,k}| = 1/n$, where $k = 1, 2, ..., n$, and $\|\mathfrak{n}_n\| = 1/n$. It is easily seen that

$$M(f, I_{n,k}) = k^2/n^2$$

and

$$m(f, I_{n,k}) = (k-1)^2/n^2,$$

where $k = 1, 2, ..., n$. Then (cf. Exercise 4 of Problem Set 2.5)

$$u(f, \mathfrak{n}_n) = \sum_{k=1}^{n} M(f, I_{n,k})|I_{n,k}|$$

$$= \sum_{k=1}^{n} (k^2/n^2)(1/n)$$

$$= (1/n^3) \sum_{k=1}^{n} k^2$$

$$= (1/n^3)(n)(n+1)(2n+1)/6$$

$$= (1 + 1/n)(2 + 1/n)/6,$$

and

$$l(f, \mathfrak{n}_n) = \sum_{k=1}^{n} m(f, I_{n,k})|I_{n,k}|$$

$$= \sum_{k=1}^{n} [(k-1)^2/n^2](1/n)$$

$$= (1/n^3) \sum_{k=1}^{n} (k-1)^2$$

$$= (1/n^3)(n-1)(n)(2n-1)/6$$

$$= (1 - 1/n)(2 - 1/n)/6.$$

It should be noted that for any natural numbers m and n,

$$l(f, \mathfrak{n}_m) < 1/3 < u(f, \mathfrak{n}_n).$$

We are now ready to prove the first theorem concerning upper and lower sums.

Theorem 6.1.1. *Let I be a compact interval, and let $f: I \to R$ such that f is bounded on I. If $\mathfrak{n}$ and $\mathfrak{n}^*$ are nets of I such that $\mathfrak{n}^*$ is a refinement of $\mathfrak{n}$, then*

$$l(f, \mathfrak{n}) \leqslant l(f, \mathfrak{n}^*) \leqslant u(f, \mathfrak{n}^*) \leqslant u(f, \mathfrak{n}).$$

Proof. We shall prove that $u(f, \mathfrak{n}^*) \leqslant u(f, \mathfrak{n})$ and leave the proof that $l(f, \mathfrak{n}) \leqslant l(f, \mathfrak{n}^*)$ as Exercise 4 of Problem Set 6.1.

Let $I_1, I_2, ..., I_n$ be the subintervals of the net $\mathfrak{n}$, and let $J_1, J_2, ..., J_m$ be the subintervals of the net $\mathfrak{n}^*$. Since $\mathfrak{n}^*$ is a refinement of $\mathfrak{n}$, each subinterval of $\mathfrak{n}$ is the union of subintervals of $\mathfrak{n}^*$. (Cf. Exercise 2 of Problem Set 6.1.) For each k, where $k = 1, 2, ..., n$, let $J_{k,1}, J_{k,2}, ..., J_{k,m_k}$ be the subintervals of the net $\mathfrak{n}^*$ that are contained in I_k, that is,

$$I_k = \bigcup_{j=1}^{m_k} J_{k,j}, \qquad k = 1, 2, ..., n.$$

Thus for each k, where $k = 1, 2, ..., n$,

$$J_{k,j} \subseteq I_k,$$

and

$$M(f, J_{k,j}) \leqslant M(f, I_k),$$

$j = 1, 2, ..., m_k$. (Cf. Exercise 11 of Problem Set 2.7.) We see that

$$u(f, \mathfrak{n}^*) = \sum_{j=1}^{m} M(f, J_j)|J_j|$$

$$= \sum_{k=1}^{n} \sum_{i=1}^{m_k} M(f, J_{k,i})|J_{k,i}|$$

$$\leqslant \sum_{k=1}^{n} M(f, I_k)|I_k| = u(f, \mathfrak{n}).$$

Therefore $u(f, \mathfrak{n}^*) \leqslant u(f, \mathfrak{n})$.

From Theorem 6.1.1 we obtain the following interesting corollary.

Corollary 6.1.1. *Let I be a closed interval, and let $f: I \to R$ such that f is bounded on I. If $\mathfrak{n}_1$ and $\mathfrak{n}_2$ are any nets of I, then*

$$l(f, \mathfrak{n}_1) \leqslant u(f, \mathfrak{n}_2).$$

Proof. Let $\mathfrak{n}$ be a common refinement of $\mathfrak{n}_1$ and $\mathfrak{n}_2$. Then Theorem 6.1.1 implies that

$$l(f, \mathfrak{n}_1) \leqslant l(f, \mathfrak{n}) \leqslant u(f, \mathfrak{n}) \leqslant u(f, \mathfrak{n}_2).$$

We next define the *upper and lower integrals* of a bounded function defined on a compact interval.

Definition 6.1.4. *Let I be a compact interval, let $f: I \to R$ such that f is bounded on I, and let $\overline{N}$ be the set of all nets of the interval I. The* **upper integral** *of f on I, denoted by $\overline{\int} f$, is defined by*

$$\overline{\int} f = \inf \{u(f, \mathfrak{n}): \mathfrak{n} \in \overline{N}\}.$$

The **lower integral** *of f on I, denoted by $\underline{\int} f$, is defined by*

$$\underline{\int} f = \sup \{l(f, \mathfrak{n}): \mathfrak{n} \in \overline{N}\}.$$

Corollary 6.1.1 now leads to the following theorem.

Theorem 6.1.2. *Let I be a closed interval, and let $f: I \to R$ such that f is bounded on I. Then*

$$\underline{\int} f \leqslant \overline{\int} f.$$

The proof is left as Exercise 6 of Problem Set 6.1.

Definition 6.1.5. *Let $I = [a, b]$ be a compact interval, and let $f: I \to R$ such that f is bounded on I. The function f is said to be* **Riemann**

integrable or **integrable** *on I if and only if the upper integral of f on I is equal to the lower integral of f on I. If f is Riemann integrable on I, then the common value of the upper and lower integrals is called the* **Riemann integral** *or the* **integral** *of f on I. The Riemann integral on the interval I = [a, b] is denoted by*

$$\int_I f, \int_a^b f, \int_a^b f, \quad \text{or} \quad \int_a^b f(x)dx.$$

The numbers a and b are called the **lower** *and* **upper limits of integration** *of the integral*

$$\int_a^b f(x)dx.$$

The Riemann integral was named for the German mathematician Georg Friedrich Bernhard Riemann (1866–1926), who first formulated the definition of the integral in this manner.

Example 6.1.3. Let f and I be the function and interval defined in Example 6.1.2. Since the upper integral is the greatest lower bound of the set of all upper sums of f on I, we have that

$$\bar{\int} f \leq u(f, \mathfrak{n}_n)$$

for each natural number n, where $\mathfrak{n}_n$ was defined in Example 6.1.2. But then

$$\bar{\int} f \leq (1 + 1/n)(2 + 1/n)/6$$

for each natural number n. Thus

$$\bar{\int} f \leq 1/3.$$

In a similar manner we have

$$\underline{\int} f \geq 1/3.$$

Theorem 6.1.2 implies that

$$\underline{\int} f \leq \bar{\int} f \leq 1/3 \leq \underline{\int} f.$$

Therefore,

$$\underline{\int} f = \bar{\int} f = \int f = 1/3,$$

and the function $f: I \to R$ defined by $f(x) = x^2$ for each $x \in I$ is Riemann integrable on I, and the Riemann integral of f has value 1/3. Thus we may write

$$\int_0^1 x^2 \, dx = 1/3.$$

Example 6.1.3 shows that there is at least one Riemann integrable function. We now present an example to show that not all bounded functions are Riemann integrable.

Example 6.1.4. Let $f: [0, 1] \to R$ such that

$$f(x) = \begin{cases} 1 & \text{if } x \text{ is rational,} \\ -1 & \text{if } x \text{ is irrational.} \end{cases}$$

Let n be any net of $[0, 1]$ with subintervals I_k, where $k = 1, 2, ..., n$. Then

$$M(f, I_k) = 1 \quad \text{and} \quad m(f, I_k) = -1,$$

$k = 1, 2, ..., n$. Thus, for the net n,

$$u(f, n) = 1 \quad \text{and} \quad l(f, n) = -1.$$

Since n was an arbitrary net of $[0, 1]$,

$$\overline{\int} f = 1 \quad \text{and} \quad \underline{\int} f = -1.$$

Therefore, $\underline{\int} f \neq \overline{\int} f$ and the function f is not Riemann integrable on $[0, 1]$.

We conclude this section by giving a necessary and sufficient condition for a function to be Riemann integrable. The proof is left as Exercise 14 of Problem Set 6.1.

Theorem 6.1.3. *Let f be a bounded function on the compact interval I. Then f is Riemann integrable on I if and only if to each $\varepsilon > 0$ there is a net n of I such that*

$$u(f, n) - l(f, n) < \varepsilon.$$

Problem Set 6.1

1. Let n be a set of the compact interval $I = [a, b]$. If the subintervals of n are I_k, where $k = 1, 2, ..., n$, show that

$$b - a = \sum_{k=1}^{n} |I_k|.$$

2. Let n and n^* be nets of the compact interval I such that n^* is a refinement of n. Show that each subinterval of n is the union of one or more subintervals of n^*.
3. Let f be a bounded function on the compact interval I, and let n be a net of I. Show that

$$l(f, n) \leqslant u(f, n).$$

4. Let f be a bounded function on the compact interval I, and let $\mathfrak{n}$ and $\mathfrak{n}^*$ be nets of I such that $\mathfrak{n}^*$ is a refinement of $\mathfrak{n}$. Show that
$$l(f, \mathfrak{n}) \leqslant l(f, \mathfrak{n}^*).$$

5. Let f be a bounded function on the compact interval I. Show that the upper and lower integrals of f on I both exist.

6. Prove Theorem 6.1.2.

7. Let $I = [a, b]$, and let $f: I \to R$ such that $f(x) = c$ for every $x \in I$, where c is a real number. Show that f is Riemann integrable on I, and find the value of the Riemann integral of f on I.

8. Let $I = [0, 1]$, and let $f: I \to R$ such that $f(x) = x$ for each $x \in I$. Let n be a natural number, and let
$$\mathfrak{n}_n = \{0 = 0/n, 1/n, 2/n, \dots, n/n = 1\}.$$

   (a) Find $u(f, \mathfrak{n}_4)$ and $l(f, \mathfrak{n}_4)$.
   (b) Find $u(f, \mathfrak{n}_n)$ and $l(f, \mathfrak{n}_n)$ for each natural number n.
   (c) Show that
$$\lim_{n \to \infty} u(f, \mathfrak{n}_n) = \lim_{n \to \infty} l(f, \mathfrak{n}_n) = 1/2.$$

   (d) Show that part (c) implies that
$$\int_0^1 x \, dx = 1/2.$$

9. Let $I = [0, 1]$, and let $f: I \to R$ such that $f(x) = \sin x$ for each $x \in I$. Let n be a natural number, and let
$$\mathfrak{n}_n = \{0 = 0/n, 1/n, 2/n, \dots, n/n = 1\}.$$

   (a) Find $u(f, \mathfrak{n}_3)$ and $l(f, \mathfrak{n}_3)$.
   (b) Find $u(f, \mathfrak{n}_n)$ and $l(f, \mathfrak{n}_n)$ for each natural number n.
   (c) Show that
$$\lim_{n \to \infty} u(f, \mathfrak{n}_n) = \lim_{n \to \infty} l(f, \mathfrak{n}_n) = 1 - \cos 1 = \int_0^1 \sin x \, dx.$$

   [*Hint*: For $\{\sin (1/n) + \sin (2/n) + \dots + \sin (n/n)\}$, multiply each term by $2 \sin (1/n)$ and use the identity $2 \sin A \sin B = \cos (A - B) - \cos (A + B)$.]

10. Let $I = [0, 1]$, and let $f: I \to R$ such that $f(x) = e^x$ for each $x \in I$. Let n be a natural number, and let
$$\mathfrak{n}_n = \{0 = 0/n, 1/n, 2/n, \dots, n/n = 1\}.$$

   (a) Find $u(f, \mathfrak{n}_4)$ and $l(f, \mathfrak{n}_4)$.
   (b) Find $u(f, \mathfrak{n}_n)$ and $l(f, \mathfrak{n}_n)$ for each natural number n.
   (c) Show that
$$\lim_{n \to \infty} u(f, \mathfrak{n}_n) = \lim_{n \to \infty} l(f, \mathfrak{n}_n) = e - 1 = \int_0^1 e^x \, dx.$$

   [*Hint*: Use the identity $(1 - e)/(1 - e^{1/n}) = 1 + e^{1/n} + e^{2/n} + \dots + e^{(n-1)/n}$.]

11. Let f be Riemann integrable on $[0, 1]$, and, for each natural number n, let

$$\mathfrak{n}_n = \{0 = 0/n, 1/n, 2/n, ..., n/n = 1\}.$$

If

$$\lim_{n \to \infty} u(f, \mathfrak{n}_n) = \lim_{n \to \infty} l(f, \mathfrak{n}_n) = A,$$

show that

$$\int_a^b f(x)dx = A.$$

12. Let f be Riemann integrable on $[0, 1]$, and let

$$s_n = (1/n) \sum_{k=1}^n f(k/n)$$

for each natural number n. Show that

$$\lim_{n \to \infty} s_n = \int_0^1 f(x)dx.$$

13. Evaluate the limits:
    (a) $\lim_{n \to \infty} (1/n)[(1/n) + (2/n) + ... + (n/n)]$;

    (b) $\lim_{n \to \infty} (1/n)[\sin(1/n) + \sin(2/n) + ... + \sin(n/n)]$;

    (c) $\lim_{n \to \infty} (1/n)[e^{1/n} + e^{2/n} + ... + e^{n/n}]$.

14. Prove Theorem 6.1.3.

15. Let n be a natural number, and let $I_n = [0, n]$. Show that the bracket function is Riemann integrable on each interval I_n, and evaluate

$$\int_0^n [x]dx.$$

16. Let f and g be bounded functions on the interval $[a, b]$. Let S be a finite subset of $[a, b]$, and let $f(x) = g(x)$ for each $x \in [a, b] \backslash S$. If f is Riemann integrable on $[a, b]$, show that g is Riemann integrable on $[a, b]$, and show that

$$\int_a^b f(x)dx = \int_a^b g(x)dx.$$

17. Let f be Riemann integrable on $[a, b]$ and let $m \leqslant f(x) \leqslant M$ for each $x \in [a, b]$. Show that

$$m(b - a) \leqslant \int_a^b f(x)dx \leqslant M(b - a).$$

18. Let $f: [0, 1] \to R$ such that

$$f(x) = \begin{cases} x & \text{if } x \text{ is rational,} \\ 1 - x & \text{if } x \text{ is irrational.} \end{cases}$$

Show that f is *not* Riemann integrable on $[0, 1]$. [*Hint*: Compute $\overline{\int} f$ and $\underline{\int} f$.]

6.2. Conditions for Riemann Integrability

In this section we shall give a necessary and sufficient condition that a bounded function on a compact interval be Riemann integrable. This condition is given in terms of *Riemann sums*. We begin by proving the following theorem.

Theorem 6.2.1. *Let f be a bounded function on the interval* $[a, b]$. *Then for each* $\varepsilon > 0$ *there is a* $\delta = \delta(\varepsilon) > 0$ *such that*

$$u(f, \mathfrak{n}) < \bar{\int} f + \varepsilon$$

and

$$l(f, \mathfrak{n}) > \underline{\int} f - \varepsilon$$

for each net $\mathfrak{n}$ *of* $[a, b]$ *with* $\|\mathfrak{n}\| < \delta$.

Proof. We prove the first of these two inequalities; the second is left as Exercise 4 of Problem Set 6.2.

Let $\varepsilon > 0$, and let M be a positive number such that $|f(x)| \leqslant M$ for each $x \in [a, b]$. Since the upper integral is the greatest lower bound of the set of all upper sums of f on $[a, b]$, there is a net $\mathfrak{n}^*$ of $[a, b]$ such that

$$u(f, \mathfrak{n}^*) < \bar{\int} f + \varepsilon/2.$$

Let I_k^*, where $k = 1, 2, ..., n$, be the subintervals of the net $\mathfrak{n}^*$, and let $\delta^* = \min\{|I_k^*| : k = 1, 2, ..., n\}$. Let $\delta = \min\{\delta^*, \varepsilon/6Mn\}$.

Let $\mathfrak{n}$ be any net with $\|\mathfrak{n}\| < \delta$. We need to show that

$$u(f, \mathfrak{n}) < \bar{\int} f + \varepsilon.$$

To do this, we let $\mathfrak{n}_1 = \mathfrak{n} \cup \mathfrak{n}^*$ be the smallest common refinement of $\mathfrak{n}$ and $\mathfrak{n}^*$. Then Theorem 6.1.1 implies that

$$u(f, \mathfrak{n}_1) \leqslant u(f, \mathfrak{n}^*) < \bar{\int} f + \varepsilon/2.$$

Let I_j, where $j = 1, 2, ..., m$, be the subintervals of the net $\mathfrak{n}$, and let J_i, where $i = 1, 2, ..., r$, be the subintervals of the net $\mathfrak{n}_1$. By the definition of the net $\mathfrak{n}$, each subinterval of $\mathfrak{n}$ can contain at most one point of the net $\mathfrak{n}^*$ as an interior point. Thus the set of subintervals of the net $\mathfrak{n}$ can be divided into two classes: (1) those subintervals of $\mathfrak{n}$ that contain a point of $\mathfrak{n}^*$ as an interior point; and (2) those subintervals of $\mathfrak{n}$ that do not contain a point of $\mathfrak{n}^*$ as an interior point. Likewise, the set of subintervals of $\mathfrak{n}_1$ can be divided into two classes: (3) those subintervals of n_1 that arise from a subinterval of n when a point from the net n^* is an interior point of the net n; and those which are in class (2) above.

We now compare the upper sums $u(f, n)$ and $u(f, n_1)$. In the difference $u(f, n) - u(f, n_1)$, all the contributions from subintervals in class (2) above cancel out. What remains are the contributions caused by subintervals of classes (1) and (3). For the net n, there are at most $(n - 1)$ subintervals in class (1), since there are only $(n - 1)$ points of n^* interior to $[a, b]$. For each of these subintervals the contribution to $u(f, n)$ is of the form $M(f, I_j)|I_j|$ and this is at most $M\delta$. Thus the contribution caused by subintervals of class (1) is at most $(n - 1)M\delta$. For the partition n_1, there are at most $2(n - 1)$ subintervals in the class (3), since there are at most $(n - 1)$ sub-intervals of n in class (1). For each of these subintervals the contribution to $u(f, n_1)$ is of the form $M(f, J_i)|J_i|$, and this is at most $M\delta$. Thus the contribution caused by subintervals of class (3) is at most $2(n - 1)M\delta$. Therefore,

$$u(f, n) - u(f, n_1) \leqslant 3(n - 1)M\delta$$
$$< 3nM\varepsilon/6Mn = \varepsilon/2,$$

and

$$u(f, n) - u(f, n_1) < \varepsilon/2.$$

Using the inequality

$$u(f, n_1) < \bar{\textstyle\int} f + \varepsilon/2,$$

we have

$$u(f, n) < \bar{\textstyle\int} f + \varepsilon,$$

which concludes the proof.

This theorem leads to the following corollary when f is integrable on $[a, b]$.

Corollary 6.2.1. *If f is integrable on $[a, b]$, then for each $\varepsilon > 0$ there is a $\delta = \delta(\varepsilon) > 0$ such that*

$$u(f, n) - \varepsilon < \int_a^b f(x)dx < l(f, n) + \varepsilon$$

for each net n of $[a, b]$ with $\|n\| < \delta$.

We are now ready to define the concept of a Riemann sum. This type of sum need not be an upper or a lower sum, but it is always greater than or equal to any lower sum and less than or equal to any upper sum.

Definition 6.2.1. *Let f be a bounded function on the compact interval I, and let $\mathfrak{n}$ be any net of I with subintervals I_k, where $k = 1, 2, ..., n$. Let ξ_k be any point in the subinterval I_k, where $k = 1, 2, ..., n$. Then the sum*

$$R(f, \mathfrak{n}, \xi) = \sum_{k=1}^{n} f(\xi_k)|I_k|,$$

*where $\xi = \{\xi_1, \xi_2, ..., \xi_n\}$, will be called a **Riemann sum** of f on I.*

Example 6.2.1. Let $f: [0, 1] \to R$ such that $f(x) = x^2$. Let $\mathfrak{n} = \{0, 1/3, 2/3, 1\}$. Then the subintervals of $\mathfrak{n}$ are $I_1 = [0, 1/3]$, $I_2 = [1/3, 2/3]$, and $I_3 = [2/3, 1]$. Let $\xi_1 = 1/4$, $\xi_2 = 1/2$, and $\xi_3 = 3/4$. Then for $\xi = \{\xi_1, \xi_2, \xi_3\} = \{1/4, 1/2, 3/4\}$, we have a Riemann sum

$$R(f, \mathfrak{n}, \xi) = \sum_{k=1}^{3} f(\xi_k)|I_k|$$

$$= (1/4)^2 (1/3) + (1/2)^2 (1/3) + (3/4)^2 (1/3)$$

$$= (1/3)(1/16 + 1/4 + 9/16) = 7/24.$$

More generally, if for each natural number n we define the net $\mathfrak{n}_n$ by

$$\mathfrak{n}_n = \{0 = 0/n, 1/n, 2/n, ..., n/n = 1\},$$

then the subintervals of $\mathfrak{n}_n$ are $I_{n,k} = [(k - 1)/n, k/n]$, $k = 1, 2, ..., n$. For each k, where $k = 1, 2, ..., n$, let

$$\xi_{n,k} = (k - 1/2)/n.$$

Then $\xi_{n,k} \in I_{n,k}$, and in fact $\xi_{n,k}$ is the mid-point of the interval $I_{n,k}$. For $\xi_n = \{\xi_{n,k} : k = 1, 2, ..., n\}$, we have a Riemann sum

$$R(f, \mathfrak{n}_n, \xi_n) = \sum_{k=1}^{n} f(\xi_{n,k})|I_{n,k}|$$

$$= \sum_{k=1}^{n} [(k - 1/2)]^2 (1/n)^3$$

$$= (1/n^3) \sum_{k=1}^{n} (k - 1/2)^2$$

$$= (1/n^3) \sum_{k=1}^{n} (k^2 - k + 1/4)$$

$$= (1 - 1/2n)(1 + 1/2n)/3.$$

Thus $R(f, \mathfrak{n}_n, \xi_n) = (1 - 1/2n)(1 + 1/2n)/3$ for each natural number n.

From Example 6.1.2 we have

$$u(f, \mathfrak{n}_n) = (1 + 1/n)(1 + 1/2n)/3,$$
$$l(f, \mathfrak{n}_n) = (1 - 1/n)(1 - 1/2n)/3,$$

and

$$l(f, \mathfrak{n}_n) < R(f, \mathfrak{n}_n, \xi_n) < u(f, \mathfrak{n}_n)$$

for each natural number n.

In Example 6.2.1 we saw that the Riemann sum was between the upper and lower sums. This fact is now put in the form of a theorem; the proof is left as Exercise 5 of Problem Set 6.2.

Theorem 6.2.2. *Let f be a bounded function on the compact interval I, and let $\mathfrak{n}$ be any net of I. Then*

$$l(f, \mathfrak{n}) \leqslant R(f, \mathfrak{n}, \xi) \leqslant u(f, \mathfrak{n}),$$

where $R(f, \mathfrak{n}, \xi)$ is any Riemann sum of f on I for the net $\mathfrak{n}$.

We now define a special kind of limit that is associated with Riemann sums.

Definition 6.2.2. *Let f be a bounded function on the compact interval I. Let $\bar{R}$ be the set of all Riemann sums of f on I. Then the set $\bar{R}$ is said to have **limit** S if and only if to each $\varepsilon > 0$ there is $\delta = \delta(\varepsilon) > 0$ such that*

$$|R(f, \mathfrak{n}, \xi) - S| < \varepsilon$$

whenever $\|\mathfrak{n}\| < \delta$. This limit is denoted by

$$\lim_{\|\mathfrak{n}\| \to 0} R(f, \mathfrak{n}, \xi) = S.$$

This definition looks quite similar to the definition of the limit of a function. However, it must be emphasized that the ε-condition must hold for each partition $\mathfrak{n}$ with $\|\mathfrak{n}\| < \delta$ as well as for every possible choice of ξ. This limit is analogous to the limit of a function in the sense that if it exists, then it is unique; the proof of this fact is left as Exercise 6 of Problem Set 6.2.

We are now ready to prove the main theorem of this section.

Theorem 6.2.3. *Let f be a bounded function on the compact interval $I = [a, b]$. Then f is integrable on I and has integral S if and only if*

$$\lim_{\|\mathfrak{n}\| \to 0} R(f, \mathfrak{n}, \xi) = S.$$

Proof. We first assume that f is integrable on I and has integral S.

Let $\varepsilon > 0$. Then Theorem 6.2.1 implies that there is a $\delta > 0$ such that

$$u(f, \mathfrak{n}) - \varepsilon < S < l(f, \mathfrak{n}) + \varepsilon,$$

when $\|\mathfrak{n}\| < \delta$. Theorem 6.2.2 implies that for any Riemann sum $R(f, \mathfrak{n}, \xi)$,

$$l(f, \mathfrak{n}) \leqslant R(f, \mathfrak{n}, \xi) \leqslant u(f, \mathfrak{n}).$$

Combining these two inequalities we have

$$S - \varepsilon < l(f, \mathfrak{n}) \leqslant R(f, \mathfrak{n}, \xi) \leqslant u(f, \mathfrak{n}) < S + \varepsilon,$$

and so

$$|R(f, \mathfrak{n}, \xi) - S| < \varepsilon$$

when $\|\mathfrak{n}\| < \delta$. Thus

$$\lim_{\|\mathfrak{n}\| \to 0} R(f, \mathfrak{n}, \xi) = S.$$

We now assume that

$$\lim_{\|\mathfrak{n}\| \to 0} R(f, \mathfrak{n}, \xi) = S.$$

Let $\varepsilon > 0$. Definition 6.2.2 implies that there is a $\delta > 0$ such that

$$S - \varepsilon/2 < R(f, \mathfrak{n}, \xi) < S + \varepsilon/2$$

whenever $\|\mathfrak{n}\| < \delta$ and for any choice of ξ.

Let $\mathfrak{n}$ be any net of I with $\|\mathfrak{n}\| < \delta$ and let I_k, where $k = 1, 2, ..., n$, be the subintervals of $\mathfrak{n}$. Since $M(f, I_k)$ is the least upper bound of the set $\{f(x) : x \in I_k\}$, let $\xi_k \in I_k$ such that

$$f(\xi_k) > M(f, I_k) - \varepsilon/(2|I|), \qquad k = 1, 2, ..., n.$$

Then

$$R(f, \mathfrak{n}, \xi) = \sum_{k=1}^{n} f(\xi_k)|I_k|$$

$$> \sum_{k=1}^{n} [M(f, I_k) - \varepsilon/(2|I|)]|I_k|$$

$$= \sum_{k=1}^{n} M(f, I_k)|I_k| - [\varepsilon/(2|I|)] \sum_{k=1}^{n} |I_k|$$

$$= u(f, \mathfrak{n}) - \varepsilon/2.$$

Using the inequality

$$R(f, \mathfrak{n}, \xi) < S + \varepsilon/2,$$

we have

$$u(f, \mathfrak{n}) - \varepsilon/2 < R(f, \mathfrak{n}, \xi) < S + \varepsilon/2$$

or

$$u(f, \mathfrak{n}) < S + \varepsilon.$$

Since the upper integral is always less than or equal to any upper sum, we have

$$\overline{\int} f \leqslant u(f, \mathfrak{n}) < S + \varepsilon.$$

Since ε was an arbitrary positive number,

$$\overline{\int} f \leqslant S.$$

By a similar argument, we have

$$S \leqslant \underline{\int} f.$$

(Cf. Exercise 7 of Problem Set 6.2.) Hence

$$\overline{\int} f \leqslant S \leqslant \underline{\int} f \leqslant \overline{\int} f$$

and

$$\overline{\int} f = \underline{\int} f = S.$$

Therefore, f is integrable on I, and its integral is S.

Corollary 6.2.3. *Let f be integrable on the compact interval $I = [a, b]$, and let f have integral S on I. If $\{\mathfrak{n}_n\}$ is any sequence of nets of I such that*

$$\lim_{n \to \infty} \|\mathfrak{n}_n\| = 0,$$

then

$$\lim_{n \to \infty} R(f, \mathfrak{n}_n, \xi) = S,$$

where $R(f, \mathfrak{n}_n, \xi)$ is any Riemann sum of f on I for the net $\mathfrak{n}_n$.

Example 6.2.2. Evaluate the limit

$$\lim_{n \to \infty} (1/n) \left[\sum_{k=1}^{n} \sin \left(\frac{2k - 1}{2n} \right) \right].$$

We shall write the limit as an integral and evaluate the integral.

If, for each positive integer n, we let $\mathfrak{n}_n$ be the net of the interval $[0, 1]$ defined by

$$\mathfrak{n}_n = \{0 = 0/n, 1/n, 2/n, ..., n/n = 1\},$$

then the subintervals of $\mathfrak{n}_n$ are

$$I_{n,k} = [(k-1)/n, k/n], \qquad k = 1, 2, ..., n,$$

and $\|\mathfrak{n}_n\| = |I_{n,k}| = 1/n$. Let

$$\xi_{n,k} = (2k-1)/(2n) \in I_{n,k}$$

for each n and k. Then

$$(1/n)\left[\sum_{k=1}^{n} \sin\left(\frac{2k-1}{2n}\right)\right] = \sum_{k=1}^{n} (\sin \xi_{n,k})|I_{n,k}|.$$

This last sum is a Riemann sum for the function $f(x) = \sin x$ on the interval $[0, 1]$ for the net $\mathfrak{n}_n$. Thus Corollary 6.2.3 implies that

$$\lim_{n \to \infty} (1/n)\left[\sum_{k=1}^{n} \sin\left(\frac{2k-1}{2n}\right)\right] = \int_0^1 \sin x \, dx.$$

Now

$$\int_0^1 \sin x \, dx = 1 - \cos 1,$$

(cf. Exercise 9 of Problem Set 6.1.), and so

$$\lim_{n \to \infty} (1/n)\left[\sum_{k=1}^{n} \sin\left(\frac{2k-1}{2n}\right)\right] = 1 - \cos 1.$$

Theorem 6.2.3 provides a necessary and sufficient condition that a bounded function f be integrable on a closed interval. This condition is very difficult to apply in actual practice, and so we shall present two sufficient conditions for a bounded function f to be integrable on a closed interval.

Theorem 6.2.4. *Let f be a monotone function on the compact interval I. Then f is integrable on I.*

Proof. We consider the case where f is increasing; the case where f is decreasing can be handled in a similar manner (cf. Exercise 8 of Problem Set 6.2).

Let $I = [a, b]$, and let $\varepsilon > 0$. We shall find a net $\mathfrak{n}$ such that

$$u(f, \mathfrak{n}) - l(f, \mathfrak{n}) < \varepsilon$$

and apply Theorem 6.1.3.

If $f(a) = f(b)$, then f is constant since f is increasing. Hence f is integrable on I. (Cf. Exercise 7 of Problem Set 6.1.) Thus we assume that $f(a) \neq f(b)$. Let $\mathfrak{n}$ be any net such that $\|\mathfrak{n}\| < \varepsilon/[f(b) - f(a)]$, and let $I_k = [x_{k-1}, x_k]$, where $k = 1, 2, ..., n$, be the subintervals of $\mathfrak{n}$. Since f is increasing, $M(f, I_k) = f(x_k)$ and $m(f, I_k) = f(x_{k-1})$. Then

$$u(f, \mathfrak{n}) - l(f, \mathfrak{n}) = \sum_{k=1}^{n} M(f, I_k)|I_k| - \sum_{k=1}^{n} m(f, I_k)|I_k|$$

$$= \sum_{k=1}^{n} [M(f, I_k) - m(f, I_k)]|I_k|$$

$$\leqslant \|\mathfrak{n}\| \sum_{k=1}^{n} [M(f, I_k) - m(f, I_k)]$$

$$= \|\mathfrak{n}\| \sum_{k=1}^{n} [f(x_k) - f(x_{k-1})]$$

$$= \|\mathfrak{n}\| [f(x_n) - f(x_0)]$$

$$= \|\mathfrak{n}\| [f(b) - f(a)]$$

$$< \{\varepsilon/[f(b) - f(a)]\} [f(b) - f(a)] = \varepsilon.$$

Thus Theorem 6.1.3 implies that f is integrable on I.

We now turn to the question of integrability of continuous functions on a compact interval. In Section 4.6 it was proved that a continuous function f on a compact interval I is uniformly continuous on I, that is, to each $\varepsilon > 0$ there is a $\delta > 0$ such that x and y in I and $|x - y| < \delta$ imply that $|f(x) - f(y)| < \varepsilon$. We use this fact to prove the following theorem.

Theorem 6.2.5. *If f is continuous on the compact interval I, then f is integrable on I.*

Proof. Let $I = [a, b]$, and let $\varepsilon > 0$. We again use Theorem 6.1.3 to show that f is integrable on I. Since f is continuous on I, Theorem 4.6.3 implies that f is uniformly continuous on I. Thus there is a $\delta > 0$ such that x and y in I and $|x - y| < \delta$ imply that $|f(x) - f(y)| < \varepsilon/(b - a)$. Let $\mathfrak{n}$ be any net with $\|\mathfrak{n}\| < \delta$, and let $I_k = [x_{k-1}, x_k]$, $k = 1, 2, ..., n$, be the subintervals of $\mathfrak{n}$. Since f is continuous on I, f is continuous on each of the subintervals I_k, $k = 1, 2, ..., n$. Theorem 4.5.2 implies that for each k, where $k = 1, 2, ..., n$, there is $\zeta_k \in I_k$ such that $M(f, I_k) = f(\zeta_k)$, and Theorem 4.5.2 implies that for each k, where $k = 1, 2, ..., n$, there is $\xi_k \in I_k$ such that $m(f, I_k) = f(\xi_k)$. Since $\|\mathfrak{n}\| < \delta$ and $\zeta_k, \xi_k \in I_k$, we have that $|\zeta_k - \xi_k| < \delta$; and, since f is uniformly continuous on I, we have that

$$f(\zeta_k) - f(\xi_k) < \varepsilon/(b - a), \qquad k = 1, 2, ..., n.$$

Then

$$u(f, n) - l(f, n) = \sum_{k=1}^{n} [M(f, I_k) - m(f, I_k)]|I_k|$$

$$= \sum_{k=1}^{n} [f(\zeta_k) - f(\xi_k)]|I_k|$$

$$\leq \sum_{k=1}^{n} [\varepsilon/(b - a)]|I_k|$$

$$= [\varepsilon/(b - a)] \sum_{k=1}^{n} |I_k|$$

$$= [\varepsilon/(b - a)](b - a) = \varepsilon.$$

Thus Theorem 6.1.3 implies that f is integrable on I

Problem Set 6.2

1. Express $\lim\limits_{n \to \infty} \sum\limits_{k=1}^{n} 1/(n + k)$ as an integral. [*Hint*: Write $\sum\limits_{k=1}^{n} 1/(n + k)$ as a lower sum and express the answer as an integral.]

2. Express $\lim\limits_{n \to \infty} \sum\limits_{k=1}^{n} k/(n^2 + k^2)$ as an integral.

3. Express $\lim\limits_{n \to \infty} n \sum\limits_{k=1}^{n} 1/(n^2 + k^2)$ as an integral.

4. Prove the remaining part of Theorem 6.2.1.

5. Prove Theorem 6.2.2.

6. Prove that if the limit defined in Definition 6.2.2 exists, then it is unique.

7. Prove the remaining part of Theorem 6.2.3.

8. Prove that a decreasing function on a compact interval is integrable.

9. Express $\lim\limits_{n \to \infty} \sum\limits_{k=1}^{n} 1/(n+k-1/2)$ as an integral. [*Hint*: Write $\sum\limits_{k=1}^{n} 1/(n+k-1/2)$ as a Riemann sum and express the answer as an integral.]

10. Express $\lim\limits_{n \to \infty} \sum\limits_{k=1}^{n} (k - 1/2)/(n + k - 1/2)^2$ as an integral.

11. Express $\lim\limits_{n \to \infty} \sum\limits_{k=1}^{n} 1/\sqrt{n} \sqrt{n + k - 1/2}$ as an integral.

12. Let $f: [a, b] \to R$, and let $c \in (a, b)$. If f is continuous and bounded on $[a, c)$ and $(c, b]$, show that f is integrable on $[a, b]$.

13. Let $f: [a, b] \to R$. If f is continuous on $[a, b]$ except for a finite number of points of $[a, b]$ and bounded on $[a, b]$, show that f is integrable on $[a, b]$.

14. Let f be a monotone function on the interval $[a, b]$. Show that

$$f(a)(b - a) \leqslant \int_a^b f(x)dx \leqslant f(b)(b - a)$$

or

$$f(b)(b - a) \leqslant \int_a^b f(x)dx \leqslant f(a)(b - a).$$

15. Show that, for each natural number n,

$$0 \leqslant \int_0^{\pi/2} \sin^{n+1} x \, dx \leqslant \int_0^{\pi/2} \sin^n x \, dx.$$

6.3. The Class of Integrable Functions

To determine which bounded functions are Riemann integrable by direct application of the definition or Theorem 6.2.3 is extremely difficult and laborious. Hence we shall derive some general properties of Riemann integrable functions and develop some methods for evaluating the integrals of Riemann integrable functions.

In the next few sections we shall use the terms "integral" and "integrable" to mean "Riemann integral" and "Riemann Integrable." For the remainder of this section, we shall let I denote the compact interval $[a, b]$ and assume that all functions discussed are bounded on I.

In order to prove that the sum of two integrable functions is integrable, we need the following lemmas.

Lemma 6.3.1. *Let f and g be any two functions with domain I. Then*

$$M(f + g, I) \leqslant M(f, I) + M(g, I)$$

and

$$m(f + g, I) \geqslant m(f, I) + m(g, I).$$

Proof. We prove the first of these inequalities; the second is left as Exercise 1 of Problem Set 6.3.

To prove that

$$M(f + g, I) \leqslant M(f, I) + M(g, I),$$

we assume to the contrary that

$$M(f + g, I) > M(f, I) + M(g, I)$$

and show that this assumption leads to a contradiction. Let x and y

be any points of I. Then $f(x) \leqslant M(f, I), g(y) \leqslant M(g, I)$, and

$$f(x) + g(y) \leqslant M(f, I) + M(g, I) < M(f + g, I).$$

Since $M(f + g, I)$ is the least upper bound of the set

$$\{(f + g)(x): x \in I\} \quad \text{and} \quad f(x) + g(y) < M(f + g, I),$$

there is $z \in I$ such that

$$f(x) + g(y) < (f + g)(z) \leqslant M(f + g, I).$$

Since the points x and y are arbitrary points of I, let $x = y = z$
Then

$$f(z) + g(z) < (f + g)(z) = f(z) + g(z),$$

which is a contradiction. Hence

$$M(f + g, I) \leqslant M(f, I) + M(g, I).$$

Lemma 6.3.2. *Let f and g be any two functions with domain I. The*

$$u(f + g, \mathfrak{n}) \leqslant u(f, \mathfrak{n}) + u(g, \mathfrak{n})$$

and

$$l(f + g, \mathfrak{n}) \geqslant l(f, \mathfrak{n}) + l(g, \mathfrak{n}).$$

Proof. We prove the first of these inequalities; the second is left a
Exercise 2 of Problem Set 6.3.

Let $\mathfrak{n}$ be any net of I, and let I_k, where $k = 1, 2, \ldots, n$, be th
subintervals of $\mathfrak{n}$. Lemma 6.3.1 implies that

$$M(f + g, I_k) \leqslant M(f, I_k) + M(g, I_k)$$

for each $k, k = 1, 2, \ldots, n$. Since $|I_k| > 0$, where $k = 1, 2, \ldots, n$,

$$M(f + g, I_k)|I_k| \leqslant \{M(f, I_k) + M(g, I_k)\}|I_k|$$
$$= M(f, I_k)|I_k| + M(g, I_k)|I_k|,$$

where $k = 1, 2, \ldots, n$. Summing from $k = 1$ to $k = n$ we have

$$u(f + g, \mathfrak{n}) = \sum_{k=1}^{n} M(f + g, I_k)|I_k|$$

$$\leqslant \sum_{k=1}^{n} M(f, I_k)|I_k| + \sum_{k=1}^{n} M(g, I_k)|I_k|$$

$$= u(f, \mathfrak{n}) + u(g, \mathfrak{n}),$$

which gives the desired result.

We now use Lemma 6.3.2 and Theorem 6.1.3 to prove that the sum of two integrable functions is integrable.

Theorem 6.3.1. *If f and g are both integrable on I, then $f + g$ is integrable on I and*

$$\int_a^b (f + g)(x)dx = \int_a^b f(x)dx + \int_a^b g(x)dx.$$

Proof. Let $\varepsilon > 0$. Since f and g are integrable on I, Theorem 6.1.3 implies that there are nets $\mathfrak{n}_1$ and $\mathfrak{n}_2$ of I such that

$$u(f, \mathfrak{n}_1) - l(f, \mathfrak{n}_1) < \varepsilon/2 \quad \text{and} \quad u(g, \mathfrak{n}_2) - l(g, \mathfrak{n}_2) < \varepsilon/2.$$

Let $\mathfrak{n}$ be a common refinement of $\mathfrak{n}_1$ and $\mathfrak{n}_2$. Then it follows from Theorem 6.1.1 that

$$u(f, \mathfrak{n}) - l(f, \mathfrak{n}) < \varepsilon/2 \quad \text{and} \quad u(g, \mathfrak{n}) - l(g, \mathfrak{n}) < \varepsilon/2.$$

Finally, Lemma 6.3.2 implies that

$$u(f + g, \mathfrak{n}) - l(f + g, \mathfrak{n}) \leqslant u(f, \mathfrak{n}) + v(g, \mathfrak{n}) - l(f, \mathfrak{n}) - l(g, \mathfrak{n})$$
$$< \varepsilon/2 + \varepsilon/2 = \varepsilon.$$

Therefore, it follows from Theorem 6.1.3 that $f + g$ is integrable on I.

To verify that the integral of the sum function is the sum of the integrals of the functions, we could show that

$$\int_a^b (f + g)(x)dx \leqslant \int_a^b f(x)dx + \int_a^b g(x)dx$$

and

$$\int_a^b (f + g)(x)dx \geqslant \int_a^b f(x)dx + \int_a^b g(x)dx.$$

We shall prove the first of these two inequalities; the second inequality follows in an analogous fashion and is left as Exercise 3 of Problem Set 6.3.

We assume that

$$\int_a^b (f + g)(x)dx > \int_a^b f(x)dx + \int_a^b g(x)dx$$

and show that this leads us to a contradiction. Let

$$\varepsilon = \int_a^b (f + g)(x)dx - \int_a^b f(x)dx - \int_a^b g(x)dx > 0.$$

Then

$$\int_a^b (f + g)(x)dx = \int_a^b f(x)dx + \int_a^b g(x)dx + \varepsilon.$$

Since f and g are integrable on I, there are nets $\mathfrak{n}_1$ and $\mathfrak{n}_2$ of I such that

$$\int_a^b f(x)dx > u(f, \mathfrak{n}_1) - \varepsilon/2$$

and

$$\int_a^b g(x)dx > u(g, \mathfrak{n}_2) - \varepsilon/2.$$

Let $\mathfrak{n}$ be a common refinement of $\mathfrak{n}_1$ and $\mathfrak{n}_2$. Then

$$\int_a^b f(x)dx > u(f, \mathfrak{n}) - \varepsilon/2$$

and

$$\int_a^b g(x)dx > u(g, \mathfrak{n}) - \varepsilon/2.$$

Lemma 6.3.2 and the above inequalities now imply that

$$u(f, \mathfrak{n}) + u(g, \mathfrak{n}) \geqslant u(f + g, \mathfrak{n})$$

$$\geqslant \int_a^b (f + g)(x)dx$$

$$= \int_a^b f(x)dx + \int_a^b g(x)dx + \varepsilon$$

$$> u(f, \mathfrak{n}) + u(g, \mathfrak{n}).$$

Since this is a contradiction, we have

$$\int_a^b (f + g)(x)dx \leqslant \int_a^b f(x)dx + \int_a^b g(x)dx.$$

Combining this with the inequality

$$\int_a^b (f + g)(x)dx \geqslant \int_a^b f(x)dx + \int_a^b g(x)dx,$$

gives the desired result.

Theorem 6.3.1 is a useful tool for evaluating integrals; we present an example to demonstrate this technique.

Example 6.3.1. Let $f: [0, 1] \to R$ be defined by $f(x) = x^2$ for each $x \in [0, 1]$, and let $g: [0, 1] \to R$ be defined by $g(x) = e^x$ for each $x \in [0, 1]$. Then Theorem 6.3.1 implies that the function $h: [0, 1] \to R$ defined by $h(x) = f(x) + g(x)$ for each $x \in [0, 1]$ is integrable on $[0, 1]$ and

$$\int_0^1 (x^2 + e^x)dx = \int_0^1 x^2 \, dx + \int_0^1 e^x \, dx$$

$$= 1/3 + e - 1 = e - 2/3.$$

(Cf. Example 6.1.3 and Exercise 10 of Problem Set 6.1.)

We use the following lemma to show that the product of a constant and an integrable function is integrable.

Lemma 6.3.3. *Let f be a bounded function on I, let $\mathfrak{n}$ be any net of I, and let c be a real number. If $c \geqslant 0$, then*

$$u(cf, \mathfrak{n}) = cu(f, \mathfrak{n}) \quad \text{and} \quad l(cf, \mathfrak{n}) = cl(f, \mathfrak{n}).$$

If $c < 0$, then

$$u(cf, \mathfrak{n}) = cl(f, \mathfrak{n}) \quad \text{and} \quad l(cf, \mathfrak{n}) = cu(f, \mathfrak{n}).$$

The proof of this lemma is left as Exercise 5 of Problem Set 6.3.

Theorem 6.3.2. *If f is integrable on I and c is any real number, then cf is integrable on I and*

$$\int_a^b cf(x)dx = c \int_a^b f(x)dx.$$

Proof. We first show that cf is integrable on I. Let $\varepsilon > 0$. Since f is integrable on I, it follows from Theorem 6.1.3 that there is a net $\mathfrak{n}$ of I such that

$$u(f, \mathfrak{n}) - l(f, \mathfrak{n}) < \varepsilon/(|c| + 1).$$

If $c \geqslant 0$, Lemma 6.3.3 implies that

$$u(cf, \mathfrak{n}) - l(cf, \mathfrak{n}) = c\{u(f, \mathfrak{n}) - l(f, \mathfrak{n})\}$$
$$< c\varepsilon/(|c| + 1) < \varepsilon,$$

while if $c < 0$, it follows from Lemma 6.3.3 that

$$u(cf, \mathfrak{n}) - l(cf, \mathfrak{n}) = |c|\{u(f, \mathfrak{n}) - l(f, \mathfrak{n})\}$$
$$< |c| \, \varepsilon/(|c| + 1) < \varepsilon.$$

In either case,

$$u(cf, \mathfrak{n}) - l(cf, \mathfrak{n}) < \varepsilon$$

and Theorem 6.1.3 implies that cf is integrable on I.

To verify that the integral of the product of a constant and an integrable function is the product of the constant and the integral of that function, we could show that

$$\int_a^b cf(x)dx \leqslant c \int_a^b f(x)dx$$

and

$$\int_a^b cf(x)dx \geqslant c \int_a^b f(x)dx.$$

Again we shall prove the first of these two inequalities; the second follows by similar arguments and is left as Exercise 6 of Problem Set 6.3. Before doing this, we point out that if $c = 0$, then $cf = 0$ and the desired result follows trivially. (Cf. Exercise 7 of Problem Set 6.1.)

We assume to the contrary that

$$\int_a^b cf(x)dx > c \int_a^b f(x)dx.$$

Since cf is integrable, there is a net $\mathfrak{n}$ of I such that

$$l(cf, \mathfrak{n}) > c \int_a^b f(x)dx.$$

If $c > 0$, then Lemma 6.3.3 implies that $l(cf, \mathfrak{n}) = cl(f, \mathfrak{n})$ and it follows that

$$cl(f, \mathfrak{n}) > c \int_a^b f(x)dx.$$

Dividing by c we get

$$l(f, \mathfrak{n}) > \int_a^b f(x)dx,$$

which contradicts the fact that any lower sum is less than or equal to the value of the integral. If $c < 0$, then Lemma 6.3.3 implies that $l(cf, \mathfrak{n}) = cu(f, \mathfrak{n})$ and it follows that

$$cu(f, \mathfrak{n}) > c \int_a^b f(x)dx.$$

Since c is negative, division by c gives

$$u(f, \mathfrak{n}) < \int_a^b f(x)dx.$$

This contradicts the fact that any upper sum is greater than or equal to the value of the integral. In either case we are lead to a contradiction and so

$$\int_a^b cf(x)dx \leqslant c \int_a^b f(x)dx.$$

Combining this with the inequality

$$\int_a^b cf(x)dx \geqslant c \int_a^b f(x)dx,$$

yields the desired result.

As a corollary to Theorem 6.3.1 and 6.3.2 we get the following result.

Corollary 6.3.2. *If f and g are integrable on I, then $f - g$ is integrable on I and*

$$\int_a^b (f - g)(x)dx = \int_a^b f(x)dx - \int_a^b g(x)dx.$$

Our next goal is to verify that the product of two integrable functions is integrable. To this end we first show that the absolute value function of an integrable function is integrable and that the square of an integrable function is integrable. The following lemma is necessary for these proofs.

Lemma 6.3.4. *Let f be a bounded function on I. Then*

$$M(|f|, I) - m(|f|, I) \leqslant M(f, I) - m(f, I).$$

Proof. We assume to the contrary that

$$M(|f|, I) - m(|f|, I) > M(f, I) - m(f, I)$$

and show that this assumption leads to a contradiction. Let

$$\varepsilon = M(|f|, I) - m(|f|, I) - [M(f, I) - m(f, I)].$$

Since $M(|f|, I)$ is the least upper bound of the set $\{|f|(x): x \in I\}$, there is $\zeta \in I$ such that $|f|(\zeta) > M(|f|, I) - \varepsilon/2$. Likewise, since $m(|f|, I)$ is the greatest lower bound of the set $\{|f|(x): x \in I\}$, there is $\xi \in I$ such that $|f|(\xi) < m(|f|, I) + \varepsilon/2$. Hence,

$$|f|(\zeta) - |f|(\xi) > M(|f|, I) - \varepsilon/2 - [m(|f|, I) + \varepsilon/2]$$
$$= M(|f|, I) - m(|f|, I) - \varepsilon$$
$$= M(f, I) - m(f, I).$$

Also for any points v and η of I, $f(v) \leqslant M(f, I)$ and $f(\eta) \geqslant m(f, I)$. It follows that

$$M(f, I) - m(f, I) \geqslant f(v) - f(\eta).$$

Using this inequality and the points ζ and ξ defined above, we have

$$M(f, I) - m(f, I) \geqslant f(\xi) - f(\zeta), \quad M(f, I) - m(f, I) \geqslant f(\zeta) - f(\xi).$$

Thus

$$M(f, I) - m(f, I) \geqslant |f(\xi) - f(\zeta)|.$$

This inequality together with the inequality

$$|f|(\zeta) - |f|(\xi) > M(f, I) - m(f, I)$$

imply that

$$|f|(\zeta) - |f|(\xi) = |f(\zeta)| - |f(\xi)| > |f(\zeta) - f(\xi)|.$$

But this inequality contradicts Corollary 2.4.6. Therefore,

$$M(|f|, I) - m(|f|, I) \leqslant M(f, I) - m(f, I).$$

To verify that the absolute value function of an integrable function is integrable, we need to define two special functions.

Definition 6.3.1. *Let* $f: D \to R$. *The functions* $f^+: D \to R$ *and* $f^-: D \to R$ *defined by the equations*

$$f^+ = (|f| + f)/2 \quad \text{and} \quad f^- = (|f| - f)/2$$

are called the **positive** *and* **negative** *parts of the function* f.

It is easily seen that if $f: D \to R$, then $f^+(x) \geqslant 0$, $f^-(x) \geqslant 0$, $f^+(x) = \max(f(x), 0)$, and $f^-(x) = \max(-f(x), 0)$ for each $x \in D$. It is left as an exercise to show that if f is integrable on the compact interval I, then f^+ and f^- are integrable on the interval I (Exercise 10 of Problem Set 6.3).

We are now ready to prove that the absolute value function of an integrable function is integrable.

Theorem 6.3.3. *Let* f *be integrable on* I. *Then* $|f|$ *is integrable on* I *and*

$$\left| \int_a^b f(x)dx \right| \leqslant \int_a^b |f|(x)dx.$$

Proof. We first show that $|f|$ is integrable on I. Let $\varepsilon > 0$. Since f is integrable on I, Theorem 6.1.3 implies that there is a net $\mathfrak{n}$ of I such that

$$u(f, \mathfrak{n}) - l(f, \mathfrak{n}) < \varepsilon.$$

We denote by I_k, where $k = 1, 2, ..., n$, the subintervals of the net $\mathfrak{n}$. Then Lemma 6.3.4 implies that

$$M(|f|, I_k) - m(|f|, I_k) \leqslant M(f, I_k) - m(f, I_k),$$

where $k = 1, 2, ..., n$. Since the length of each subinterval of the net $\mathfrak{n}$ is greater than zero, we have

$$[M(|f|, I_k) - m(|f|, I_k)]|I_k| \leqslant [M(f, I_k) - m(f, I_k)]|I_k|,$$

where $k = 1, 2, ..., n$. Summing from $k = 1$ to $k = n$ we get

$$u(|f|, \mathfrak{n}) - l(|f|, \mathfrak{n}) = \sum_{k=1}^{n} [M(|f|, I_k) - m(|f|, I_k)]|I_k|$$

$$\leqslant \sum_{k=1}^{n} [M(f, I_k) - m(f, I_k)]|I_k|$$

$$= u(f, \mathfrak{n}) - l(f, \mathfrak{n}) < \varepsilon.$$

Thus there is a net $\mathfrak{n}$ of I such that

$$u(|f|, I) - l(|f|, I) < \varepsilon$$

and Theorem 6.1.3 implies that $|f|$ is integrable on I.

Since f is integrable on I, f^+ and f^- are integrable on I (cf. Exercise 10 of Problem Set 6.3) and it follows that

$$\left| \int_a^b f(x)dx \right| = \left| \int_a^b [f^+(x) - f^-(x)]dx \right|$$

$$= \left| \int_a^b f^+(x)dx - \int_a^b f^-(x)dx \right|$$

$$\leqslant \left| \int_a^b f^+(x)dx \right| + \left| \int_a^b f^-(x)dx \right|$$

$$= \int_a^b f^+(x)dx + \int_a^b f^-(x)dx$$

$$= \int_a^b [f^+(x) + f^-(x)]dx$$

$$= \int_a^b |f|(x)dx,$$

which completes the proof.

INTEGRATION

Before proving that the square of an integrable function is integrable, we present an example to show that the converse of Theorem 6.3.3 is not necessarily true.

Example 6.3.2. Let $f: [0, 1] \to R$ such that

$$f(x) = \begin{cases} 1 & \text{if } x \text{ is rational,} \\ -1 & \text{if } x \text{ is irrational.} \end{cases}$$

Then $|f|(x) = 1$ for $x \in [0, 1]$ and $|f|$ is integrable on $[0, 1]$. (Cf. Exercise 7 of Problem Set 6.1.) In Example 6.1.4 it was shown that f is not integrable on $[0, 1]$. Thus f is an example of a function whose absolute value function is integrable on $[0, 1]$, but the function itself is not integrable on $[0, 1]$.

We now apply Theorem 6.3.3 to prove that the square of an integrable function is integrable.

Theorem 6.3.4. *If f is integrable on I, then $f^2 = f \cdot f$ is integrable on I.*

Proof. Since f is integrable on I, f is bounded on I and there is a real number $M > 0$ such that $|f(x)| \leqslant M$ for each $x \in I$. Also, Theorem 6.3.3 implies that $|f|$ is integrable on I. From Theorem 6.1.3 it follows that there is a net $\mathfrak{n}$ of I such that

$$u(|f|, \mathfrak{n}) - l(|f|, \mathfrak{n}) < \varepsilon/(2M).$$

Let I_k, where $k = 1, 2, ..., n$, be the subintervals of $\mathfrak{n}$, then

$$M(f^2, I_k) = [M(|f|, I_k)]^2$$

and

$$m(f^2, I_k) = [m(|f|, I_k)]^2,$$

where $k = 1, 2, ..., n$. (Cf. Exercise 13 of Problem Set 6.3.) Thus

$$\begin{aligned} M(f^2, I_k) &- m(f^2, I_k) \\ &= [M(|f|, I_k)]^2 - [m(|f|, I_k)]^2 \\ &= [M(|f|, I_k) + m(|f|, I_k)][M(|f|, I_k) - m(|f|, I_k)] \\ &\leqslant 2M[M(|f|, I_k) - m(|f|, I_k)], \end{aligned}$$

where $k = 1, 2, ..., n$. Since the length of each subinterval of $\mathfrak{n}$ is greater than zero,

$$[M(f^2, I_k) - m(f^2, I_k)]|I_k| \leqslant 2M[M(|f|, I_k) - m(|f|, I_k)]|I_k|,$$

where $k = 1, 2, ..., n$. Summing from $k = 1$ to $k = n$, we have

$$u(f^2, \mathfrak{n}) - l(f^2, \mathfrak{n}) = \sum_{k=1}^{n} [M(f^2, I_k) - m(f^2, I_k)]|I_k|$$

$$\leqslant \sum_{k=1}^{n} 2M[M(|f|, I_k) - m(|f|, I_k)]|I_k|$$

$$= 2M \sum_{k=1}^{n} [M(|f|, I_k) - m(|f|, I_k)]|I_k|$$

$$= 2M[u(|f|, \mathfrak{n}) - l(|f|, \mathfrak{n})]$$

$$< 2M \varepsilon/(2M) = \varepsilon.$$

Thus there is a net $\mathfrak{n}$ of I such that

$$u(f^2, \mathfrak{n}) - l(f^2, \mathfrak{n}) < \varepsilon,$$

and it follows from Theorem 6.1.3 that f^2 is integrable on I.

Theorem 6.3.4 now allows us to prove easily that the product of two integrable functions is integrable.

Theorem 6.3.5. *If f and g are integrable on I, then fg is integrable on I.*

Proof. Write

$$fg = [(f + g)^2 - f^2 - g^2]/2.$$

Since f and g are integrable on I, it follows from Theorems 6.3.1, 6.3.2, and 6.3.4 that fg is integrable on I.

Problem Set 6.3

1. Let f and g be any bounded functions with domain I. Show that

$$m(f + g, I) \geqslant m(f, I) + m(g, I).$$

2. Let f and g be any bounded function with domain I, and let $\mathfrak{n}$ be any net of I. Show that

$$l(f + g, \mathfrak{n}) \geqslant l(f, \mathfrak{n}) + l(g, \mathfrak{n}).$$

3. If f and g are integrable on the interval $I = [a, b]$, show that

$$\int_a^b (f + g)(x)dx \geqslant \int_a^b f(x)dx + \int_a^b g(x)dx.$$

4. Evaluate the following integrals using the theorems of this section, Exercises 8–11 of Problem Set 6.1, and Example 6.1.3.

(a) $\displaystyle\int_0^1 (3 \sin x + 2x)dx.$

(b) $\displaystyle\int_0^1 (e^x + 2 \sin x)dx.$

(c) $\displaystyle\int_0^1 (a + bx - cx^2)dx$, where a, b, and c are constants.

5. Prove Lemma 6.3.3.

6. Let f be integrable on $[a, b]$, and let c be a constant. Show that

$$\int_a^b cf(x)dx \geqslant c \int_a^b f(x)dx.$$

7. Prove Corollary 6.3.2.

8. Let α and β be constants. If f and g are integrable on $[a, b]$, prove that the function $\alpha f + \beta g$ is integrable on $[a, b]$ and that

$$\int_a^b (\alpha f + \beta g)(x)dx = \alpha \int_a^b f(x)dx + \beta \int_a^b g(x)dx.$$

9. Let $f_1, f_2, ..., f_n$ be integrable on $[a, b]$, and let $c_1, c_2, ..., c_n$ be constants. Prove that the function

$$\sum_{k=1}^n c_k f_k$$

is integrable on $[a, b]$ and that

$$\int_a^b \left(\sum_{k=1}^n c_k f_k\right)(x)dx = \sum_{k=1}^n c_k \int_a^b f_k(x)dx.$$

10. Let f be integrable on $[a, b]$, and let f^+ and f^- be the positive and negative parts of f. Show that f^+ and f^- are integrable on $[a, b]$ and that

$$\int_a^b f^+(x)dx \geqslant 0 \quad \text{and} \quad \int_a^b f^-(x)dx \geqslant 0.$$

11. Give an example to show that the converse of Theorem 6.3.4 is not necessarily true; that is, give an example of a function f such that f^2 is integrable on an interval I, but f is not integrable on I.

12. Give an example to show that the converse of Theorem 6.3.5 is not necessarily true; that is, give an example of two functions f and g, such that fg is integrable on an interval I, but f and g are not integrable on I.

13. Let f be a bounded function with domain I. Show that

$$M(f^2, I) = [M(|f|, I)]^2 \quad \text{and} \quad m(f^2, I) = [m(|f|, I)]^2.$$

4. Give an example of a bounded function f with domain I such that

$$M(f^2, I) \neq [M(f, I)]^2 \quad \text{and} \quad m(f^2, I) \neq [m(f, I)]^2.$$

5. Let f be integrable on I, and let $1/f$ be bounded on I. Show that $1/f$ is integrable on I.

6. If f is integrable on $[a, b]$ and $|f(x)| \leq M$ for each $x \in [a, b]$, show that

$$\left| \int_a^b f(x)dx \right| \leq M(b - a).$$

7. If f and g are integrable on $[a, b]$, and if $f(x) \leq g(x)$ for each $x \in [a, b]$, show that

$$\int_a^b f(x)dx \leq \int_a^b g(x)dx,$$

8. Let f be integrable on $[a, b]$, and let $f(x) \geq 0$ for each $x \in [a, b]$. Prove that

$$\int_a^b f(x)dx \geq 0.$$

9. Let f be continuous on $[a, b]$, let $f(x) \geq 0$ for each $x \in [a, b]$, and let $f(c) > 0$ for some $c \in [a, b]$. Prove that

$$\int_a^b f(x)dx > 0.$$

0. Let f be continuous on $[a, b]$, let $f(x) \geq 0$ for each $x \in [a, b]$, and let

$$\int_a^b f(x)dx = 0.$$

Prove that f is identically zero on $[a, b]$.

1. (a) If $0 \leq x \leq 1$ show that $\dfrac{x^2}{\sqrt{2}} \leq \dfrac{x^2}{\sqrt{1 + x}} \leq x^2$.

(b) Show that $\dfrac{1}{3\sqrt{2}} \leq \displaystyle\int_0^1 \dfrac{x^2}{\sqrt{1 + x}} \, dx \leq \dfrac{1}{3}$.

2. (a) If $\pi/6 \leq x \leq \pi/2$ show that $\dfrac{2x}{\sin x} \leq 4x$.

(b) Show that $\dfrac{2\pi^2}{9} \leq \displaystyle\int_{\pi/6}^{\pi/2} \dfrac{2xdx}{\sin x} \leq \dfrac{4\pi^2}{9}$.

23. Let $f: D \rightarrow R$, where D is symmetric with respect to 0, that is, if $x \in D$, then $-x \in D$. Then f is said to be an **even function** if and only if $f(-x) = f(x)$ for each $x \in D$. Let f be an even function on the closed interval $[-a, a]$, and let f

be integrable on $[0, a]$. Show that f is integrable on $[-a, a]$ and that

$$\int_{-a}^{a} f(x)dx = 2\int_{0}^{a} f(x)dx.$$

24. Let f be an even function on $[-a, a]$, and let $[c, d] \subseteq [0, a]$. If f is integrabl
on $[c, d]$, show that f is integrable on $[-d, -c]$ and that

$$\int_{-d}^{-c} f(x)dx = \int_{c}^{d} f(x)dx.$$

25. Let $f: D \rightarrow R$, where D is symmetric with respect to 0. Then f is said to be a
odd function if and only if $f(-x) = -f(x)$ for each $x \in D$. Let f be an od
function on the closed interval $[-a, a]$, and let f be integrable on $[0, a]$. Shov
that f is integrable on $[-a, a]$ and that

$$\int_{-a}^{a} f(x)dx = 0.$$

26. Let f be an odd function on $[-a, a]$, and let $[c, d] \subseteq [0, a]$. If f is integrabl
on $[c, d]$, show that f is integrable on $[-d, -c]$ and that

$$\int_{-d}^{-c} f(x)dx = -\int_{c}^{d} f(x)dx.$$

27. Let $f: D \rightarrow R$ and $g: D \rightarrow R$.
   (a) If f and g are both even functions, show that $f + g$ is an even function.
   (b) If f and g are both odd functions, show that $f + g$ is an odd function.
   (c) If f and g are both even functions, show that fg is an even function.
   (d) If f and g are both odd functions, show that fg is an even function.
   (e) If f is an even function and g is an odd function, show that fg is an od
   function.
   (f) Show that parts (a), (b), and (c) can be extended to any finite sum or pro
   duct.

28. Let $f: D \rightarrow R$. If f is both an even function and an odd function, show tha
$f \equiv 0$.

29. Let $f: D \rightarrow R$ be differentiable on D.
   (a) If f is an even function, show that f' is an odd function.
   (b) If f is an odd function, show that f' is an even function.

30. Let D be a set such that $-x \in D$ whenever $x \in D$, and let $f: D \rightarrow R$.
   (a) Show that the function $g: D \rightarrow R$ defined by $g(x) = f(x) + f(-x)$ is a
   even function.
   (b) Show that the function $h: D \rightarrow R$ defined by $h(x) = f(x) - f(-x)$ is a
   odd function.
   (c) Show that the function f can be uniquely represented as the sum of an eve
   function and an odd function. [*Hint*: Let $f = g/2 + h/2$.]

6.4. Properties of Integrable Functions

We have defined the integral only for intervals, that is, integrals for which the lower limit of integration is less than the upper limit. We can remove this restriction by the following convention. If f is defined at a, then

$$\int_a^a f(x)dx = 0,$$

and if f is integrable on $[a, b]$, then

$$\int_b^a f(x)dx = -\int_a^b f(x)dx.$$

We now establish some further properties of integrals and integrable functions.

Theorem 6.4.1. *If f is integrable on $I = [a, b]$ and $J = [c, d] \subseteq I$, then f is integrable on J.*

Proof. If $c = d$, then the result follows from the above convention, and so we assume $c \neq d$. Let $\varepsilon > 0$ and assume that f is integrable on I. It follows from Theorem 6.1.3 that there is a net $\mathfrak{n}$ of I such that

$$u(f, \mathfrak{n}) - l(f, \mathfrak{n}) < \varepsilon.$$

Let $\mathfrak{n}'$ be the refinement of $\mathfrak{n}$ that contains the points c and d (if c and d are both in $\mathfrak{n}$, then let $\mathfrak{n}' = \mathfrak{n}$). Theorem 6.1.1 implies that

$$u(f, \mathfrak{n}') - l(f, \mathfrak{n}') \leqslant u(f, \mathfrak{n}) - l(f, \mathfrak{n}) < \varepsilon.$$

Let $I_k = [x_{k-1}, x_k]$, $k = 1, 2, ..., n$, be the subintervals of $\mathfrak{n}'$ where $c = x_\alpha$, $d = x_\beta$, and $0 \leqslant \alpha < \beta \leqslant n$. Let

$$\mathfrak{n}^* = \{c = x_\alpha, x_{\alpha+1}, ..., x_\beta = d\}.$$

Then $\mathfrak{n}^*$ is a net of J with subintervals I_k, where

$$k = \alpha + 1, \alpha + 2, ..., \beta.$$

Since $M(f, I_k) - m(f, I_k) \geqslant 0$ and $|I_k| > 0$ for $k = 1, 2, ..., n$, we have

$$u(f, \mathfrak{n}^*) - l(f, \mathfrak{n}^*) = \sum_{k=\alpha+1}^{\beta} [M(f, I_k) - m(f, I_k)]|I_k|$$

$$\leqslant \sum_{k=1}^{n} [M(f, I_k) - m(f, I_k)]|I_k|$$

$$= u(f, \mathfrak{n}') - l(f, \mathfrak{n}') < \varepsilon.$$

Thus there is a net $\mathfrak{n}^*$ of J such that

$$u(f, \mathfrak{n}^*) - l(f, \mathfrak{n}^*) < \varepsilon,$$

and Theorem 6.1.3 implies that f is integrable on J.

If f is integrable on $[a, b]$ and $c \in (a, b)$, then Theorem 6.4.1 implies that f is integrable on $[a, c]$ and that f is integrable on $[c, b]$. We now obtain a relation between the three integrals

$$\int_a^b f(x)dx, \quad \int_a^c f(x)dx, \quad \text{and} \quad \int_c^b f(x)dx.$$

Theorem 6.4.2. *If f is integrable on $[a, b]$ and $c \in [a, b]$, then*

$$\int_a^b f(x)dx = \int_a^c f(x)dx + \int_c^b f(x)dx.$$

Proof. Assume that f is integrable on $[a, b]$; then from the above remarks it follows that f is integrable on the intervals $[a, c]$ and $[c, b]$. If $a = c$ or $b = c$, then the result follows from the convention that

$$\int_c^c f(x)dx = 0.$$

Thus we assume that $a \neq c$ and $c \neq b$. Let $\mathfrak{n}_1$ be any net of $[a, c]$, and let $\mathfrak{n}_2$ be any net of $[c, b]$. Then $\mathfrak{n} = \mathfrak{n}_1 \cup \mathfrak{n}_2$ is a net of $[a, b]$. Since f is integrable on $[a, b]$, we have

$$u(f, \mathfrak{n}_1) + u(f, \mathfrak{n}_2) = u(f, \mathfrak{n}) \geqslant \int_a^b f(x)dx.$$

Hence

$$u(f, \mathfrak{n}_1) \geqslant \int_a^b f(x)dx - u(f, \mathfrak{n}_2),$$

that is, the right-hand side of this inequality is a lower bound for the set S of all upper sums of f on $[a, c]$. Thus

$$\inf S \geqslant \int_a^b f(x)dx - u(f, \mathfrak{n}_2).$$

Since f is integrable on $[a, c]$, we have

$$\int_a^c f(x)dx = \inf S \geqslant \int_a^b f(x)dx - u(f, \mathfrak{n}_2).$$

We can now write the above inequality as

$$u(f, n_2) \geq \int_a^b f(x)dx - \int_a^c f(x)dx.$$

Since n_2 is an arbitrary net of $[c, b]$, the right-hand side of this inequality is a lower bound for the set T of all upper sums of f on $[c, b]$. Thus

$$\inf T \geq \int_a^b f(x)dx - \int_a^c f(x)dx.$$

Since f is integrable on $[c, b]$, it follows that

$$\int_c^b f(x)dx = \inf T \geq \int_a^b f(x)dx - \int_a^c f(x)dx.$$

Hence we have the inequality

$$\int_a^c f(x)dx + \int_c^b f(x)dx \geq \int_a^b f(x)dx.$$

By using lower sums in a manner similar to that above, we have

$$\int_a^c f(x)dx + \int_c^b f(x)dx \leq \int_a^b f(x)dx.$$

These two inequalities imply that

$$\int_a^b f(x)dx = \int_a^c f(x)dx + \int_c^b f(x)dx.$$

Corollary 6.4.2a. *Let a, b, c be any real numbers. Then*

$$\int_a^b f(x)dx = \int_a^c f(x)dx + \int_c^b f(x)dx$$

whenever these three integrals exist.

Proof. The case where $a < c < b$ is covered in Theorem 6.4.2. We use Theorem 6.4.2 to show that the result holds when $b < a < c$ and leave the proof of the remaining cases as Exercise 13 of Problem Set 6.4.

Assume $b < a < c$, then Theorem 6.4.2 implies that

$$\int_b^c f(x)dx = \int_b^a f(x)dx + \int_a^c f(x)dx.$$

We may write this equality as

$$-\int_c^b f(x)dx = -\int_a^b f(x)dx + \int_a^c f(x)dx$$

or

$$\int_a^b f(x)dx = \int_a^c f(x)dx + \int_c^b f(x)dx.$$

Corollary 6.4.2b. *Let* $a_0, a_1, \ldots, a_n$ *be any* $n+1$ *real numbers. Then*

$$\int_{a_0}^{a_n} f(x)dx = \sum_{k=1}^n \int_{a_{k-1}}^{a_k} f(x)dx$$

whenever these $n+1$ *integrals exists.*

The proof of Corollary 6.4.2b follows from Corollary 6.4.2a by using Mathematical Induction on the number of integrals. (Cf. Exercise 14 of Problem Set 6.4.)

A result related to Theorem 6.4.1 and 6.4.2 is the following theorem, whose proof is left as Exercise 15 of Problem Set 6.4.

Theorem 6.4.3. *If the function* f *is integrable on the intervals* $[a, c]$ *and* $[c, b]$, *then* f *is integrable on the interval* $[a, b]$, *and*

$$\int_a^b f(x)dx = \int_a^c f(x)dx + \int_c^b f(x)dx.$$

We now turn our attention to functions defined in terms of integrals. Let f be integrable on $[a, b]$, and let

$$F(x) = \int_a^x f(\zeta)d\zeta$$

for each $x \in [a, b]$. Does the function F have any special properties? We can show that F is continuous on $[a, b]$ and differentiable at each point c of $[a, b]$ where f is continuous, and that $F'(c) = f(c)$ at each point c of $[a, b]$ where f is continuous.

Theorem 6.4.4. *Let* f *be integrable on* $[a, b]$, *and let*

$$F(x) = \int_a^x f(\zeta)d\zeta$$

for each $x \in [a, b]$. *Then* F *is continuous on* $[a, b]$, *and, at each point* $c \in [a, b]$ *where* f *is continuous,* F *is differentiable and* $F'(c) = f(c)$.

Proof.   Since f is integrable on $[a, b]$, f is bounded on $[a, b]$ and there is $M > 0$ such that $|f(x)| \leqslant M$ for each x in $[a, b]$. For any point y of $[a, b]$ we shall show that F is continuous at y. Let $\varepsilon > 0$, and let $\delta = \varepsilon/(M + 1)$. Then, for any x in $[a, b]$ with $|x - y| < \delta$, we consider

$$F(x) - F(y) = \int_a^x f(\zeta)d\zeta - \int_a^y f(\zeta)d\zeta.$$

Since f is integrable on $[a, b]$, Corollary 6.4.2a allows us to write

$$\int_a^x f(\zeta)d\zeta - \int_a^y f(\zeta)d\zeta = \int_a^x f(\zeta)d\zeta + \int_y^a f(\zeta)d\zeta$$

$$= \int_y^x f(\zeta)d\zeta.$$

Then

$$|F(x) - F(y)| = \left| \int_y^x f(\zeta)d\zeta \right| \leqslant M|x - y|$$

$$\leqslant M\delta = M\varepsilon/(M + 1) < \varepsilon.$$

(Cf. Exercise 16 of Problem Set 6.3.) Thus, for each ε, there is a δ (where $\delta = \varepsilon/(M + 1)$) such that for any x in $[a, b]$ with $|x - y| < \delta$, we have $|F(x) - F(y)| < \varepsilon$. Hence F is continuous at y. Since y was an arbitrary point of $[a, b]$, F is continuous on $[a, b]$.

We now assume that f is continuous at $c \in [a, b]$ and show that F is differentiable at c and that $F'(c) = f(c)$. Let $\varepsilon > 0$. Since f is continuous at c, there is a $\delta > 0$ such that $x \in [a, b]$ and $|x - c| < \delta$ imply that $|f(x) - f(c)| < \varepsilon/2$. If $c < x < c + \delta$ and $x \in [a, b]$, then

$$\left| \frac{F(x) - F(c)}{x - c} - f(c) \right| = \frac{1}{x - c} \left| \int_a^x f(\zeta)d\zeta - \int_a^c f(\zeta)d\zeta - (x - c)f(c) \right|$$

$$= \frac{1}{x - c} \left| \int_c^x f(\zeta)d\zeta - \int_c^x f(c)d\zeta \right|$$

$$= \frac{1}{x - c} \left| \int_c^x [f(\zeta) - f(c)]d\zeta \right|$$

$$\leqslant \frac{1}{x - c} \cdot \frac{\varepsilon}{2} \cdot (x - c) < \varepsilon,$$

where $f(c)$ denotes the constant function with domain $[a, b]$ and value $f(c)$. If $c - \delta < x < c$ and $x \in [a, b]$, then, in a similar manner, we get

$$\left| \frac{F(x) - F(c)}{x - c} - f(c) \right| < \varepsilon.$$

Hence to each $\varepsilon > 0$ there is a $\delta > 0$ such that $x \in [a, b]$ and $|x - c| < \delta$ imply that

$$\left| \frac{F(x) - F(c)}{x - c} - f(c) \right| < \varepsilon.$$

Therefore,
$$\lim_{x \to c} \left[\frac{F(x) - F(c)}{x - c} \right] = f(c),$$

that is, F is differentiable at c and $F'(c) = f(c)$.

Corollary 6.4.4. *Let f be continuous on $[a, b]$, and let*

$$F(x) = \int_a^x f(\zeta) d\zeta$$

for each $x \in [a, b]$. Then F is differentiable on $[a, b]$ and $F'(x) = f(x)$ for each $x \in [a, b]$.

Example 6.4.1. Let $f: R \to R$ such that

$$f(x) = \int_0^x \frac{d\zeta}{\sqrt{16 + \zeta^2}}$$

for each $x \in R$. Then

$$f'(x) = \frac{1}{\sqrt{16 + x^2}}$$

for each $x \in R$. In particular, if $x = 3$ we have

$$f'(3) = \frac{1}{\sqrt{16 + 3^2}} = 1/5.$$

We are now ready to discuss the important *Fundamental Theorem of Integral Calculus*, which provides a method for evaluating integrals.

Theorem 6.4.5. (*The Fundamental Theorem of Integral Calculus*) *Let f be continuous on $[a, b]$, and let $G: [a, b] \to R$ such that $G'(x) = f(x)$ for each $x \in [a, b]$. Then*

$$\int_a^b f(x) dx = G(b) - G(a).$$

Proof. Let $F: [a, b] \to R$ such that

$$F(x) = \int_a^x f(\zeta)d\zeta$$

for each $x \in [a, b]$. Since f is continuous on $[a, b]$, Corollary 6.4.4 implies that $F'(x) = f(x)$ for each $x \in [a, b]$. Since $G'(x) = f(x)$ for each $x \in [a, b]$, we have that $F'(x) = G'(x)$ for each $x \in [a, b]$. By Theorem 5.3.6 there is a real number C such that $F(x) = G(x) + C$ for each $x \in [a, b]$. Thus,

$$F(b) - F(a) = G(b) + C - [G(a) + C] = G(b) - G(a).$$

But

$$F(a) = \int_a^a f(\zeta)d\zeta = 0,$$

and so

$$F(b) = F(b) - F(a) = G(b) - G(a).$$

Since

$$F(b) = \int_a^b f(\zeta)d\zeta,$$

we have

$$G(b) - G(a) = \int_a^b f(\zeta)d\zeta,$$

which completes the proof.

For notation purposes, the value $G(b) - G(a)$ is usually written

$$G(b) - G(a) = G(x)\Big|_a^b.$$

Then, under the conditions of Theorem 6.4.5, we may write

$$\int_a^b f(x)dx = G(x)\Big|_a^b.$$

Example 6.4.2. Evaluate the integral

$$\int_2^6 x^3 dx.$$

If we let $G(x) = x^4/4$ for each $x \in [2, 6]$, then $G'(x) = x^3$ and Theorem 6.4.5 implies that

$$\int_2^6 x^3 dx = x^4/4 \Big|_2^6 = 6^4/4 - 2^4/4$$

$$= 324 - 4 = 320.$$

From Theorem 6.4.5 we observe that if F' is continuous on $[a, b]$, then

$$F(x) = \int_a^x F'(\zeta)d\zeta,$$

where $x \in [a, b]$, that is, if the derivative of a function is continuous then the function itself is the integral of its derivative. This leads one to the following question. If F is differentiable on $[a, b]$, does

$$F(x) = \int_a^x F'(\zeta)d\zeta ?$$

The answer is *no*, since a derivative of a differentiable function need not be integrable. The following example presents such a function.

Example 6.4.3. Let $f: [0, 1] \to R$ such that

$$f(x) = \begin{cases} x^2 \sin(1/x^2) & \text{if } x \neq 0, \\ 0 & \text{if } x = 0. \end{cases}$$

Then f is differentiable on $[0, 1]$ and f' is given by

$$f'(x) = \begin{cases} 2x \sin(1/x^2) - (2/x) \cos(1/x^2) & \text{if } x \neq 0, \\ 0 & \text{if } x = 0. \end{cases}$$

Thus, f' is not bounded in any neighborhood of 0, and so f' is not an integrable function on $[0, 1]$. Therefore

$$f(x) \neq \int_0^x f'(\zeta)d\zeta,$$

since the integral does not exist.

The following theorem assures us that if f' is integrable on $[a, b]$, then its integral is the function f. This theorem provides a useful method for evaluating integrals.

Theorem 6.4.6. *If f is differentiable on $[a, b]$, and if f' is integrable on $[a, b]$, then*

$$\int_a^b f'(x)dx = f(b) - f(a).$$

Proof. Let $\mathfrak{n}$ be any net of $[a, b]$, and let $I_k = [x_{k-1}, x_k]$, $k = 1, 2, \ldots, n$, be the subintervals of $\mathfrak{n}$. Then, for each subinterval

I_k, where $k = 1, 2, ..., n$, the Mean Value Theorem implies that there is a $\xi_k \in I_k$ such that

$$f(x_k) - f(x_{k-1}) = f'(\xi_k)|I_k|.$$

Thus

$$f(b) - f(a) = \sum_{k=1}^{n} \{f(x_k) - f(x_{k-1})\}$$

$$= \sum_{k=1}^{n} f'(\xi_k)|I_k|.$$

Since $m(f', I_k) \leqslant f'(\xi_k) \leqslant M(f', I_k)$, where $k = 1, 2, ..., r$, we have

$$l(f', n) = \sum_{k=1}^{n} m(f', I_k)|I_k|$$

$$\leqslant f(b) - f(a)$$

$$\leqslant \sum_{k=1}^{n} M(f', I_k)|I_k| = u(f', n).$$

Since n was an arbitrary net of $[a, b]$, the number $f(b) - f(a)$ is an upper bound for the set of all lower sums of f' on $[a, b]$ and $f(b) - f(a)$ is a lower bound for the set of all upper sums of f' on $[a, b]$. Since f' is integrable on $[a, b]$,

$$\int_a^b f'(x)dx \leqslant f(b) - f(a) \leqslant \int_a^b f'(x)dx.$$

Therefore,

$$\int_a^b f'(x)dx = f(b) - f(a),$$

which concludes the proof.

Problem Set 6.4

Evaluate the integrals given in Exercises 1–6.

1. $\int_0^1 (x^2 - 2x)dx.$

2. $\int_1^2 \dfrac{dx}{x^2}.$

3. $\int_0^2 \sqrt{1 + x^3}\, x^2 dx$. [*Hint*: If $f(x) = (1 + x^3)^{3/2}$, then $f'(x) = 9(1 + x^3)^{1/2}\, x^2/2$.]

4. $\int_0^{\pi/2} \sin x \, dx$.

5. $\int_0^\pi \cos x \, dx$.

6. $\int_0^\pi \sin x \cos x \, dx$. [*Hint*: If $f(x) = \sin^2 x$, then $f'(x) = 2 \sin x \cos x$.]

For each of the functions F defined below in Exercises 7–12, find F'.

7. $F(x) = \int_a^b \sin(\zeta^2)d\zeta$.

8. $F(x) = \int_0^x \cos(\zeta^2)d\zeta$.

9. $F(x) = \int_x^\pi \sin(\zeta^2)d\zeta$.

10. $F(x) = \int_0^{x^2} \cos(\zeta^2)d\zeta$.

11. $F(x) = \int_{x^2}^1 \dfrac{d\zeta}{1 + \zeta^4}$.

12. $F(x) = \int_{x^3}^{x^5} \sqrt{1 + \zeta^2} \, d\zeta$.

13. Prove the remaining parts of Corollary 6.4.2a.

14. Prove Corollary 6.4.2b.

15. Prove Theorem 6.4.3.

16. Let f be continuous on $[a, b]$, and let

$$F(x) = \int_x^b f(\zeta)d\zeta$$

for each $x \in [a, b]$. Find F'.

17. Let f be continuous on the interval $[a, b]$, and let $v: D \to [a, b]$ such that v is differentiable on D. Let $F: D \to R$ such that

$$F(x) = \int_a^{v(x)} f(\zeta)d\zeta$$

for each $x \in D$. Show that

$$F'(x) = f(v(x))v'(x) = (f \circ v)(x)v'(x)$$

for each $x \in D$. [*Hint*: Define $\phi(v) = \int_a^v f(\zeta)d\zeta$, where $v = v(x)$. Then $F(x) = (\phi \circ v)(x)$.]

18. Let f be continuous on the interval $I = [a, b]$, and let $v: D \to I$ and $u: D \to I$ such that u and v are differentiable on D. Let $F: D \to R$ such that

$$F(x) = \int_{u(x)}^{v(x)} f(\zeta)d\zeta$$

for each $x \in D$. Show that

$$F'(x) = f(v(x))v'(x) - f(u(x))u'(x)$$

$$= (f \circ v)(x)v'(x) - (f \circ u)(x)u'(x)$$

for each $x \in D$. [*Hint*: Write $F(x) = \int_a^{v(x)} f(\zeta)d\zeta - \int_a^{u(x)} f(\zeta)d\zeta.$]

19. Let f be continuous on $[0, 1]$ such that, for each $x \in [0, 1]$,

$$\int_0^x f(\zeta)d\zeta = \int_x^1 f(\zeta)d\zeta.$$

Find f.

20. What is wrong with

$$\int_{-1}^1 \frac{dx}{x^2} = \frac{-1}{x} \bigg|_{-1}^1 = -2?$$

21. Let f be integrable on $[a, b]$, and let $f(x) \geqslant 0$ for each $x \in [a, b]$. Show that $\sqrt{f}$ is integrable on $[a, b]$. Let f be continuous on $[a, b]$. Show that f^+ and f^- are continuous on $[a, b]$.

22. Let f have a continuous derivative on $[a, b]$. Show that f can be written as the difference of two increasing functions.

$$[\textit{Hint}: \text{Consider } \int_a^x (f')^+(\zeta)d\zeta - \int_a^x (f')^-(\zeta)d\zeta.]$$

23. The function in Example 6.4.3 has a derivative which is not integrable in any neighborhood of 0. Show that, for any $b > 0$,

$$\lim_{a \to 0+} \int_a^b f'(x)dx = f(b)$$

24. Let u and v be continuous on $[a, b]$, and let c be a positive constant. If

$$|u(x) - v(x)| \leqslant c \int_a^x |u(\zeta) - v(\zeta)|d\zeta$$

for each $x \in [a, b]$, show that $u \equiv v$ on $[a, b]$.

25. Let f be continuous on $[a, b]$, and let $f(x) \leqslant M$ for each $x \in [a, b]$. If

$$\int_a^b f(x)dx \geqslant M(b - a),$$

show that $f(x) \equiv M$ for each $x \in [a, b]$.

24. Let P be a polynomial of degree at most 3. Show that

$$\int_a^b P(x)dx = [P(a) + P([a + b]/2) + P(b)](b - a)/6.$$

27. If f and g are integrable on $[a, b]$, use Exercise 4 of Problem Set 4.4 to show that the function max (f, g) and min (f, g) are integrable on $[a, b]$.

28. Let $f: [0, 1] \to R$ such that

$$f(x) = \begin{cases} 1 & \text{if } x \text{ is rational,} \\ 0 & \text{if } x \text{ is irrational.} \end{cases}$$

Use the Intermediate Value Theorem for Derivatives to show that there exist *no* function $F:[0, 1] \to R$ such that $F'(x) = f(x)$ for each $x \in [0, 1]$.

29. Let f be the bracket function with domain $[0, 2]$. Show that there exists no function $f: [0, 2] \to R$ such that $F'(x) = f(x)$ for each $x \in [0, 2]$.

6.5. Additional Properties of Integrable Functions

We conclude this chapter by proving some properties about integrable functions that the reader should be familiar with from elementary calculus.

We begin with the formula for *integration by parts*. If f and g are differentiable on $[a, b]$, then Theorem 5.2.1 implies that fg is differentiable on $[a, b]$ and that $(fg)' = fg' + f'g$. If f' and g' are also integrable, then $fg', f'g$, and $(fg)'$ are integrable and

$$\int_a^b (fg)'(x)dx = \int_a^b (fg')(x)dx + \int_a^b (f'g)(x)dx$$

and we have the following theorem.

Theorem 6.5.1. (*Integration by Parts*) *If f and g are differentiable on $[a, b]$ and f' and g' are integrable on $[a, b]$, then*

$$\int_a^b (fg')(x)dx = (fg)(x)\Big|_a^b - \int_a^b (f'g)(x)dx.$$

Example 6.5.1. Evaluate the integral

$$\int_0^1 xe^x dx.$$

Let $f(x) = x$ and $g'(x) = e^x$ for each $x \in [0, 1]$. Then $f'(x) = 1$ and $g(x) = e^x$ for each $x \in [0, 1]$. Using Theorem 6.5.1 we have

$$\int_0^1 xe^x dx = xe^x \Big|_0^1 - \int_0^1 e^x dx$$

$$= e - (e - 1) = 1.$$

(Cf. Exercises 8 and 10 of Problem Set 6.1.)

The next theorem of this section deals with the problem of changing variables when integrating. If $f: D \to E$ and $g: E \to F$ such that f is differentiable on D and g is differentiable on E, then Theorem 5.2.3 implies that

$$(g \circ f)'(x) = g'(f(x)) \cdot f'(x) = (g' \circ f)(x) \cdot f'(x)$$

for each $x \in D$. This formula is the basis for the following theorem.

Theorem 6.5.2. *Let f be continuous on $[a, b]$, and let v have a continuous derivative on $[c, d]$. If $v(c) = a$, $v(d) = b$, and $v(x) \in [a, b]$ for each $x \in [c, d]$, then*

$$\int_a^b f(y) dy = \int_c^d f[v(x)] v'(x) dx.$$

Proof.   Let

$$g(y) = \int_a^y f(\zeta) d\zeta,$$

then

$$g[v(x)] = \int_a^{v(x)} f(\zeta) d\zeta$$

and

$$\frac{d\{g[v(x)]\}}{dx} = f[v(x)] v'(x)$$

for each $x \in [c, d]$. (Cf. Exercise 17 of Problem Set 6.4.) Theorem 5.2.3 implies that

$$\frac{d\{g[v(x)]\}}{dx} = g'[v(x)] \cdot v'(x)$$

for each $x \in [c, d]$. Thus

$$g'[v(x)] v'(x) = f[v(x)] v'(x).$$

From the Fundamental Theorem of Integral Calculus (Theorem 6.4.4) it follows that

$$\int_c^d f[v(x)]v'(x)dx = g[v(x)]\Big|_c^d$$

$$= g[v(d)] - g[v(c)]$$

$$= g(b) - g(a)$$

$$= \int_a^b f(y)dy.$$

Example 6.5.2. Evaluate the integral

$$\int_0^1 \sqrt{1 - x^2}\, dx.$$

If we define f by

$$f(x) = \sqrt{1 - x^2}$$

for $x \in [0, 1]$ and let $g: [0, \pi/2] \to R$ such that $g(x) = \sin x$, then $g(0) = 0$, $g(\pi/2) = 1$, and

$$(f \circ g)(x) = \cos x = g'(x).$$

Thus

$$\int_0^1 \sqrt{1 - x^2}\, dx = \int_0^{\pi/2} \cos^2 x\, dx$$

$$= \int_0^{\pi/2} \left(\frac{1}{2} + \frac{\cos 2x}{2}\right) dx = \left[\frac{x}{2} - \frac{\sin 2x}{4}\right]_0^{\pi/2} = \pi/4.$$

We now apply the Mean Value Theorem for Derivatives (Theorem 5.3.4) to prove a *Mean Value Theorem for Integrals*.

Theorem 6.5.3. (*Mean Value Theorem for Integrals*) *If f is continuous on $[a, b]$, then there is $c \in (a, b)$ such that*

$$f(c)(b - a) = \int_a^b f(x)dx.$$

Proof. Let $F(x) = \int_a^x f(\zeta)d\zeta$ for each $x \in [a, b]$. Since f is continuous on $[a, b]$, the function F is differentiable on $[a, b]$, and $F'(x) = f(x)$ for each $x \in [a, b]$. The Mean Value Theorem (Theorem 5.3.4) implies that there is $c \in (a, b)$ such that

$$F'(c)(b - a) = F(b) - F(a).$$

But

$$F(b) - F(a) = \int_a^b f(x)dx$$

and

$$F'(c) = f(c).$$

Thus,

$$f(c)(b - a) = \int_a^b f(x)dx.$$

Problem Set 6.5

For Exercises 1–6, evaluate the given integrals.

1. $\int_0^1 e^{ax}dx$, where a is a constant.

2. $\int_0^{\pi/2} \cos(2x)\,dx$.

3. $\int_0^{\pi} \sin(ax)\,dx$, where a is a constant.

4. $\int_0^{\pi} \sin^2 x\,dx$.

5. $\int_0^{\pi} x\cos x\,dx$.

6. $\int_0^1 xe^x dx$.

7. By using Integration by Parts twice and collecting like terms, evaluate the integral.

$$\int_0^{\pi/2} e^x \sin x\,dx.$$

8. If f and g are continuous on $[a, b]$ and $g(x) \geqslant 0$ for each $x \in [a, b]$, show that there exists $c \in (a, b)$ such that

$$\int_a^b f(x)g(x)dx = f(c)\int_a^b g(x)dx.$$

Also consider the case where $a > b$. [This result is called the **Generalized Form of the Mean Value Theorem for Integrals.**]

9. If f, ϕ, and ϕ' are continuous on $[a, b]$ and $\phi'(x) \geqslant 0$ for each $x \in [a, b]$, show that there exists $c \in (a, b)$ such that

$$\int_a^b f(x)\phi(x)dx = \phi(a)\int_a^c f(x)dx + \phi(b)\int_c^b f(x)dx.$$

[This result is called the **Second Mean Value Theorem for Integrals.**] [*Hint*: Use Integration by Parts, letting

$$F(x) = \int_a^x f(\zeta)d\zeta$$

and apply the Generalized Form of the Mean Value Theorem for Integrals.]

10. If f and g are integrable on $[a, b]$ show that

$$\left[\int_a^b f(x)g(x)dx\right]^2 \leqslant \int_a^b [f(x)]^2 dx \int_a^b [g(x)]^2 dx.$$

[This inequality is called the **Schwarz (or Cauchy) inequality.**] [*Hint*: First show that $[f(x) + \lambda g(x)]^2$ is integrable on $[a, b]$ for all real λ. Let

$$A = \int_a^b [f(x)]^2 dx, \quad B = \int_a^b f(x)g(x)dx, \text{ and } \quad C = \int_a^b [g(x)]^2 dx.$$

Show that $A + 2B\lambda + C\lambda^2 \geqslant 0$ for all real λ. If $C = 0$ then $B = 0$, and if $C \neq 0$ the discriminant of $A + 2B\lambda + C\lambda^2$ must be nonpositive.]

11. If f and g are integrable on $[a, b]$, show that

$$\left\{\int_a^b [f(x) + g(x)]^2 dx\right\}^{1/2} \leqslant \left\{\int_a^b [f(x)]^2 dx\right\}^{1/2} + \left\{\int_a^b [g(x)]^2 dx\right\}^{1/2}.$$

[This inequality is called the **Minkowski inequality.**] [*Hint*: Square both members of this inequality, expand each term on the left, cancel identical terms that result, divide by 2, use the Schwarz inequality, and reverse the steps.]

12. If f is continuous on $[a, b]$ and

$$\int_a^b f(x)\phi(x)dx = 0$$

for every integrable function ϕ on $[a, b]$, show that $f(x) = 0$ for every $x \in [a, b]$.

13. Let f be integrable on $[a, b]$, let ϕ continuous on $[a, b]$, and let

$$\int_a^b f(x)\phi(x)dx = 0.$$

Show that $f(x) = 0$ at each point x of $[a, b]$ where f is continuous.

The remaining exercises of this section deal with the *logarithmic function*. The **natural logarithm function**, ln, is the function $\ln: R^+ \to R$ such that

$$\ln x \equiv \int_1^x \frac{1}{\zeta} d\zeta.$$

If x is a positive number, then the number $\ln x$ is called the **natural logarithm** of x. The natural logarithm of 1 is zero, that is, $\ln 1 = 0$. Define the number x such that $\ln x = 1$ to be $x = e$, that is, $\ln e = 1$.

14. Show that the function $\ln x$ is strictly increasing on R^+.

15. Show that the function $\ln x$ is continuous on R^+.

16. Show that the function $\ln x$ is differentiable on R^+, and find $(\ln x)'$.

For Exercises 17–21, let x and y be any positive real numbers, let n be an integer, and let r be a rational number. Prove the given statements.

17. $\ln(xy) = \ln x + \ln y$.

18. $\ln(x/y) = \ln x - \ln y$.

19. $\ln(x^n) = n(\ln x)$.

20. $\ln \sqrt[n]{x} = (\ln x)/n$.

21. $\ln x^r = r(\ln x)$.

22. Show that $\lim\limits_{x \to 0^-} \ln x = -\infty$.

23. Show that $\lim\limits_{x \to +\infty} \ln x = +\infty$.

24. If $f: R^+ \to R$ such that $f(1) = 0$ and $f'(x) = 1/x$ for each $x \in R^+$, show that $f(x) = \ln x$ for each $x \in R^+$.

In Exercises 25–29 find the derivative of the function f, assuming a domain for which the composite function is defined.

25. $f(x) = \ln(x^2 + 1)$.

26. $f(x) = \ln \sin(3x)$.

27. $f(x) = \dfrac{\ln x^2}{x}$.

28. $f(x) = \ln \cos^2(x^2)$.

29. $f(x) = \ln(x^3 + x^2)$.

For Exercises 30–38, evaluate the given integrals.

30. $\displaystyle\int_1^x \frac{1}{\zeta}\, d\zeta$.

31. $\displaystyle\int_{\pi/4}^{\pi/2} \frac{\cos x}{\sin x}\, dx$.

32. $\displaystyle\int_1^e (\ln x)\, dx$.

33. $\displaystyle\int_1^e \frac{(\ln x)}{\sqrt{x}}$.

34. $\displaystyle\int_1^e \frac{\cos(\ln x)}{x}\, dx.$

35. $\displaystyle\int_1^e x \ln x\, dx.$

36. $\displaystyle\int_1^e \frac{(\ln x)^2}{x}\, dx.$

37. $\displaystyle\int_e^{e^2} \frac{1}{x \ln x}\, dx.$

Chapter Seven

INFINITE SERIES

7.1. Convergent Series

An important concept in analysis is that of *infinite series*. The *convergence* of an infinite series is defined in terms of its associated *sequence of partial sums*. Thus a thorough knowledge of sequences is necessary in order to understand infinite series. However, a large percentage of the students completing calculus do not have a good working knowledge of the concepts of series. It is hoped that this chapter will help the student to obtain a thorough understanding of these concepts.

Definition 7.1.1. *Let* $\{a_n\}$ *be a sequence of real numbers. The expression*

$$\sum_{n=1}^{\infty} a_n = \sum_{n=1}^{\infty} a_n = a_1 + a_2 + \dots + a_n + \dots$$

*is called an **infinite series** or a **series**. The real number* a_n *is called the* **nth term** *of the series.*

An infinite series is a sum of the terms of a sequence. It would be useful if each infinite series had a real number associated with it, similar to the addition of a finite number of real numbers. One of the purposes of this chapter is to decide when a real number can be assigned as the *sum* of an infinite series.

We shall use the notation $\Sigma\, a_n$ for the infinite series $\sum_{n=1}^{\infty} a_n$ when there is no possibility of confusion.

In order to give meaning to the term "sum" of an infinite series, we define its *sequence of partial sums.*

Definition 7.1.2. *Let* $\sum\limits_{k=1}^{\infty} a_k$ *be an infinite series. For each positive integer* n, *let*

$$s_n = \sum_{k=1}^{n} a_k = a_1 + a_2 + \ldots + a_n.$$

Then s_n *is called the* **nth partial sum** *of the series, and* $\{s_n\}$ *is called the* **sequence of partial sums** *of the series.*

We give two examples of series and their sequences of partial sums.

Examples 7.1.1. For each positive integer n, assume that $a_n = 1$. The series so defined is

$$\sum_{n=1}^{\infty} 1 = 1 + 1 + \ldots + 1 + \ldots .$$

Then $s_1 = 1$, $s_2 = 2$, and in general $s_n = n$. The sequence of partial sums is the sequence $\{n\}$.

Example 7.1.2. As a second example, let $a_n = ar^n$, for each positive integer n, where $r \neq 1$ and a is a constant. This results in the series

$$\sum_{n=1}^{\infty} ar^n = ar + ar^2 + \ldots + ar^n + \ldots .$$

Thus $s_1 = ar$, $s_2 = ar + ar^2 = ar(1 + r)$, and in general we have

$$s_n = \sum_{k=1}^{n} ar^k = ar + ar^2 + \ldots + ar^n$$

$$= ar(1 + r + \ldots + r^{n-1}).$$

Since $r \neq 1$, s_n may be rewritten as

$$s_n = ar(1 - r^n)/(1 - r).$$

The *convergence* of an infinite series is now defined in terms of its sequence of partial sums.

Definition 7.1.3. *Let* $\Sigma\, a_n$ *be an infinite series with sequence of partial sums* $\{s_n\}$. *The series* $\Sigma\, a_n$ *is said to* **converge** *if and only if its sequence of partial sums* $\{s_n\}$ *converges. If* $\Sigma\, a_n$ *converges, then it is said to have* **sum** *equal to the limit of its sequence of partial sums. If* $A \equiv \lim\limits_{n} s_n$, *then we say that the series* $\Sigma\, a_n$ **converges** *to* A, *and we write*

$$\Sigma\, a_n = a_1 + a_2 + \ldots = A.$$

*A series is said to **diverge** if and only if its sequence of partial sums diverges.*

We note that the sigma notation $\Sigma \, a_n$ is used to represent the series whose terms are the terms of the sequence $\{a_n\}$. If the series $\Sigma \, a_n$ converges, we also use the symbol $\Sigma \, a_n$ to represent the sum of the series $\Sigma \, a_n$.

Example 7.1.3. The series defined in Example 7.1.1 diverges since its sequence of partial sums is increasing and not bounded (that is, its sequence of partial sums is divergent). Since $\lim_n r^n = 0$ for $-1 < r < 1$ (cf. Example 3.1.3), the series defined in Example 7.1.2 converges and has sum

$$\lim_n s_n = \lim_n \frac{ar(1 - r^n)}{1 - r} = ar/(1 - r),$$

and we write

$$\sum_{n=1}^{\infty} ar^n = ar/(1 - r).$$

This is an important series and is called a **geometric series**.

Let $a_n \geqslant 0$ for each positive integer n. Then the series $\Sigma \, a_n$ is called a **nonnegative term series** or a **nonnegative series**. If $a_n > 0$ for each positive integer n, then the series $\Sigma \, a_n$ is called a **positive term series** or a **positive series**. In either case the sequence of partial sums of $\Sigma \, a_n$ is increasing, and so the series $\Sigma \, a_n$ converges if and only if its sequence of partial sums is bounded. We use this fact in the following example.

Example 7.1.4. Let $a_n = 1/n$ for each positive integer n. The resulting series

$$\sum_{n=1}^{\infty} 1/n = 1 + 1/2 + ... + 1/n + ...$$

is called the **harmonic series**. We shall show that the sequence of partial sums of this series is unbounded. For each positive integer k, let $n = 2^k$. Then

$$\begin{aligned}
s_n &= 1 + 1/2 + ... + 1/n = 1 + 1/2 + ... + 1/2^k \\
&= 1 + 1/2 + (1/3 + 1/4) + (1/5 + 1/6 + 1/7 + 1/8) \\
&\quad + ... + (1/[2^{k-1} + 1] + 1/[2^{k-1} + 2] + ... + 1/2^k) \\
&\geqslant 1 + 1/2 + 2(1/4) + 4(1/8) + ... + 2^{k-1}(1/2^k) = 1 + k/2.
\end{aligned}$$

The above shows that the sequence $\{s_{2^k}\}$ is unbounded, and so the sequence $\{s_n\}$ is unbounded and the harmonic series diverges.

It was proved in Section 3.4 that a sequence of real numbers is convergent if and only if it is a Cauchy sequence. This condition can be restated for series.

Theorem 7.1.1. *(Cauchy Condition) The series $\Sigma\, a_n$ is convergent if and only if to each $\varepsilon > 0$ there is an $N = N(\varepsilon)$ such that*

$$|a_{m+1} + a_{m+2} + \ldots + a_n| < \varepsilon$$

for all positive integers m and n such that $n > m \geqslant N$.

Proof. This is a direct consequence of the Cauchy convergence condition for sequences (Theorem 3.3.3). Write

$$s_n = a_1 + a_2 + \ldots + a_n,$$

then

$$s_n - s_m = a_{m+1} + a_{m+2} + \ldots + a_n.$$

The condition given in this theorem is now equivalent to $|s_n - s_m| < \varepsilon$, and the theorem is merely the statement of Cauchy's condition for the sequence of partial sums $\{s_n\}$.

We now examine the implications of the above theorem. It follows from the theorem that changing finitely many terms of a series does not alter its convergence, although for convergent series the sum may be altered. Also, the convergence of an infinite series is independent of whether the terms are indexed beginning with $n = 1$ or with some integer other than 1.

By taking $n = m + 1$ in Theorem 7.1.1, we see that a *necessary condition for convergence* is that for each $\varepsilon > 0$ there is an N such that $n > N$ implies that $|a_n| < \varepsilon$. That is, the sequence $\{a_n\}$ is a null sequence; thus $\{a_n\} \to 0$.

Corollary 7.1.1. *If $\Sigma\, a_n$ converges, then $\lim_n a_n = 0$.*

Corollary 7.1.1 implies that if $\lim_n a_n \neq 0$ (or does not exist), then the series $\Sigma\, a_n$ diverges. However, although this condition is necessary it is not sufficient; for example, consider $\Sigma\,(1/n)$ (cf. Example 7.1.4). It would be very nice to have a simple criterion for deciding the convergence or divergence of a series by considering its nth term—but there is no such criterion. However, we shall develop various tests for convergence for certain types of series.

We next define a stronger type of convergence, called *absolute convergence*.

Definition 7.1.4. *The infinite series $\Sigma\, a_n$ is said to **converge absolutely** (or be **absolutely convergent**) if and only if the series $\Sigma\, |a_n|$ is convergent. The series $\Sigma\, a_n$ is said to **converge conditionally** (or be **conditionally convergent**) if and only if $\Sigma\, a_n$ converges but is not absolutely convergent.*

If the terms of a series are all positive or all negative, then convergence and absolute convergence are equivalent.

Example 7.1.5. The series $\sum_{n=1}^{\infty} (-1)^n ar^n$, where $-1 < r < 1$, is absolutely convergent, while the series $\sum_{n=1}^{\infty} (-1)^n/n$ can be at most conditionally convergent since $\sum_{n=1}^{\infty} 1/n$ diverges (we shall see later that this series is conditionally convergent (cf. Example 7.3.1)).

The following theorem establishes that absolute convergence implies ordinary convergence.

Theorem 7.1.2. *If the series $\Sigma\, a_n$ converges absolutely, then it converges.*

Proof. Let $\varepsilon > 0$. Since $\Sigma\, a_n$ converges absolutely, $\Sigma\, |a_n|$ converges. By Theorem 7.1.1, there is an N such that

$$||a_{m+1}| + |a_{m+2}| + \dots + |a_n|| < \varepsilon$$

for all positive integers m and n such that $n > m > N$. But

$$|a_{m+1} + a_{m+2} + \dots + a_n| \leqslant |a_{m+1}| + |a_{m+2}| + \dots + |a_n| < \varepsilon,$$

and, by Theorem 7.1.1, $\Sigma\, a_n$ converges.

Problem Set 7.1

1. Prove Corollary 7.1.1 using only the sequence of partial sums. [*Hint*: Use the fact that $a_n = s_n - s_{n-1}$.]
2. If the series $\Sigma\, a_n$ converges, then prove that $\Sigma\, ca_n$ converges, where c is a constant.
3. Let $\{a_n\}$ be a sequence of real numbers. Prove that the series

$$\sum_{n=1}^{\infty} (a_n - a_{n+1})$$

converges if and only if the sequence $\{a_n\}$ converges. If this series converges, what is its sum?
4. Which of the following series are convergent and which are divergent? If a series converges, find its sum.

(a) $\sum_{n=1}^{\infty} 1/n(n+1)$;

(b) $\sum_{n=1}^{\infty} 1/(2n-1)(2n+1)$;

(c) $\sum_{n=1}^{\infty} \ln(1 + 1/n)$;

(d) $\sum_{n=1}^{\infty} n/(n + 1)$;

(e) $\sum_{n=1}^{\infty} 2^{n-1}/(3n)$.

5. Assume that the series $\Sigma\, a_n$ converges absolutely, and assume that the sequenc $\{b_n\}$ is bounded. Prove that $\Sigma\, a_n b_n$ converges absolutely.

6. Give an example of a convergent series $\Sigma\, a_n$ and a bounded sequence $\{b_n$ such that $\Sigma\, a_n b_n$ diverges. [*Hint*: See Example 7.1.5.]

7. Let a and b be any real numbers. Prove that the series $\Sigma(a + nb)$ diverge unless $a = b = 0$.

8. Assume that $\Sigma\, a_n$ converges to A. Prove that $\Sigma(a_n + a_{n+1})$ converges. What i its sum?

9, Assume that $\Sigma\, a_n$ converges to A. Prove that

$$a_1 + 0 + a_2 + 0 + a_3 + 0 + \ldots$$

also converges to A. More generally, prove that any number of 0 terms may b inserted (or deleted) anywhere in a convergent series without affecting its con vergence or sum.

10. Prove that altering a finite number of terms of a series does not alter its con vergence or divergence. [*Hint*: See Theorem 3.1.2.]

11. Give an example of a series $\Sigma\, a_n$ such that the series

$$(a_1 + a_2) + (a_3 + a_4) + (a_5 + a_6) + \ldots$$

converges, but the series $\Sigma\, a_n$ diverges.

12. Assume $\lim_n a_n = 0$. Prove that the series $\Sigma\, a_n$ converges if and only if the serie $\Sigma(a_n + 2a_{n+1})$ converges.

13. If $\{a_n\}$ is a decreasing sequence of positive numbers such that the series $\sum_{n=1}^{\infty} a$ converges, prove that $\lim_n na_n = 0$. [*Hint*: Let $s_n = a_1 + a_2 + \ldots + a_n$. If

$$\sum_{n=1}^{\infty} a_n = A,$$

then

$$\lim_n s_{2n} = A = \lim_n s_n \quad \text{and} \quad \lim_n (s_{2n} - s_n) = 0.$$

Show that

$$s_{2n} - s_n \geqslant na_{2n} \geqslant 0.$$

Also

$$a_{2n+1} \leqslant a_{2n} \quad \text{and} \quad (2n + 1)a_{2n+1} \leqslant \left(\frac{2n + 1}{2n}\right) 2na_{2n}.]$$

14. Give an example to show that Exercise 13 is no longer true if the hypothesi that $\{a_n\}$ is decreasing is dropped.

15. Let $\{p_n\}$ be an unbounded increasing positive term sequence.

(a) Prove that $\sum\limits_{n=1}^{\infty} (p_{n+1} - p_n)$ diverges.

(b) Prove that $\sum\limits_{n=1}^{\infty} (1/p_n - 1/p_{n+1})$ converges. Find its sum.

7.2. Tests for Convergence

      This section discusses a number of tests that can be used to determine whether a series is convergent or divergent. It would be useful if there were one test that could always be used, but unfortunately this is not the case. We shall prove some of the results and leave the others as exercises.

The first test we give is called the *Comparison Test*.

> **Theorem 7.2.1.** (*Comparison Test*) *Let $\Sigma\, a_n$ be a series, and let $\Sigma\, b_n$ be a positive term series.*
> (1) *If $\Sigma\, b_n$ converges and there are positive numbers N and M such that $|a_n| \leqslant Mb_n$ for each positive integer $n \geqslant N$, then $\Sigma\, a_n$ converges absolutely.*
> (2) *If $\Sigma\, b_n$ diverges and there are positive numbers N and M such that $a_n \geqslant Mb_n$ for each positive integer $n \geqslant N$, then $\Sigma\, a_n$ diverges.*

We prove Part (1) and leave Part (2) as Exercise 29 of Problem Set 7.2.

> *Proof.* Assume that the hypotheses of (1) hold, that is, $\Sigma\, b_n$ converges and there are numbers N and M such that $|a_n| \leqslant Mb_n$ for each positive integer $n \geqslant N$. Since $\Sigma\, b_n$ converges, for $\varepsilon > 0$ there is an N_1 such that
>
> $$b_{m+1} + b_{m+2} + \ldots + b_n < \varepsilon/M$$
>
> when $n > m \geqslant N_1$. Let $N_2 = \max(N, N_1)$. Then
>
> $$|a_{m+1}| + |a_{m+2}| + \ldots + |a_n| \leqslant Mb_{m+1} + Mb_{m+2} + \ldots + Mb_n < \varepsilon$$
>
> when $n > m \geqslant N_2$. Thus, $\Sigma\, |a_n|$ converges, and $\Sigma\, a_n$ converges absolutely.

> **Example 7.2.1.** As an example of Part (1) of this theorem we show that the series
>
> $$\sum_{n=1}^{\infty} (-1)^n/(n+1)^2$$

is absolutely convergent. To do this, we compare this series with the series

$$\sum_{n=1}^{\infty} 1/n(n+1).$$

Exercise 4(a) of Problem Set 7.1 implies that this series converges. For each positive integer n,

$$|(-1)^n/(n+1)^2| = 1/(n+1)^2 < 1/n(n+1).$$

Thus, the series

$$\sum_{n=1}^{\infty} (-1)^n/(n+1)^2$$

converges absolutely. From this we conclude that the series

$$\sum_{n=1}^{\infty} 1/(n+1)^2 = \sum_{n=2}^{\infty} 1/n^2$$

converges. Therefore, the series

$$\sum_{n=1}^{\infty} 1/n^2$$

also converges.

Example 7.2.2. The series

$$\sum_{n=1}^{\infty} 1/\sqrt{n}$$

can be compared with the divergent harmonic series

$$\sum_{n=1}^{\infty} 1/n$$

($1/\sqrt{n} \geq 1/n$ for every positive integer n) and Part (2) of Theorem 7.2.1 implies that

$$\sum_{n=1}^{\infty} 1/\sqrt{n}$$

diverges.

The second result we prove is called the *Ratio Test*.

Theorem 7.2.2. (*Ratio Test*) *Let Σa_n be an infinite series of nonzero terms.*

(1) *If there is a positive integer N such that*

$$\left| \frac{a_{n+1}}{a_n} \right| \leq r < 1$$

for each positive integer $n \geqslant N$, then Σa_n converges absolutely.

(2) *If there is an N such that*

$$\left| \frac{a_{n+1}}{a_n} \right| \geqslant 1$$

for each positive integer $n \geqslant N$, then Σa_n diverges.

Proof. Assume that the hypotheses of Part (1) hold. Then

$$|a_{N+1}| \leqslant r\,|a_N|,$$
$$|a_{N+2}| \leqslant r\,|a_{N+1}| \leqslant r^2\,|a_N|, \dots,$$
$$|a_{N+p}| \leqslant r\,|a_{N+p-1}| \leqslant \dots \leqslant r^p\,|a_N|.$$

Thus, for $n \geqslant N$, $|a_n| \leqslant |a_N|\,r^{n-N}$. Since $0 < r < 1$, the series

$$\sum_{n=1}^{\infty} r^n$$

converges, and Part (1) of Theorem 7.2.1 implies that the series

$$\sum_{n=1}^{\infty} a_n$$

converges absolutely.

Assume that the hypotheses of Part (2) hold. Since $a_n \neq 0$ for every positive integer n and $|a_{n+1}| \geqslant |a_n|$ for $n \geqslant N$, the sequence $\{|a_n|\}$ is increasing for $n \geqslant N$ and does not converge to zero. Hence

$$\sum_{n=1}^{\infty} a_n$$

diverges.

If in the above theorem the sequence $\{|a_{n+1}/a_n|\}$ converges, we have the following corollary.

Corollary 7.2.2. *Let Σa_n be an infinite series of nonzero terms, and let*

$$r = \lim_{n} \left| \frac{a_{n+1}}{a_n} \right|.$$

(1) *If $r < 1$, the series converges absolutely.*

(2) *If $r > 1$, the series diverges.*

(3) *If $r = 1$, the test fails; that is, no conclusion can be made about the convergence of the series.*

Part (3) of this corollary is proved by giving an example of a divergent series with $r = 1$ and an example of a convergent series with $r = 1$. (Cf. Example 7.2.5.)

Example 7.2.3. As an example of Part (1) of Theorem 7.2.2 and Corollary 7.2.2, consider the series

$$\sum_{n=1}^{\infty} n/e^n.$$

Then

$$\left|\frac{a_{n+1}}{a_n}\right| = \frac{(n+1)}{e^{n+1}} \Bigg/ \frac{n}{e^n} = \frac{1}{e}\left(\frac{n+1}{n}\right)$$

and

$$r = \lim_n \left|\frac{a_{n+1}}{a_n}\right| = \frac{1}{e} < 1.$$

Thus, the series

$$\sum_{n=1}^{\infty} n/e^n$$

converges.

Example 7.2.4. An example of Part (2) of Theorem 7.2.2 and Corollary 7.2.2 is the series

$$\sum_{n=1}^{\infty} e^n/n^2.$$

Then

$$\left|\frac{a_{n+1}}{a_n}\right| = \frac{e^{n+1}}{(n+1)^2} \Bigg/ \frac{e^n}{n^2} = \frac{en^2}{(n+1)^2}$$

and

$$r = \lim_n \left|\frac{a_{n+1}}{a_n}\right| = e > 1.$$

Hence, the series

$$\sum_{n=1}^{\infty} e^n/n^2$$

diverges.

Example 7.2.5. Lastly, we give examples to show that, when the limit r in Corollary 7.2.2 is 1, we can have either convergence or divergence. We first consider the series

$$\sum_{n=1}^{\infty} 1/n.$$

Then

$$\lim_n \left| \frac{a_{n+1}}{a_n} \right| = \lim_n \frac{1/(n+1)}{1/n} = \lim_n \frac{n}{n+1} = 1,$$

and the series

$$\sum_{n=1}^{\infty} 1/n$$

diverges. The series

$$\sum_{n=1}^{\infty} 1/n^2$$

converges, while

$$\lim_n \left| \frac{a_{n+1}}{a_n} \right| = \lim_n \frac{1/(n+1)^2}{1/n^2} = \lim_n \frac{n^2}{(n+1)^2} = 1.$$

The *Root Test*, which we develop next, is better than the Ratio Test in the sense that if the Root Test fails, then so does the Ratio Test. However, the Root Test is usually more difficult to apply than the Ratio Test.

Theorem 7.2.3. *(Root Test) Let $\Sigma\, a_n$ be an infinite series.*
(1) *If there is an N such that*

$$|a_n|^{1/n} \leqslant r < 1$$

for each positive integer $n \geqslant N$, then $\Sigma\, a_n$ converges absolutely
(2) *If*

$$|a_n|^{1/n} \geqslant 1$$

for infinitely many n, then $\Sigma\, a_n$ diverges.

Proof. If for $n \geqslant N$, $|a_n|^{1/n} \leqslant r < 1$, then for $n \geqslant N$, $|a_n| \leqslant r^n$. By the Comparison Test,

$$\sum_{n=1}^{\infty} a_n$$

converges absolutely.

If $|a_n|^{1/n} \geqslant 1$ for infinitely many values of n, then $\lim_n a_n \neq 0$ and the series diverges.

If, in the above theorem, the sequence $\{|a_n|^{1/n}\}$ converges, then we have the following corollary.

Corollary 7.2.3. *Let* $\Sigma\, a_n$ *be an infinite series of nonzero terms, and let*

$$r = \lim_n |a_n|^{1/n}.$$

(1) *If* $r < 1$, *then the series converges absolutely.*

(2) *If* $r > 1$, *then the series diverges.*

(3) *If* $r = 1$, *then the test fails, that is, no conclusion can be made about the convergence of the series.*

Part (3) of this corollary is proved by giving an example of a divergent series with $r = 1$ and an example of a convergent series with $r = 1$. The series given in Example 7.2.5 serve as such examples.

We now show that whenever the Ratio Test (Theorem 7.2.2.) does give a check on the convergence or divergence of a series, then the Root Test (Theorem 7.2.3) provides the same results. Example 7.2.6 illustrates that the converse is *not* true.

We begin by assuming that there is an N such that

$$\left|\frac{a_{n+1}}{a_n}\right| \leqslant r < 1$$

for each $n \geqslant N$. Then by the Ratio Test, the series $\Sigma\, a_n$ converges absolutely. We see that

$$|a_n| \leqslant r^{n-N}|a_N| = r^n(|a_N|/r^N)$$

for $n > N$. Thus

$$|a_n|^{1/n} \leqslant r(|a_N|/r^N)^{1/n}$$

for $n > N$, and so

$$\lim_n \sup |a_n|^{1/n} \leqslant \lim_n \sup r(|a_N|/r^N)^{1/n} = r.$$

It follows from the definition of limit superior that there is an M such that

$$|a_n|^{1/n} \leqslant (r + 1)/2 < 1$$

for each positive integer $n > M$. Therefore the Root Test implies that the series $\Sigma\, a_n$ converges absolutely. That is, when ever the Ratio Test implies that the series $\Sigma\, a_n$ converges absolutely, the Root Test also implies that the series $\Sigma\, a_n$ converges absolutely.

On the other hand, if there is an N such that

$$\left|\frac{a_{n+1}}{a_n}\right| \geqslant 1$$

for each positive integer $n \geqslant N$, then the Ratio Test implies that the series $\Sigma\, a_n$ diverges. In fact, if K is any nonzero constant, then the Ratio Test implies that the series $\Sigma\, Ka_n$ diverges. We see that

$$|a_n| \geqslant |a_N|$$

for each $n > N$. Let K be a positive constant such that

$$K|a_N| \geq 1.$$

Then

$$|Ka_n| \geq K|a_N| \geq 1$$

for $n > N$. Hence, for $n > N$,

$$|Ka_n|^{1/n} \geq 1,$$

and the Root Test implies that the series $\Sigma\, Ka_n$ diverges. Since K is a constant, the series $\Sigma\, a_n$ also diverges. Therefore, whenever the Ratio Test shows that a series diverges, we can also show that the series diverges by using the Root Test. It is left as an exercise to show that if Corollary 7.2.2 can be used to check convergence or divergence of a series, then Corollary 7.2.3 produces the same result (Exercise 30 of Problem Set 7.2.).

The following example gives a series for which the Ratio Test fails, but the Root Test proves convergence. Thus the Root Test is a stronger test than the Ratio Test. In most cases, however, the Root Test is more difficult to use.

Example 7.2.6. Let

$$a_n = \begin{cases} 1/3^n & \text{if } n \text{ is odd,} \\ 1/2^n & \text{if } n \text{ is even.} \end{cases}$$

Then

$$\frac{a_{n+1}}{a_n} = \begin{cases} (3/2)^n/2 & \text{if } n \text{ is odd,} \\ (2/3)^n/3 & \text{if } n \text{ is even.} \end{cases}$$

Thus, the Ratio Test does not apply. But

$$|a_n|^{1/n} = \begin{cases} 1/3 & \text{if } n \text{ is odd,} \\ 1/2 & \text{if } n \text{ is even.} \end{cases}$$

Hence, $|a_n|^{1/n} \leq 1/2$ for each positive integer n, and the series $\Sigma\, a_n$ converges.

We conclude this section with the *Integral Test*.

A function f is **decreasing** on its domain D if and only if $f(x) \leq f(y)$ whenever $x, y \in D$ and $y < x$. If $f(x)$ is defined for all $x \geq a > 0$, then we write

$$\int_a^{\infty} f(x)dx = \lim_{n} \int_a^{n} f(x)dx,$$

if the limit exists. The integral is said to **converge** if the limit exists (that is, is finite).

Theorem 7.2.4. (*Maclaurin–Cauchy Integral Test*) *Let f be a positive decreasing continuous function for all $x \geq N$ such that $f(n) = a_n$ for $n \geq N$. Then the series $\Sigma\, a_n$ converges if and only if*

$$\int_N^\infty f(x)dx$$

converges.

Proof. If $m \leqslant x \leqslant m + 1$, we have $f(m + 1) \leqslant f(x) \leqslant f(m)$, and so

$$a_{m+1} = f(m + 1) \leqslant \int_m^{m+1} f(x)dx \leqslant f(m) = a_m.$$

Then

$$a_{N+1} + \ldots + a_{n+1} \leqslant \int_N^{n+1} f(x)dx \leqslant a_N + \ldots + a_n.$$

If $\int_N^\infty f(x)dx = \lim_n \int_N^n f(x)dx = K$, then

$$S_{n+1} = a_1 + a_2 + \ldots + a_N + a_{N+1} + \ldots + a_{n+1} \leqslant S_N + K$$

for each positive integer $n \geqslant N$. Thus the series $\Sigma\, a_n$ converges. On the other hand if $\Sigma\, a_n$ converges to A, then

$$\int_N^{n+1} f(x)dx \leqslant a_N + \ldots + a_n \leqslant A.$$

Hence $\int_N^\infty f(x)dx = \lim_n \int_N^n f(x)dx$ exists and is less than or equal to A. Therefore, the series $\Sigma\, a_n$ converges if and only if the integral $\int_N^\infty f(x)dx$ converges.

Example 7.2.7. As an example of Theorem 7.2.4, consider the series

$$\sum_{n=1}^{\infty} 1/n^p,$$

where p is any positive real number. Let $f(x) = 1/x^p$. Then $f(x)$ is defined for $x > 0$. If $p > 1$, then

$$\int_1^n x^{-p}dx = \frac{x^{-p+1}}{-p+1}\Big|_1^n = \frac{1}{1-p}\left(\frac{1}{n^{p-1}} - 1\right)$$

and

$$\lim_n \int_1^n x^{-p}dx = \frac{1}{p-1}.$$

Hence, for $p > 1$,

$$\sum_{n=1}^{\infty} 1/n^p$$

converges. It is easy to verify that, for $p \leqslant 1$, the series

$$\sum_{n=1}^{\infty} 1/n^p$$

diverges.

Problem Set 7.2

For Exercises 1–13, check the convergence or divergence of the given series.

1. $\displaystyle\sum_{n=2} 1/\ln n.$

2. $\displaystyle\sum_{n=1}^{\infty} n/(n+1)2^n.$

3. $\displaystyle\sum_{n=1}^{\infty} (n+1)/n!.$ [*Hint*: Show that $(n+1)/n! < 8/2^n.$]

4. $\displaystyle\sum_{n=1}^{\infty} (3 \cdot 5 \cdot 7 \cdot \ldots \cdot (2n+1))/(5 \cdot 10 \cdot 15 \cdot \ldots \cdot 5n).$ [*Hint*: Compare to a suitable multiple of the series $\displaystyle\sum_{n=1}^{\infty} 1/2^{n-1}.$]

5. $\displaystyle\sum_{n=1}^{\infty} n^2/(n^3 + 1).$

6. $\displaystyle\sum_{n=1}^{\infty} n!/2^n.$

7. $\displaystyle\sum_{n=1}^{\infty} n\,(1/3)^n.$

8. $\displaystyle\sum_{n=1}^{\infty} n^n/n!.$

9. $\displaystyle\sum_{n=1}^{\infty} 1/n \ln n.$

10. $\displaystyle\sum_{n=2}^{\infty} 1/n(\ln n)^2.$

11. $\displaystyle\sum_{n=6}^{\infty} 1/\{n \ln n[\ln(\ln n)]\}.$

12. $\sum_{n=6}^{\infty} 1/\{n \ln n[\ln(\ln n)]^2\}$.

13. $\sum_{n=2}^{\infty} (\ln n)^2/n^3$.

14. Let $\Sigma\, a_n$ and $\Sigma\, b_n$ be infinite series with $b_n > 0$ for each positive integer. Prove that if

$$\lim_{n} \left|\frac{a_n}{b_n}\right| = L,$$

where L is a real number, and $\Sigma\, b_n$ converges, then $\Sigma\, a_n$ converges absolutely. Prove that if $\Sigma\, b_n$ diverges and $L \neq 0$, then $\Sigma\, |a_n|$ diverges.

15. Let $\Sigma\, a_n$ be a convergent positive term series. Let $\{n_k\}$ be a subsequence of the sequence of positive integers. Prove that the series $\sum_{k=1}^{\infty} a_{n_k}$ is convergent.

16. Let $\Sigma\, a_n$ and $\Sigma\, b_n$ be two series such that $\Sigma\, b_n$ is a positive term series. Prove that if the inequality

$$\left|\frac{a_{n+1}}{a_n}\right| \leqslant \frac{b_{n+1}}{b_n}$$

holds for each positive integer n and $\Sigma\, b_n$ converges, then $\Sigma\, a_n$ converges absolutely.

17. Let $\Sigma\, a_n$ be an infinite series with nonzero terms. Let

$$\lim_{n} \inf \left|\frac{a_{n+1}}{a_n}\right| = r \quad \text{and} \quad \lim_{n} \sup \left|\frac{a_{n+1}}{a_n}\right| = R.$$

Prove that, if $R < 1$, the series converges absolutely; if $r > 1$, the series diverges; and, if either $R = 1$ or $r = 1$, the test fails.

18. Let $\Sigma\, a_n$ be an infinite series. Let

$$r = \lim_{n} \sup |a_n|^{1/n}.$$

Prove that, if $r < 1$, the series converges absolutely; if $r > 1$, the series diverges; and, if $r = 1$, the test fails.

19. For what values of p does the series

$$\sum_{n=2}^{\infty} 1/n(\ln n)^p$$

converges, and for what values of p does the series diverge?

20. For what values of p does the series

$$\sum_{n=4}^{\infty} 1\{n \ln n[\ln(\ln n)]^p\}$$

converge, and for what values of p does the series diverge?

21. Prove that the series

$$\sum_{n=2}^{\infty} 1/(\ln n)^p$$

diverges for all values of p.

22. For what values of p and q does the series

$$\sum_{n=2}^{\infty} (\ln n)^q/n^p$$

converge, and for what values of p and q does the series diverge?

23. Let $C_n = (1 + 1/2 + \ldots + 1/n) - \ln n$ for each positive integer n. Prove that the sequence $\{C_n\}$ converges. The limit of this sequence is called **Euler's Constant**. [*Hint*: Use the fact that

$$\ln x = \int_1^x t^{-1}\, dt$$

for each $x > 0$.]

24. Let $D_n = (1 + 1/\sqrt{2} + \ldots + 1/\sqrt{n}) - 2(\sqrt{n} - 1)$ for each positive integer n. Prove that the sequence $\{D_n\}$ converges.

25. Let $\Sigma\, a_n$ be a convergent positive term series. Give an example of a positive continuous function f such that $f(n) = a_n$ for each positive integer n, but the integral

$$\int_1^{\infty} f(x)dx$$

diverges.

26. Let $\Sigma\, a_n$ be a divergent positive term series. Give an example of a positive continuous function f such that $f(n) = a_n$ for each positive integer n, but the integral

$$\int_1^{\infty} f(x)dx$$

converges.

27. For what values of p does the series

$$\sum_{n=N}^{\infty} 1/[n(\ln n)(\ln \ln n) \ldots (\ln \ln \ldots \ln n)^p]$$

converge, where N is sufficiently large for $(\ln \ln \ldots \ln n)$ to be defined. For what values of p does the series diverge?

28. Give an example of a series such that Theorem 7.2.2 can be used to show convergence, while Corollary 7.2.2 fails.

29. Prove Part (2) of Theorem 7.2.1.

30. If

$$\lim_n \left| \frac{a_{n+1}}{a_n} \right| = L,$$

then prove that $\lim_n |a_n|^{1/n} = L$. [*Hint*: For $0 < L < +\infty$. Let α and β b

arbitrary numbers such that $0 < \alpha < L < \beta$. Then, for n greater than some N

$$\alpha^n(a_N/\alpha^N) < a_n < \beta^n(a_N/\beta^N),$$

and

$$\alpha \leqslant \begin{Bmatrix} \lim_n \sup |a_n|^{1/n} \\ \lim_n \inf |a_n|^{1/n} \end{Bmatrix} \leqslant \beta.]$$

31. If the series $\Sigma \, a_n$ converges absolutely, then prove that the series $\Sigma \, a_n^2$ conver
ges. Give an example to show that this statement is false if $\Sigma \, a_n$ converge
conditionally. [*Hint*: See Example 7.1.5.]

7.3. Series of Arbitrary Terms

In the last section we studied positive term series and absolutely
convergent series. In this section we shall discuss conditionally convergen
series. We first consider a special kind of series called an *alternating series*.

Let $\{a_n\}$ be a sequence such that $a_n > 0$ for each positive integer n. Then th
series

$$\sum_{n=1}^{\infty} (-1)^{n+1} a_n = a_1 - a_2 + a_3 - \ldots$$

is called an **alternating series**.

Theorem 7.3.1. *If the sequence $\{a_n\}$ is a decreasing null sequence, then
the alternating series*

$$\sum_{n=1}^{\infty} (-1)^{n+1} a_n$$

*is convergent. Let s_n denote the nth partial sum of the alternating series
If the alternating series converges to S, then, for each positive integer n
$|S - s_n| \leqslant a_{n+1}$ If the sequence $\{a_n\}$ is strictly decreasing then
$|S - s_n| < a_{n+1}$.*

Proof. We first consider the two sequences $\{s_{2n}\}$ and $\{s_{2n-1}\}$
For each positive integer n,

$$s_{2n+2} = s_{2n} + (a_{2n+1} - a_{2n+2}) \geqslant s_{2n},$$

$$s_{2n+1} = s_{2n-1} - (a_{2n} - a_{2n+1}) \leqslant s_{2n-1},$$

since $\{a_n\}$ is decreasing. Hence $\{s_{2n}\}$ is an increasing sequence, while $\{s_{2n-1}\}$ is a decreasing sequence. Furthermore,

$$s_{2n} = s_{2n-1} - a_{2n} \leqslant s_{2n-1} \leqslant s_1$$

and

$$s_{2n-1} = s_{2n-2} + a_{2n-1} \geqslant s_{2n-2} \geqslant s_2.$$

Thus, the sequences $\{s_{2n}\}$ and $\{s_{2n-1}\}$ both converge. We write $s_{2n} - s_{2n-1} = -a_{2n}$, and we see that

$$\lim_n (s_{2n} - s_{2n-1}) = \lim_n (-a_{2n}),$$

or

$$\lim_n s_{2n} - \lim_n s_{2n-1} = 0.$$

Therefore, $\lim_n s_{2n} = \lim_n s_{2n-1}$ and the sequence of partial sums converges, say, to S. Since $\{s_{2n}\}$ is increasing and $\{s_{2n-1}\}$ is decreasing, we have

$$s_{2n} \leqslant S \leqslant s_{2n-1}$$

and

$$s_{2n} \leqslant S \leqslant s_{2n+1}.$$

We may rewrite these two inequalities as

$$-a_{2n} = s_{2n} - s_{2n-1} \leqslant S - s_{2n-1} \leqslant 0$$

and

$$0 \leqslant S - s_{2n} \leqslant s_{2n+1} - s_{2n} = a_{2n+1}.$$

In either case, for each positive integer n,

$$|S - s_n| \leqslant a_{n+1}.$$

We present two examples of convergent alternating series.

Example 7.3.1. The **alternating harmonic series,**

$$\sum_{n=1}^{\infty} (-1)^{n+1}/n = 1 - 1/2 + 1/3 - \ldots,$$

is convergent since $\{1/n\}$ is a decreasing null sequence.

This series gives an example to show that the converse of Theorem 7.1.2 is not true. That is, there are series that are convergent but not absolutely convergent.

Example 7.3.2. As a second example of a convergent alternating series, we consider

$$\sum_{n=1}^{\infty} (-1)^{n+1} [\ln (n + 2)]/(n + 2).$$

To apply Theorem 7.3.1, we need to show that the sequence $\{(\ln n)/n\}$ is a decreasing null sequence. It follows from Exercise 7 of Problem Set 3.1 that $\lim_n \{(\ln n)/n\} = 0$, so we need only show that the sequence is decreasing. The simplest way to establish this is to prove that the function

$$f(x) = \frac{\ln x}{x}$$

is decreasing. We have

$$f'(x) = \frac{1 - \ln x}{x^2} < 0, \quad \text{for } x \geqslant 1.$$

Thus $f(x)$ is decreasing and $f(n + 1) \leqslant f(n)$, for each positive integer $n \geqslant 3$.

Let Σa_n be a series with terms of arbitrary sign. For each positive integer n, let

$$p_n = \max(a_n, 0), \quad q_n = \max(-a_n, 0).$$

In the case where $a_n = 0$, we have $p_n = q_n = 0$. The numbers p_n and q_n are nonnegative with at least one of them being zero. The numbers satisfy the two equations

$$a_n = p_n - q_n, \quad |a_n| = p_n + q_n$$

and the two inequalities

$$0 \leqslant p_n \leqslant |a_n|, \quad 0 \leqslant q_n \leqslant |a_n|.$$

The two positive term series

$$\sum_{n=1}^{\infty} p_n \quad \text{and} \quad \sum_{n=1}^{\infty} q_n$$

are called the **positive (or nonnegative) part** and the **negative (or nonpositive) part**, respectively, of the series Σa_n.

Example 7.3.3. To illustrate the positive part and the negative part of a series, we consider the alternating harmonic series

$$\sum_{n=1}^{\infty} (-1)^{n+1}/n.$$

Then

$$p_n = \begin{cases} 0 & \text{if } n \text{ is even,} \\ 1/n & \text{if } n \text{ is odd,} \end{cases}$$

$$q_n = \begin{cases} 0 & \text{if } n \text{ is odd,} \\ 1/n & \text{if } n \text{ is even,} \end{cases}$$

and

$$\sum_{n=1}^{\infty} p_n = \sum_{k=1}^{\infty} 1/(2k-1), \quad \sum_{n=1}^{\infty} q_n = \sum_{k=1}^{\infty} 1/(2k).$$

In considering a series of arbitrary terms, there are four series that we may speak of: the series itself, the series of absolute values, the positive part series, and the negative part series. We shall use the following notation to indicate the nth partial sum of each of these series:

$$s_n = a_1 + \dots + a_n, \quad A_n = |a_1| + \dots + |a_n|,$$

$$P_n = p_1 + \dots + p_n, \quad Q_n = q_1 + \dots + q_n.$$

We see that

$$s_n = P_n - Q_n \quad \text{and} \quad A_n = P_n + Q_n.$$

From these two equations, it follows that

$$P_n = Q_n + s_n \quad \text{and} \quad Q_n = P_n - s_n.$$

These two equations together with $s_n = P_n - Q_n$ yield the fact that the convergence of any two of the three series

$$\Sigma a_n, \quad \Sigma p_n, \quad \Sigma q_n$$

implies the convergence of the third series. This means that if both the positive part series and the negative part series converge, then the series itself converges (actually, absolutely). It also means that if a series converges, then the positive part series and the negative part series either both converge or both diverge. If they both diverge the series is conditionally convergent, and if they both converge the series is absolutely convergent. In fact we can make the stronger statement: *the series Σa_n converges absolutely if and only if the series Σp_n and Σq_n both converge.*

Example 7.3.4. For the alternating harmonic series

$$\sum_{n=1}^{\infty} (-1)^{n+1}/n,$$

$$p_n = \begin{cases} 1/n & \text{if } n \text{ is odd,} \\ 0 & \text{if } n \text{ is even,} \end{cases}$$

while

$$q_n = \begin{cases} 1/n & \text{if } n \text{ is even,} \\ 0 & \text{if } n \text{ is odd.} \end{cases}$$

Since both the series

$$\sum_{n=1}^{\infty} 1/(2n), \quad \sum_{n=1}^{\infty} 1/(2n-1)$$

diverge, the alternating harmonic series is only conditionally convergent. (This we already knew since the harmonic series is divergent.)

In considering finite sums, the order in which the terms appear does not effect the sum. As we shall see, the situation is quite different for infinite series.

Definition 7.3.1. *Let ϕ be a* 1–1 *mapping of the set of positive integers onto the set of positive integers. Then the series*

$$\sum_{n=1}^{\infty} a_{\phi(n)}$$

is called a **rearrangement** *of the series*

$$\sum_{n=1}^{\infty} a_n.$$

We present two examples to illustrate the concept of rearrangement of series.

Example 7.3.5. First, let

$$\phi(n) = \begin{cases} n - 1 & \text{if } n \text{ is a positive even integer,} \\ n + 1 & \text{if } n \text{ is a positive odd integer.} \end{cases}$$

Then ϕ is a 1–1 mapping of the set of positive integers onto the set of positive integers, and the series

$$\sum_{n=1}^{\infty} 1/\phi(n) = 1/2 + 1/1 + 1/4 + 1/3 + \ldots$$

is a rearrangement of the series

$$\sum_{n=1}^{\infty} 1/n.$$

Example 7.3.6. As our second example, let

$$\phi(n) = (k - 1)^2 + (k^2 - n) + 1 \quad \text{if} \quad (k - 1)^2 < n \leqslant k^2.$$

Then ϕ is a 1–1 mapping of the set of positive integers onto the set of positive integers, and the series

$$\sum_{n=1}^{\infty} a_{\phi(n)} = a_1 + a_4 + a_3 + a_2 + \ldots$$

$$+ \, a_{n^2} + a_{n^2-1} + \ldots + a_{(n-1)^2+1} + \ldots$$

is a rearrangement of the series

$$\sum_{n=1}^{\alpha} a_n.$$

Theorem 7.3.2. *Let $\Sigma\, a_n$ be an absolutely convergent series with sum A. Then any rearrangement of this series also converges to A.*

Proof. Let $\Sigma\, a_{\phi(n)}$ be any rearrangement of $\Sigma\, a_n$. For each positive integer n, let

$$s_n = \sum_{k=1}^{n} a_k, \quad t_n = \sum_{k=1}^{n} a_{\phi(k)}.$$

By hypotheses, $\{s_n\} \to A$. We need to show that $\{t_n\} \to A$. Since the given series converges absolutely, for each $\varepsilon > 0$ there is an N such that

$$\sum_{k=n+1}^{\infty} |a_k| < \varepsilon/2$$

when $n \geqslant N$. Also, for $n \geqslant N$,

$$|s_n - A| = \left| \sum_{k=n+1}^{\infty} a_k \right| \leqslant \sum_{k=n+1}^{\infty} |a_k| < \varepsilon/2.$$

Since ϕ is a 1–1 mapping of the set of positive integers onto the set of positive integers, there is an $M \geqslant N$ such that $\{1, 2, ..., N\} \subseteq \{\phi(1), \phi(2), ..., \phi(M)\}$. Then, for $n \geqslant M$,

$$|t_n - A| = |t_n - s_N + s_N - A|$$

$$\leqslant |t_n - s_N| + |s_N - A|$$

$$< \left| \sum_{k=1}^{n} a_{\phi(k)} - \sum_{k=1}^{N} a_k \right| + \varepsilon/2.$$

Since

$$\sum_{k=1}^{n} a_{\phi(k)} - \sum_{k=1}^{N} a_k$$

can contain only terms of the series $\Sigma\, a_n$ with subscripts greater than N, we have

$$\left| \sum_{k=1}^{n} a_{\phi(k)} - \sum_{k=1}^{N} a_k \right| < \varepsilon/2.$$

Thus,

$$|t_n - A| < \varepsilon/2 + \varepsilon/2 = \varepsilon$$

when $n \geqslant M$, and so $\{t_n\} \to A$.

Theorem 7.3.3. *Let Σa_n be a conditionally convergent series, and let c be any real number. Then there is a rearrangement of Σa_n that converges to c.*

Proof. Let Σp_n be the positive part of Σa_n, and let Σq_n be the negative part of Σa_n. Since Σa_n is conditionally convergent, both Σp_n and Σq_n diverge. Let r_1 be the least positive integer such that

$$p_1 + p_2 + \ldots + p_{r_1} > c.$$

We know that such a positive integer exists, for otherwise the sequence of partial sums of Σp_n would be bounded and Σp_n would be convergent. Let s_1 be the least positive integer such that

$$p_1 + p_2 + \ldots + p_{r_1} - q_1 - q_2 - \ldots - q_{s_1} < c.$$

We know that such a positive integer exists, for otherwise the sequence of partial sums of Σq_n would be bounded and Σq_n would be convergent. Let r_2 be the least positive integer greater than r_1 such that

$$p_1 + p_2 + \ldots + p_{r_1} - q_1 - q_2 - \ldots - q_{s_1} + p_{r_1+1} + \ldots + p_{r_2} > c.$$

Continuing in this manner, we have a series that converges to c, since $\lim_n a_n = 0$.

For each term a_n in the series Σa_n, two terms, p_n and q_n, appear in the series just constructed, one of which is zero. If $a_n \geqslant 0$, we delete the term q_n; if $a_n < 0$, we delete the term p_n. The resulting series is a rearrangement of the series Σa_n, and it converges to the number c.

Problem Set 7.3

In Exercises 1–8, test for absolute convergence, conditional convergence, or divergence.

1. $\displaystyle\sum_{n=1}^{\infty} (-1)^n/\sqrt{n}.$

2. $\displaystyle\sum_{n=1}^{\infty} (-1)^{n+1}/\ln(n+1).$

3. $\displaystyle\sum_{n=1}^{\infty} (-1)^{n+1}(\sqrt{n+1} - \sqrt{n}).$

4. $\displaystyle\sum_{n=1}^{\infty} (-1)^{n+1}/(n^{1+1/n}).$

5. $\sum_{n=1}^{\infty} (-1)^{n+1} n/(n^2 - 5n + 1)$.

6. $\sum_{n=1}^{\infty} (-1)^{n+1} n^3/e^{2n}$.

7. $\sum_{n=1}^{\infty} (-1)^{n+1} n^4/(n+1)!$.

8. $\sum_{n=1}^{\infty} (1)^{n+1} (n \ln n)/e^n$.

9. Show by three counterexamples that each of the three conditions of the alternating series test (Theorem 7.3.1) is needed to assure convergence (that is, the alternating signs, the decreasing nature of the terms, and the sequence of terms is a null sequence).

10. Let $\{n_k\}$ be a subsequence of the sequence of positive integers $\{n\}$. Then

$$\sum_{k=1}^{\infty} a_{n_k}$$

is called a **subseries** of the series

$$\sum_{n=1}^{\infty} a_n.$$

Give an example of a convergent series with at least one divergent subseries.

11. If every subseries of a series converges, show that the series converges absolutely.

12. Show that, if $\Sigma\, a_n$ is a conditionally convergent series, there is a rearrangement whose sequence of partial sums tends to $+\infty$ (that is, if $\{s_n\}$ is the sequence of partial sums, then for each real number M there is an N such that $n > N$ implies that $s_n > M$).

13. Construct a rearrangement of $\sum_{n=1}^{\infty} (-1)^n/n$ that converges to zero.

7.4. More Delicate Tests

The proofs of the Ratio Test and the Root Test both depend on a comparison with the geometric series. These tests are inconclusive when the "limit" is 1. In order to obtain sharper tests, we use for comparison the telescopic series (cf. Exercise 3 of Problem Set 7.1). We begin with *Kummer's Test*. This test is named for the German mathematician Ernst E. Kummer (1810–1893.).

Theorem 7.4.1. (*Kummer's Test*) *Let* $\{c_n\}$ *be a positive term sequence and let* $\{a_n\}$ *be a nonzero term sequence. If there are positive numbers h and N such that*

$$c_n - c_{n+1}|a_{n+1}/a_n| \geqslant h > 0$$

for each $n \geqslant N$, *then the series* $\Sigma \, a_n$ *converges absolutely.*

Proof. Assume that

$$c_n - c_{n+1}|a_{n+1}/a_n| \geqslant h > 0$$

for each $n \geqslant N$. Then

$$0 < h|a_n| \leqslant c_n|a_n| - c_{n+1}|a_{n+1}|$$

for each $n \geqslant N$. Hence the positive sequence $\{c_{n+N}|a_{n+N}|\}$ is a decreasing sequence and Theorem 3.1.6 implies that this sequence is convergent. Suppose that $\{c_{n+N}|a_{n+N}|\}$ converges to L. Exercise 3 of Problem Set 7.1 implies that the telescopic series

$$\sum_{n=N}^{\infty} (c_n|a_n| - c_{n+1}|a_{n+1}|)$$

converges to $c_N|a_N| - L$. The Comparison Test (Theorem 7.2.1) implies that the series $\Sigma \, h|a_n|$ converges. Thus the series $\Sigma \, |a_n|$ converges, and the series $\Sigma \, a_n$ converges absolutely.

Kummer's Test is a test for convergence of a given series. The corresponding test for divergence of a given series is called *Jensen's Test*. This test is named for the Danish telephone engineer and mathematician J. L. W. V. Jensen (1859–1925).

Theorem 7.4.2. (*Jensen's Test*) *Let* $\Sigma \, 1/c_n$ *be a divergent positive term series, and let* $\{a_n\}$ *be a positive term sequence. If there is a positive number N such that*

$$c_n - c_{n+1} \, a_{n+1}/a_n \leqslant 0$$

for each $n \geqslant N$, *then the positive term series* $\Sigma \, a_n$ *is divergent.*

The proof of this test is left as Exercise 9 of Problem Set 7.4.

As a corollary to Theorems 7.4.1 and 7.4.2 we have the following *limit form* of these two tests.

Corollary 7.4.2. (*Kummer–Jensen Test*) *Let* $\{a_n\}$ *and* $\{c_n\}$ *be positive term sequences, let*

$$K = \lim_n \sup [c_n - c_{n+1} \, a_{n+1}/a_n],$$

and let

$$k = \lim_n \inf [c_n - c_{n+1} \, a_{n+1}/a_n].$$

(1) *If $k > 0$, then the series $\Sigma \, a_n$ converges.*

(2) *If $K < 0$ and $\Sigma \, 1/c_n$ diverges, then $\Sigma \, a_n$ diverges.*

Proof.   (1) If $k > 0$, let $h = k/2$. Then the definition of limit inferior implies that there is an N such that

$$c_n - c_{n+1} \, a_{n+1}/a_n \geqslant h$$

for each $n \geqslant N$. Theorem 7.4.1 implies that $\Sigma \, a_n$ converges.

(2) If $K < 0$, let $h = K/2$. Then the definition of limit superior implies that there is an N such that

$$c_n - c_{n+1} \, a_{n+1}/a_n \leqslant h < 0$$

for each $n \geqslant N$. Theorem 7.4.2 implies that $\Sigma \, a_n$ diverges if $\Sigma \, 1/c_n$ diverges.

By using special sequences $\{c_n\}$, we are now able to derive various other tests. We begin by using Theorems 7.4.1 and 7.4.2 to give a new proof of the Ratio Test.

Example 7.4.1. If in Theorem 7.4.1 we let $c_n = 1$ for each positive integer n, then we have

$$c_n - c_{n+1}|a_{n+1}/a_n| = 1 - |a_{n+1}/a_n| \geqslant h > 0$$

or

$$|a_{n+1}/a_n| \leqslant 1 - h < 1$$

for $n \geqslant N$. This is precisely Part (1) of the Ratio Test (Theorem 7.2.2).

If in Theorem 7.4.2 we let $c_n = 1$ for each positive integer n, then we have

$$c_n - c_{n+1} \, a_{n+1}/a_n = 1 - a_{n+1}/a_n \leqslant 0$$

or

$$a_{n+1}/a_n \geqslant 1$$

for $n \geqslant N$. This is precisely Part (2) of the Ratio Test (Theorem 7.2.2).

If in Corollary 7.4.2 we let $c_n = 1$ for each positive integer n, then we have

$$K = \lim_n \sup \, [1 - a_{n+1}/a_n]$$

and

$$k = \lim_n \inf \, [1 - a_{n+1}/a_n].$$

If $k > 0$, then we write

$$\lim_n \inf \, [1 - a_{n+1}/a_n] = -\lim_n \sup \, [a_{n+1}/a_n - 1]$$
$$= -\lim_n \sup \, [a_{n+1}/a_n] + 1$$
$$> 0,$$

or
$$1 > \lim_n \sup [a_{n+1}/a_n].$$

This is precisely the convergence part of the "lim sup" form of the Ratio Test (cf. Exercise 17 of Problem Set 7.2).

If $K < 0$, then we write

$$\begin{aligned} \lim_n \sup [1 - a_{n+1}/a_n] &= -\lim_n \inf [a_{n+1}/a_n - 1] \\ &= -\lim_n \inf [a_{n+1}/a_n] + 1 \\ &< 0, \end{aligned}$$

or
$$1 < \lim_n \inf [a_{n+1}/a_n].$$

This is precisely the divergence part of the "lim inf" form of the Ratio Test (cf. Exercise 17 of Problem Set 7.2).

If in Corollary 7.4.2 we let $c_n = n - 1$ for each n, then we have *Raabe's Test*. This test is named for the mathematician Joseph Ludwig Raabe (1800–1859). Raabe was born in Brody, Ukraine, and he taught at Zurich, Switzerland.

Theorem 7.4.3. (*Raabe's Test*) *Let*

$$P = \lim_n \sup [n(1 - a_{n+1}/a_n)]$$

and

$$\rho = \lim_n \inf [n(1 - a_{n+1}/a_n)],$$

where $\{a_n\}$ is a positive term sequence.

(1) *If $\rho > 1$, then the series Σa_n converges.*

(2) *If $P < 1$, then the series Σa_n diverges.*

The proof of this test follows from Corollary 7.4.2 by letting

$$c_n = n - 1.$$

The proof is left as Exercise 10 of Problem Set 7.4.

We now present two examples to demonstrate the use of Raabe's Test.

Example 7.4.2. Let α be a real number. Check the series

$$\sum_{n=0}^{\infty} a_n = 1 + \frac{\alpha}{1} + \frac{\alpha(\alpha + 1)}{2!} + \frac{\alpha(\alpha + 1)(\alpha + 2)}{3!} + \dots$$

for convergence or divergence. We first note that

$$a_{n+1}/a_n = (\alpha + n)/(n + 1)$$

and

$$\lim_n \{a_{n+1}/a_n\} = 1.$$

Thus the Ratio Test fails.

It is easily seen that

$$n(1 - a_{n+1}/a_n) = n\{1 - (\alpha + n)/(n + 1)\} = (1 - \alpha)\,n/(n + 1),$$

and that

$$\lim_n \{n(1 - a_{n+1}/a_n)\} = 1 - \alpha.$$

When $n > \alpha$ the terms $a_{n+1}/a_n > 0$ and the terms of the series $\Sigma\, a_n$ ultimately have the same sign. Thus we can apply Raabe's Test. If $1 - \alpha > 1$ ($\alpha < 0$), then we have convergence; if $1 - \alpha < 1$ ($\alpha > 0$), then we have divergence.

Example 7.4.3. Check the series

$$\sum_{n=1}^{\infty} [2^n(n!)]^2/(2n + 1)!$$

for convergence or divergence. We first note that

$$a_{n+1}/a_n = 4(n + 1)^2/(2n + 2)(2n + 3)$$

and

$$\lim_n \{a_{n+1}/a_n\} = 1.$$

Thus the Ratio Test fails. We see that

$$n(1 - a_{n+1}/a_n) = n/(2n + 3)$$

and

$$\lim_n \{n(1 - a_{n+1}/a_n)\} = 1/2.$$

Applying Raabe's Test, we have that this series diverges.

In Raabe's Test if either $\rho = 1$ or $P = 1$, the test fails. We again use Corollary 7.4.2 to derive a test that sometimes works when Raabe's Test fails. This test is called *Bertrand's Test* and is named for the Parisian mathematician Joseph Bertrand (1822–1900).

Theorem 7.4.4. (*Bertrand's Test*) *Let*

$$B = \lim_n \sup \{\ln n\, [n\{\frac{a_n - a_{n+1}}{a_n}\} - 1]\}$$

and

$$b = \lim_n \inf \{\ln n\, [n\{\frac{a_n - a_{n+1}}{a_n}\} - 1]\},$$

where $\{a_n\}$ is a positive term sequence.

(1) *If $b > 1$, then the series $\Sigma\, a_n$ converges.*

(2) *If $B < 1$, then the series $\Sigma\, a_n$ diverges.*

The proof of this test follows from Corollary 7.4.2 by letting

$$c_n = (n - 1) \ln (n - 1).$$

The proof is left as Exercise 11 of Problem Set 7.4.

We conclude this section by proving a test named for the great German mathematician Karl Friedrich Gauss (1777–1855).

Theorem 7.4.5. (*Gauss's Test*) *If it is possible to write*

$$n \left(\frac{a_n - a_{n+1}}{a_n} \right) = h + \frac{k_n}{n^p},$$

where $p > 1$, h is a constant, and the sequence $\{k_n\}$ is bounded, then the positive term series $\Sigma \, a_n$ converges if $h > 1$ and diverges if $h \leqslant 1$.

Proof. Since

$$\lim_n \left\{ n \left(\frac{a_n - a_{n+1}}{a_n} \right) \right\} = h,$$

Raabe's Test implies that the series $\Sigma \, a_n$ converges if $h > 1$ and diverges if $h < 1$. Thus we need only consider the case where $h = 1$. We use Bertrand's Test in this case.

Write

$$\ln n \left[n \left(\frac{a_n - a_{n+1}}{a_n} \right) - 1 \right] = \ln n \left[1 + \frac{k_n}{n^p} - 1 \right]$$

$$= \ln n \left[\frac{k_n}{n^p} \right].$$

Since

$$\lim_n \left\{ \ln n \left[\frac{k_n}{n^p} \right] \right\} = 0,$$

Bertrand's Test implies that $\Sigma \, a_n$ diverges.

We shall consider a special case of Gauss' Test where the ratio a_{n+1}/a_n can be written as the quotient of two polynomials with the same term of highest degree; namely

$$\frac{a_{n+1}}{a_n} = \frac{n^k + an^{k-1} + P(n)}{n^k + bn^{k-1} + Q(n)},$$

where P and Q are polynomials of degree at most $k - 2$. Since

$$\lim_n \{a_{n+1}/a_n\} = 1,$$

the Ratio Test fails. We see that

$$n\left(1 - \frac{a_{n+1}}{a_n}\right) = \frac{(b-a)n^k + n[Q(n) - P(n)]}{n^k + bn^{k-1} + Q(n)}$$

and

$$\lim_n \{n(1 - a_{n+1}/a_n)\} = b - a.$$

Then Raabe's Test implies that the positive term series Σa_n converges if $b - a > 1$ and diverges if $b - a < 1$. Thus we need only consider the case where $b - a = 1$. For this case we use Bertrand's Test.

We see that

$$\ln n[n(1 - a_{n+1}/a_n) - 1] = \ln n \left\{\frac{n[Q(n) - P(n)] - bn^{k-1} - Q(n)}{n^k + bn^{k-1} + Q(n)}\right\}$$

and since the degree of the numerator is at most $k - 1$,

$$\lim \{\ln n[n(1 - a_{n+1}/a_n) - 1]\} = 0.$$

Thus the Bertrand Test implies that the series Σa_n diverges for $b - a = 1$. We have proved the following useful result.

Theorem 7.4.6. *If it is possible to write*

$$\frac{a_{n+1}}{a_n} = \frac{n^k + an^{k-1} + P(n)}{n^k + bn^{k-1} + Q(n)},$$

where k is a positive integer and P and Q are polynomials of degree at most $k - 2$, then the positive term series Σa_n converges if $b - a > 1$ and diverges if $b - a \leqslant 1$.

Example 7.4.4. Let α, β, γ be real numbers such that none of them is zero or a negative integer. Test the **hypergeometric series**

$$\sum_{n=0}^{\infty} a_n = 1 + \frac{\alpha\beta}{1 \cdot \gamma} + \frac{\alpha(\alpha+1) \cdot \beta(\beta+1)}{1 \cdot 2 \cdot \gamma(\gamma+1)} + \dots$$

$$+ \frac{\alpha(\alpha+1) \dots (\alpha+n-1) \cdot \beta(\beta+1) \dots (\beta+n-1)}{n! \cdot \gamma(\gamma+1) \dots (\gamma+n-1)} + \dots$$

for convergence.

The conditions on α and β imply that the series does not terminate after a finite number of terms and conditions on γ imply that there is no zero in the denominator. The terms of the series Σa_n ultimately have a constant sign, and the ratio a_{n+1}/a_n may be written

$$\frac{a_{n+1}}{a_n} = \frac{(\alpha+n)(\beta+n)}{(1+n)(\gamma+n)} = \frac{n^2 + (\alpha+\beta)n + \alpha\beta}{n^2 + (\gamma+1)n + \gamma}.$$

Theorem 7.4.6 implies that the series $\Sigma\, a_n$ converges for $\gamma > \alpha + \beta$ and diverges for $\gamma \leqslant \alpha + \beta$.

Problem Set 7.4

In Exercises 1–4, test the given series for convergence or divergence.

1. $\sum\limits_{n=1}^{\infty} n!/(\alpha + 1)(\alpha + 2) \ldots (\alpha + n)$, where $\alpha > 0$.

2. $\sum\limits_{n=1}^{\infty} (2n)!/[2^n(n!)]^2(2n + 1)$.

3. $\sum\limits_{n=1}^{\infty} (2n)!(4n + 3)/[2^n(n!)]^2(2n + 2)$.

4. $\sum\limits_{n=1}^{\infty} \{(2n)!/[2^n(n!)]^2\}^k$, where $k = 1, 2, 3$.

5. For $k = 2$ in Exercise 4, use Gauss' Test to test for convergence.

6. For which values of α and β will the series

$$1 + \frac{\alpha}{\beta} + \frac{\alpha(\alpha + 1)}{\beta(\beta + 1)} + \frac{\alpha(\alpha + 1)(\alpha + 2)}{\beta(\beta + 1)(\beta + 2)} + \cdots$$

converge? Diverge?

7. For which values of α, β, γ, and δ will the series

$$1 + \frac{\alpha\beta}{\gamma\delta} + \frac{\alpha(\alpha + 1)\beta(\beta + 1)}{\gamma(\gamma + 1)\delta(\delta + 1)} + \cdots$$

converge? Diverge?

8. Discuss the convergence of the **binomial series**

$$1 + \alpha x + \frac{\alpha(\alpha - 1)}{2!} x^2 + \frac{\alpha(\alpha - 1)(\alpha - 2)}{3!} x^3 + \cdots$$

for all values of x and α.

9. Prove Theorem 7.4.2.

10. Prove Theorem 7.4.3.

11. Prove Theorem 7.4.4.

12. Using Corollary 7.4.2, derive a test for convergence by letting

$$c_n = (n - 1)\ln(n - 1)\ln \ln(n - 1).$$

7.5 Operations with Series

We now define and prove some results about sums and products of series.

The following theorem is a consequence of results for sequences, and the proof is left as Exercise 1 of Problem Set 7.5.

Theorem 7.5.1. *Let $\Sigma\, a_n$ and $\Sigma\, b_n$ be convergent series, and let α and β be any two real numbers. Then the series $\Sigma\, (\alpha a_n + \beta b_n)$ converges and*

$$\Sigma\, (\alpha a_n + \beta b_n) = \alpha\, \Sigma\, a_n + \beta\, \Sigma\, b_n.$$

Example 7.5.1. As an application of Theorem 7.5.1, we evaluate the series

$$\sum_{n=1}^{\infty} (5 \cdot 3^n + 8 \cdot 2^n)/6^n.$$

We see that

$$(5 \cdot 3^n + 8 \cdot 2^n)/6^n = 5/2^n + 8/3^n.$$

Since

$$\sum_{n=1}^{\infty} 1/2^n = 1 \quad \text{and} \quad \sum_{n=1}^{\infty} 1/3^n = 1/2,$$

we have

$$\sum_{n=1}^{\infty} (5 \cdot 3^n + 8 \cdot 2^n)/6^n = \sum_{n=1}^{\infty} (5/2^n + 8/3^n)$$

$$= 5 \sum_{n=1}^{\infty} 1/2^n + 8 \sum_{n=1}^{\infty} 1/3^n$$

$$= 5 \cdot 1 + 8(1/2) = 9.$$

The following very useful result is called **Abel's summation** or **summation by parts**.

Theorem 7.5.2. *Let $\{a_n\}$ and $\{b_n\}$ be two sequences, and let $A_n = a_1 + a_2 + \ldots + a_n$. Then, for each positive integer n,*

$$\sum_{k=1}^{n} a_k b_k = A_n b_{n+1} + \sum_{k=1}^{n} A_k(b_k - b_{k+1}).$$

Proof. Define $A_0 = 0$. Then $a_k = A_k - A_{k-1}$, and we can write

$$\sum_{k=1}^{n} a_k b_k = \sum_{k=1}^{n} (A_k - A_{k-1}) b_k$$

$$= (A_1 - A_0) b_1 + (A_2 - A_1) b_2 + \ldots$$
$$+ (A_{n-1} - A_{n-2}) b_{n-1} + (A_n - A_{n-1}) b_n$$

$$= A_1(b_1 - b_2) + A_2(b_2 - b_3) + \ldots$$
$$+ A_{n-1}(b_{n-1} - b_n) + A_n b_n$$

$$= \sum_{k=1}^{n-1} A_k(b_k - b_{k+1}) + A_n b_n$$

$$= \sum_{k=1}^{n-1} A_k(b_k - b_{k+1}) + A_n b_n - A_n b_{n+1} + A_n b_{n+1}$$

$$= \sum_{k=1}^{n} A_k(b_k - b_{k+1}) + A_n b_{n+1}.$$

This completes the proof.

Note that we can write the difference of two partial sums of the series $\Sigma\, a_n b_n$ in the following way:

$$\sum_{k=m+1}^{n} a_k b_k = \sum_{k=m+1}^{n} A_k(b_k - b_{k+1}) + A_n b_{n+1} - A_m b_{m+1}.$$

Another consequence of Abel's summation is the following result, called *Abel's Lemma*.

Lemma 7.5.1. *(Abel's Lemma) Let $\{a_n\}$ be a sequence whose sequence of partial sums $\{A_n\}$ satisfies the relation*

$$m \leqslant A_n \leqslant M$$

for each positive integer n, where m and M are real numbers. If $\{b_n\}$ is a decreasing sequence of nonnegative real numbers, then

$$mb_1 \leqslant \sum_{k=1}^{n} a_k b_k \leqslant Mb_1$$

for each positive integer n.

Proof. From Abel's summation (Theorem 7.5.2), we have

$$\sum_{k=1}^{n} a_k b_k = \sum_{k=1}^{n} A_k(b_k - b_{k+1}) + A_n b_{n+1}.$$

Since $b_k - b_{k+1} \geqslant 0$ and $A_k \geqslant m$ for each positive integer k, we obtain

$$\sum_{k=1}^{n} a_k b_k \geqslant m \sum_{k=1}^{n} (b_k - b_{k+1}) + mb_{n+1}$$
$$= m(b_1 - b_{n+1}) + mb_{n+1}$$
$$= mb_1.$$

This proves the left-hand side of the inequality. The right-hand side is proved in a similar manner, and the proof is left to the reader.

We notice that, in Abel's Lemma, if

$$\left| \sum_{k=1}^{n} a_k \right| \leqslant M$$

for each positive integer n, then

$$\left| \sum_{k=1}^{n} a_k b_k \right| \leqslant Mb_1$$

for each positive integer n. We now use Abel's Lemma to prove the following useful theorem, called the *Dirichlet Test*. This theorem is named in honor of the German mathematician Peter Gustav Lejeune Dirichlet (1805–1859).

Theorem 7.5.3. (*Dirichlet Test*) *Let $\Sigma\, a_n$ be a series whose sequence of partial sums is $\{A_n\}$. If the sequence $\{A_n\}$ is bounded, and if $\{b_n\}$ is a decreasing null sequence, then the series $\Sigma\, a_n b_n$ converges.*

Proof. We shall show that the sequence of partial sums of the series $\Sigma\, a_n b_n$ is a Cauchy sequence. That is, if $\varepsilon > 0$, then we must find a positive number N such that

$$\left| \sum_{k=m+1}^{n} a_k b_k \right| < \varepsilon$$

whenever $n > m > N$.

Since the sequence of partial sums $\{A_n\}$ is bounded, there exists a positive number M such that $|A_n| \leqslant M$ for each positive integer n. Hence, for any positive integers m and n with $n > m$, we have

$$\left| \sum_{k=m+1}^{n} a_k \right| = |A_n - A_m| \leqslant |A_n| + |A_m| \leqslant 2M.$$

Applying the remark at the end of Abel's Lemma to the sequences $\{a_n\}_{n=m+1}^{+\infty}$ and $\{b_n\}_{n=m+1}^{+\infty}$ we have

$$\left| \sum_{k=m+1}^{n} a_k b_k \right| \leqslant 2Mb_m$$

for all positive integers m and n with $n > m$. Since $\{b_n\}$ is a decreasing null sequence, there exists a positive number N such that

$$b_n < \varepsilon/2M$$

for all positive integers $n > N$. Thus

$$\left| \sum_{k=m+1}^{n} a_k b_k \right| \leqslant 2M b_m < 2M\varepsilon/2M = \varepsilon$$

whenever $n > m > N$. Therefore, the sequence of partial sums of the series $\Sigma\, a_n b_n$ is a Cauchy sequence and thus converges.

Example 7.5.2. As an application of this result, we present a second proof of the alternating series test (Theorem 7.3.1). Let $a_n = (-1)^{n+1}$. Then

$$A_n = \sum_{k=1}^{n} a_n = \sum_{k=1}^{n} (-1)^{n+1} = \begin{cases} 0 & \text{if } k \text{ is even,} \\ 1 & \text{if } k \text{ is odd,} \end{cases}$$

and $\{A_n\}$ is bounded. If $\{b_n\}$ is a decreasing null sequence, then Theorem 7.5.3 implies that

$$\sum_{n=1}^{\infty} a_n b_n = \sum_{n=1}^{\infty} (-1)^{n+1} b_n$$

converges.

Example 7.5.3. From the identity

$2 \sin (x/2)(\sin x + \sin 2x + \dots + \sin nx)$

$$= \cos (x/2) - \cos ([2n + 1]\, x/2),$$

it follows that if $\sin (x/2) \neq 0$ then the partial sums

$$S_n (x) = \sum_{k=1}^{n} \sin kx$$

satisfy the relation

$$|S_n (x)| \leqslant 1/|\sin (x/2)|$$

for each positive integer n. If $\sin (x/2) = 0$, then $S_n (x) = 0$ for each positive integer n. Thus, for any real number x, the sequence of partial sums $\{S_n (x)\}$ of the series $\Sigma \sin nx$ is bounded. If $\{b_n\}$ is any decreasing null sequence, then the Dirichlet Test (Theorem 7.5.3) implies that the series

$$\sum_{n=1}^{\infty} b_n \sin nx$$

converges for each real x. In particular, the series

$$\sum_{n=1}^{\infty} (\sin nx)/n$$

converges for each real x.

Let $p(x) = \sum_{n=0}^{N} a_n x^n$ and $q(x) = \sum_{n=0}^{M} b_n x^n$ be two polynomials in x. The product of these two polynomials is given by

$$p(x)q(x) = \left(\sum_{n=0}^{N} a_n x^n\right)\left(\sum_{n=0}^{M} b_n x^n\right)$$

$$= (a_0 + a_1 x + \ldots + a_N x^N)(b_0 + b_1 x + \ldots + b_M x^M)$$

$$= a_0 b_0 + (a_0 b_1 + a_1 b_0) x + \ldots$$

$$+ (a_0 b_n + a_1 b_{n-1} + \ldots + a_n b_0) x^n$$

$$+ \ldots + a_N b_M x^{N+M}.$$

If we think of $a_n = 0$ for $n > N$ and $b_n = 0$ for $n > M$, then we can write

$$p(x)q(x) = c_0 + c_1 x + c_2 x^2 + \ldots + c_{N+M} x^{N+M},$$

where $c_n = \sum_{k=0}^{n} a_k b_{n-k}$. In the special case where $x = 1$, we have

$$(a_0 + a_1 + \ldots + a_N)(b_0 + b_1 + \ldots + b_M) = c_0 + c_1 + \ldots + c_{N+M}.$$

We generalize this product of finite sums to a product of series.

Definition 7.5.1. *Let $\sum_{n=0}^{\infty} a_n$ and $\sum_{n=0}^{\infty} b_n$ be two series. For each positive integer n, let*

$$c_n = \sum_{k=0}^{n} a_k b_{n-k}.$$

Then the series

$$\sum_{n=0}^{\infty} c_n$$

*is called the **Cauchy product** of $\sum_{n=0}^{\infty} a_n$ and $\sum_{n=0}^{\infty} b_n$.*

From the above discussion it might be hoped that if

$$\sum_{n=0}^{\infty} a_n = A \quad \text{and} \quad \sum_{n=0}^{\infty} b_n = B,$$

then

$$\sum_{n=0}^{\infty} c_n = AB,$$

that is, the Cauchy product of two convergent series converges to the product of their sums. Unfortunately, this is not the case, as can be seen from the following example.

Example 7.5.4. Let

$$a_n = b_n = \frac{(-1)^n}{\sqrt{n+1}}$$

for each nonnegative integer n. Then

$$\sum_{n=0}^{\infty} a_n = \sum_{n=0}^{\infty} b_n = \sum_{n=0}^{\infty} (-1)^n/\sqrt{n+1}.$$

This is an alternating series, and it is easily seen to be conditionally convergent. The nth term, c_n, of the Cauchy product of this series with itself is given by

$$c_n = \sum_{k=0}^{n} a_k b_{n-k} = (-1)^n \sum_{k=0}^{n} 1/\sqrt{(n-k+1)(k+1)}.$$

If $k \leqslant n$, then we may write

$$(n-k+1)(k+1) = \left(\frac{n}{2}+1\right)^2 - \left(\frac{n}{2}-k\right)^2 \leqslant \left(\frac{n}{2}+1\right)^2;$$

thus

$$\frac{1}{\sqrt{(n-k+1)(k+1)}} \geqslant \frac{1}{\frac{1}{2}n+1} = \frac{2}{n+2}.$$

Therefore

$$|c_n| \geqslant \sum_{k=0}^{n} 2/(n+2) = 2(n+1)/(n+2) \geqslant 1,$$

and the series $\sum_{n=0}^{\infty} c_n$ does not converge.

Since Example 7.5.4 shows that conditional convergence of two series is not sufficient to guarantee the convergence of their Cauchy product, one might ask what conditions must be put on the two series so that their Cauchy product is convergent. The following theorem gives such a set of conditions.

Theorem 7.5.4. *If Σa_n converges absolutely to A and Σb_n converges to B, then their Cauchy product series Σc_n converges to AB.*

Proof.   Let

$$A_n = \sum_{k=0}^{n} a_n, \quad B_n = \sum_{k=0}^{n} b_n,$$

$$C_n = \sum_{k=0}^{n} c_n, \quad S = \sum_{k=0}^{\infty} |a_n|.$$

We need to show that $\{C_n\} \to AB$. For each positive integer n, we write

$$
\begin{aligned}
C_n &= c_0 + c_1 + \ldots + c_n \\
&= a_0 b_0 + (a_0 b_1 + a_1 b_0) + (a_0 b_2 + a_1 b_1 + a_2 b_0) \\
&\quad + \ldots + (a_0 b_n + a_1 b_{n-1} + \ldots + a_n b_0) \\
&= a_0(b_0 + b_1 + \ldots + b_n) + a_1(b_0 + b_1 + \ldots + b_{n-1}) \\
&\quad\quad\quad\quad\quad + \ldots + a_n b_0 \\
&= a_0 B_n + a_1 B_{n-1} + \ldots + a_n B_0 \\
&= a_0[B + (B_n - B)] + a_1[B + (B_{n-1} - B)] \\
&\quad\quad\quad\quad + \ldots + a_n[B + (B_0 - B)] \\
&= B(a_0 + a_1 + \ldots + a_n) + a_0(B_n - B) + a_1(B_{n-1} - B) \\
&\quad\quad\quad\quad + \ldots + a_n(B_0 - B) \\
&= BA_n + a_0(B_n - B) + a_1(B_{n-1} - B) + \ldots + a_n(B_0 - B).
\end{aligned}
$$

Since the sequence $\{BA_n\} \to AB$, we need only show that the sequence

$$\{a_0(B_n - B) + a_1(B_{n-1} - B) + \ldots + a_n(B_0 - B)\} \to 0.$$

Assume that $\varepsilon > 0$ is given. Since $\{B_n\} \to B$, the sequence $\{B_n - B\}$ is bounded, say, by M, and there is an N_1 such that

$$|B_n - B| < \varepsilon/2S$$

when $n > N_1$. Since $\sum_{n=0}^{\infty} |a_n|$ converges to S, there is an N_2 such that

$$\sum_{k=p+1}^{q} |a_k| < \varepsilon/2M$$

for each $q > p > N_2$. Let $N = \max(N_1, N_2)$. Then, for $n > 2N$,

$$|a_0(B_n - B) + a_1(B_{n-1} - B) + \ldots + a_n(B_0 - B)|$$

$$\leqslant \sum_{k=0}^{n} |a_k(B_{n-k} - B)|$$

$$= \sum_{k=0}^{N} |a_k||B_{n-k} - B| + \sum_{k=N+1}^{n} |a_k||B_{n-k} - B|$$

$$\leqslant (\varepsilon/2S) \sum_{k=0}^{N} |a_k| + M \sum_{k=N+1}^{n} |a_k|$$

$$< (\varepsilon/2S) \cdot S + M(\varepsilon/2M) = \varepsilon.$$

Thus $\{C_n\} \to AB$.

By adding the condition that the series $\Sigma\, b_n$ also converges absolutely, we guarantee that the Cauchy product series converges absolutely. The proof is left as Exercise 6 of Problem Set 7.5.

Theorem 7.5.5. *If $\Sigma\, a_n$ converges absolutely to A and $\Sigma\, b_n$ converges to B, then their Cauchy product series converges absolutely to AB.*

Problem Set 7.5

1. Prove Theorem 7.5.1.
2. Prove the following **Abel test**: If $\Sigma\, a_n$ converges and $\{b_n\}$ is a bounded decreasing (increasing) sequence, then $\Sigma\, a_n b_n$ converges. [*Hint*: Assume that $\{b_n\}$ is decreasing and converges to b. Write

$$a_n b_n = a_n(b_n - b) + a_n b$$

and use Theorem 7.5.2.]
3. Prove the **Schwarz** (or **Cauchy**) and **Minkowski inequalities** for series: If $\Sigma\, a_n^2$ and $\Sigma\, b_n^2$ both converge, then so do $\Sigma\, a_n b_n$ and $\Sigma(a_n + b_n)^2$, and

$$\{\Sigma a_n b_n\}^2 \leqslant (\Sigma\, a_n^2)(\Sigma\, b_n^2),$$

$$\{\Sigma(a_n + b_n)^2\}^{\frac{1}{2}} \leqslant (\Sigma\, a_n^2)^{\frac{1}{2}}(\Sigma\, b_n^2)^{\frac{1}{2}}.$$

[Cf. Exercises 10 and 11 of Problems Set 6.5.]

4. Give an example of two divergent series whose Cauchy product series is absolutely convergent. [*Hint*: Let $a_0 = 2$, $a_n = 2^n$ for $n \geqslant 1$, and let $b_0 = -1$, $b_n = 1$ for $n \geqslant 1$.]
5. Prove the commutative law for Cauchy product series: Let $\Sigma\, a_n$, $\Sigma\, b_n$ be two series. If the Cauchy product of $\Sigma\, a_n$ and $\Sigma\, b_n$ is convergent, then it is equal to the Cauchy product of $\Sigma\, b_n$ and $\Sigma\, a_n$.
6. Prove Theorem 7.5.5.
7. Let $\Sigma\, a_n$ converge and $\{n_k\}$ be a subsequence of the sequence of positive integers. For each k, define

$$b_k = a_{n_{k-1}+1} + a_{n_{k-1}+2} + \dots + a_{n_k}.$$

Prove that $\sum_{k=1}^{\infty} b_k$ converges and that $\sum_{k=1}^{\infty} b_k = \sum_{n=1}^{\infty} a_n$.

Chapter Eight

SEQUENCES AND SERIES OF FUNCTIONS

8.1. Convergent Sequences and Series

Sequences and series of functions play an important role in analysis. They form a powerful tool for constructing examples and counter-examples involving functions. In Chapters 3 and 7, sequences and series of constants were discussed. Our discussion of *convergence of sequences and series of functions* will be based on the convergence of sequences and series of constants.

Definition 8.1.1. *Let* $f_n: D \to R$ *for each positive integer n. The sequence* $\{f_n\}$ *is called a* **sequence of functions.** *The set D is called the* **common domain** *of the functions of the sequence* $\{f_n\}$.

We give some examples of sequences of functions.

Example 8.1.1. If $f_n(x) = x^n$ and $g_n(x) = x^n/n$ for each positive integer n, then $\{x^n\}$ and $\{x^n/n\}$ are sequences of functions whose common domain is the set of real numbers R. The sequence $\{1/(1 - x^2)^n\}$ has common domain $R - \{-1, 1\}$.

If $\{f_n\}$ is a sequence of functions with common domain D, then to each $\in D$ there is associated a sequence of constants, $\{f_n(a)\}$. By employing the properties of sequences of constants, we may check the convergence of the sequence $\{f_n(a)\}$, that is, check whether the sequence of functions $\{f_n\}$ *converges* or *diverges* at a. This idea leads to the following definition of *convergence of sequences of functions.*

Definition 8.1.2. *Let* $\{f_n\}$ *be a sequence of functions with commo* *domain D. The sequence* $\{f_n\}$ *is said to* **converge** *at* $a \in D$ *if and onl* *if the sequence of constants* $\{f_n(a)\}$ *converges. If the sequence o functions* $\{f_n\}$ *does not converge at a, then it is said to* **diverge** *at a.*

Example 8.1.2. In Example 8.1.1, the sequence of functions $\{x^n$ converges for each real number x such that $-1 < x \leqslant 1$ and diverge for all other real numbers. The sequence of functions $\{x^n/n\}$ con verges for each real number x with $-1 \leqslant x \leqslant 1$, while it diverge for all other real numbers. Finally, the sequence of function $\{1/(1 - x^2)^n\}$ converges for $x = 0$ and for all real numbers x suc that $|x| > \sqrt{2}$ and diverges for all real numbers x with $0 < |x| <$ or $1 < |x| \leqslant \sqrt{2}$.

From Example 8.1.2 it can be seen that a sequence of functions need no converge at each point of the common domain of the sequence. If a sequenc of functions converges at each point of some subset S of its common domain then the sequence of functions is said to *converge pointwise on S.*

Definition 8.1.3. *Let* $\{f_n\}$ *be a sequence of functions with commo domain D. The sequence* $\{f_n\}$ *is said to* **converge pointwise** *on* $S \subseteq L$ *if and only if the sequence of functions converges at each point* $a \in S$ *If* $\{f_n\}$ *converges pointwise on its common domain D, then the sequenc* $\{f_n\}$ *is said to* **converge pointwise***. If* $\{f_n\}$ *converges pointwise o* $S \subseteq D$, *then define* $f: S \to R$ *by*

$$f(x) = \lim_n f_n(x)$$

for each $x \in S$. *The function f is called the* **limit function** *of* $\{f_n\}$ *on S If* $D = S$, *then f is called the* **limit function** *of the sequence* $\{f_n\}$.

Example 8.1.3. Define $f_n(x) = x^n$ for each $x \in (-1, 1)$. Then, fo each $a \in (-1, 1)$, the sequence $\{a^n\}$ converges, and so the sequenc of functions $\{x^n\}$ converges pointwise on $(-1, 1)$. The limit functio is $f \equiv 0$.

We now define the concept of a *series of functions.*

Definition 8.1.4. *Let* $\{f_n\}$ *be a sequence of functions with commo domain D. The expression*

$$\sum f_n = \sum_{n=1}^{\infty} f_n = f_1 + f_2 + \ldots + f_n + \ldots$$

is called **a series of functions***. The set D is called the* **domain** *of th series.*

A series of functions is the sum of the terms of a sequence of functions. The domain of the series of functions Σf_n is the common domain of the sequence of functions $\{f_n\}$. As in the case of series of constants, we define the *sequence of partial sums*.

Definition 8.1.5. *Let* $\sum_{n=1}^{\infty} f_n$ *be a series of functions with domain D. Let*

$$S_n = \sum_{k=1}^{n} f_k = f_1 + f_2 + \dots + f_n.$$

Then S_n *is called the **nth partial sum** of the series of functions,* Σf_n. *The sequence* $\{S_n\}$ *is called the **sequence of partial sums** of the series of functions.*

Convergence of a series of functions is defined in terms of the convergence of its sequence of partial sums.

Definition 8.1.6. *Let* Σf_n *be a series of functions with domain D and sequence of partial sums* $\{S_n\}$. *The series* Σf_n *is said to* **converge** *at* $a \in D$ *if and only if the sequence of partial sums* $\{S_n\}$ *converges at* a. *If the series* Σf_n *converges at* $a \in D$, *then* $\lim_{n} S_n(a)$ *is called its* **sum** *at* a. *If* Σf_n *does not converge at* $a \in D$, *then it is said to* **diverge** *at* a.

Example 8.1.4. Let $f_n(x) = x^{n-1}$ for each $x \in (-1, 1)$. Then

$$S_n(x) = 1 + x + x^2 + \dots + x^{n-1} = (1 - x^n)/(1 - x)$$

for each $x \in (-1, 1)$. Since $\{S_n(x)\}$ converges to $1/(1 - x)$ for each $x \in (-1, 1)$, we see that the series

$$\sum_{n=1}^{\infty} x^{n-1}$$

converges for each $x \in (-1, 1)$.

In Example 8.1.4, the series converges for each point in the domain of the series of functions and the series of functions is said to *converge pointwise*.

Definition 8.1.7. *Let* Σf_n *be a series of functions with domain D. If the series of functions* Σf_n *converges for each point* $x \in S \subseteq D$, *then the series of functions* Σf_n *is said to* **converge pointwise** *on S. If* $S = D$, *then the series of functions* Σf_n *is said to* **converge pointwise**.

Problem Set 8.1

1. Let $f_n(x) = x^n/n!$ for each real number x and for each positive integer n. Show that the sequence $\{f_n\}$ converges pointwise on the set R of real numbers. Show that the series Σf_n converges pointwise on R.

2. Let $f_n(x) = nx/(1 + n^2x^2)$ for each real number x and for each positive integer n. Show that the sequence $\{f_n\}$ converges pointwise on R. For what values of x does the series Σf_n converge?

3. Let $f_n(x) = 1/(1 + n^2x^2)$ for each real number x and for each positive integer n. Show that the sequence $\{f_n\}$ converges pointwise on R. For what values of x does the series Σf_n converge?

4. Let $f_n(x) = (\sin x)^n$ for each $x \in [0, \pi]$ and for each positive integer n. Show that the sequence $\{f_n\}$ converges pointwise on $[0, \pi]$.

5. Let $\{r_n\}$ be a sequence that contains each rational number in $[0, 1]$ exactly once. Define, for each positive integer n, the function $f_n: [0, 1] \to R$ by

$$f_n(x) = \begin{cases} 0 & \text{if } x \text{ is irrational,} \\ 0 & \text{if } x = r_i \text{ and } i > n, \\ 1 & \text{if } x = r_i \text{ and } i \leqslant n. \end{cases}$$

Show that the sequence $\{f_n\}$ converges pointwise on $[0, 1]$. For what values of x does the series Σf_n converge?

6. Define the functions $f_n: [0, 1] \to R$ by

$$f_n(x) = \begin{cases} 1 & \text{if } 1/n \leqslant x < 1, \\ n^2x & \text{if } 0 \leqslant x < 1/n, \end{cases}$$

for each positive integer n. Show that the sequence $\{f_n\}$ converges pointwise on $[0, 1]$.

7. Let $\{f_n\}$ be a sequence of functions with domain D. The sequence $\{f_n\}$ is said to be **uniformly bounded** on D if and only if there exists a positive number M such that

$$|f_n(x)| \leqslant M$$

for each positive integer n and for each $x \in D$. Give an example of a sequence of continuous functions which is *not* uniformly bounded but converges pointwise to the zero function.

8. For each positive integer n, define the function f_n by the equation

$$f_n(x) = \frac{2^{n-1}}{x^{2^{n-1}} + 1},$$

for each real number $x > 1$. Define the function f by the equation $f(x) = 1/(1 - x)$, for each real number $x > 1$. Show that the series Σf_n converges to f for each real number $x > 1$.

8.2. Uniform Convergence

In Chapter 4 we discussed operations that preserve continuity. If $\{f_n\}$ is a sequence of continuous functions which converges pointwise, then is the limit function continuous? We now present an example to show that the answer to this question is no.

Example 8.2.1. Let $f_n(x) = x^n$ for each $x \in [0, 1]$ and for each positive integer n. Then f_n is continuous for each positive integer n. But the limit function f is given by

$$f(x) = \begin{cases} 0 & \text{for } 0 \leqslant x < 1, \\ 1 & \text{for } x = 1, \end{cases}$$

which is not continuous at 1.

Since convergence of a sequence of continuous functions is not enough to guarantee that the limit function is continuous, is there a stronger type of convergence that will assure that the limit function is continuous? The answer to this question is *yes*. *Uniform convergence* is one such stronger type of convergence.

Definition 8.2.1. *Let $\{f_n\}$ be a sequence of functions with common domain D. Let $S \subseteq D$. The sequence $\{f_n\}$ is said to **converge uniformly** to $f: S \to R$ or **converge uniformly** if and only if to each $\varepsilon > 0$ there is an $N = N(\varepsilon) > 0$ such that $x \in S$ and $n > N$ imply that $|f_n(x) - f(x)| < \varepsilon$.*

We see in the case of uniform convergence that a single $N = N(\varepsilon)$ works for all $x \in S$. If $\{f_n\}$ is a sequence of functions which converges uniformly on S, and if $x_0 \in S$, then the definition of uniform convergence guarantees that, to each $\varepsilon > 0$, there is an N such that $|f(x_0) - f_n(x_0)| < \varepsilon$ for $n > N$. That is, the sequence $\{f_n\}$ converges at x_0 for each $x_0 \in S$. Thus, *if the sequence of functions $\{f_n\}$ with common domain D converges uniformly on $S \subseteq D$, then the sequence $\{f_n\}$ converges pointwise on S.* If the converse were true, then uniform convergence and convergence would be equivalent. Examples will be given of sequences of functions that converge pointwise on their common domain but do not converge uniformly on their common domain.

Before presenting examples of sequences of functions that are convergent but not uniformly convergent, we present an example of a sequence of functions that is uniformly convergent.

Example 8.2.2. For each positive integer n, let $f_n(x) = x^n/n$ for each $x \in [-1, 1]$. We show that the sequence $\{x^n/n\}$ converges uniformly on $[-1, 1]$. From our earlier remarks, the sequence $\{x^n/n\}$ converges pointwise to the zero function on $[-1, 1]$. Let $\varepsilon > 0$ be given. Choose a positive integer N such that $n > N$ implies that $1/n < \varepsilon$. Then, for all $x \in [-1, 1]$, we have

$$|x^n/n - 0| = |x^n/n| \leqslant 1/n < \varepsilon$$

when $n > N$. Thus, the sequence $\{x^n/n\}$ converges uniformly to the zero function on $[-1, 1]$.

One method for showing that a sequence of functions does not converge uniformly on a set is called *negation of uniform convergence.*

Negation of uniform convergence. *Let $\{f_n\}$ be a sequence of functions with common domain D, and let $S \subseteq D$. The sequence $\{f_n\}$ fails to converge uniformly on S to a function f with domain S if and only if there is a positive number ε having the property that for any number N there is a positive integer $n > N$ and an $x \in S$ such that*

$$|f_n(x) - f(x)| \geqslant \varepsilon.$$

The proof is left as Exercise 10 of Problem Set 8.2.

The following two examples illustrate the difference between convergence and uniform convergence.

Example 8.2.3. For each positive integer n, let $f_n(x) = x^n$, for each $x \in [0, 1]$. As pointed out at the beginning of this section, the sequence $\{x^n\}$ converges pointwise to the function f defined by

$$f(x) = \begin{cases} 0 & \text{if } 0 \leqslant x < 1, \\ 1 & \text{if } x = 1. \end{cases}$$

Let $\varepsilon = 1/2$ and, for each positive integer n, let $x_n = 1/2^{1/n}$. Then

$$|f(x_n) - f_n(x_n)| = |0 - (1/2^{1/n})^n| = 1/2 = \varepsilon.$$

Thus, the sequence of functions $\{x^n\}$ does *not* converge uniformly on $[0, 1]$.

Example 8.2.3 displays a sequence of continuous functions whose limit function is *not* continuous and whose convergence is *not* uniform. The next example is of a sequence of continuous functions whose convergence is not uniform but whose limit function is continuous.

Example 8.2.4. For each positive integer n, let

$$f_n(x) = \begin{cases} nx & \text{for } 0 \leqslant x \leqslant 1/n, \\ -nx + 2 & \text{for } 1/n < x \leqslant 2/n, \\ 0 & \text{for } 2/n < x \leqslant 1. \end{cases}$$

Since $f_n(0) = 0$, we see that the sequence $\{f_n\}$ converges to 0 at $x = 0$. If $0 < x \leqslant 1$, then there is an N such that $2/N < x$. Hence $f_n(x) = 0$ for $n > N$ and the sequence $\{f_n\}$ converges to 0 at x. Thus, the sequence $\{f_n\}$ converges pointwise on $[0, 1]$ to the zero function. Let $\varepsilon = 1$ and $x_n = 1/n$ for each positive integer n. Then

$$|f(x_n) - f_n(x_n)| = |0 - 1| = 1 = \varepsilon$$

and the sequence of functions does *not* converge uniformly on $[0, 1]$.

Our first theorem of this chapter presents the *Cauchy Condition for Uniform Convergence of a Sequence of Functions.*

Theorem 8.2.1. (*Cauchy Criterion for Sequences of Functions*) *Let* $\{f_n\}$ *be a sequence of functions with common domain D. Then the sequence* $\{f_n\}$ *converges uniformly on* $S \subseteq D$ *if and only if to each* $\varepsilon > 0$ *there is an* $N = N(\varepsilon) > 0$ *such that* $n > m > N$ *implies that*

$$|f_n(x) - f_m(x)| < \varepsilon$$

for every $x \in S$.

The proof is left as Exercise 11 of Problem Set 8.2.

We now define *uniform convergence of series of functions* and present two results about series of functions.

Definition 8.2.2. *Let* Σf_n *be a series of functions with domain D. The series* Σf_n *is said to* **converge uniformly** *on* $S \subseteq D$ *if and only if its sequence of partial sums* $\{S_n\}$ *converges uniformly on S.*

The *Cauchy Condition for Series of Functions* now follows from Theorem 8.2.1.

Corollary 8.2.1. (*Cauchy Criterion for Series of Functions*) *The series of functions* Σf_n *with domain D is uniformly convergent on* $S \subseteq D$ *if and only if to each* $\varepsilon > 0$ *there is an* $N = N(\varepsilon) > 0$ *such that* $n > m > N$ *implies that*

$$|f_{m+1}(x) + \ldots + f_n(x)| < \varepsilon$$

for every $x \in S$.

Example 8.2.5. We consider the series of functions $\sum_{n=0}^{\infty} x^n/n!$ with domain the set of real numbers R. For each real x_0,

$$\lim_n \left[|\{x_0^{n+1}/(n+1)!\}/\{x_0^n/n!\}|\right] = \lim_n \left[|x_0|/(n+1)\right] = 0.$$

Thus, Theorem 7.2.2, the Ratio Test, implies that the series $\sum_{n=0}^{\infty} x^n/n!$ converges at x_0, that is, the series $\sum_{n=0}^{\infty} x^n/n!$ converges pointwise on R.

We show that the convergence of the series $\sum_{n=0}^{\infty} x^n/n!$ is not uniform on R. Let $\varepsilon = 1$ and $x_n = n$ for each integer $n > 1$. Then

$$(x_n)^n/n! = n^n/n! > 1,$$

and Corollary 8.2.1 implies that the convergence is *not* uniform on R.

Let M be any positive real number, and let $S = \{x : |x| \leqslant M\}$. We show that the series $\sum_{n=0}^{\infty} x^n/n!$ converges uniformly on S. Select $\varepsilon > 0$. Since the series $\sum_{n=0}^{\infty} x^n/n!$ converges at M, Theorem 7.1.1 implies that there is an N such that

$$M^{m+1}/(m+1)! + \ldots + M^n/n!$$
$$= |M^{m+1}/(m+1)! + \ldots + M^n/n!| < \varepsilon$$

whenever $n > m > N$. Then, for any $x \in S$, we have $|x| \leqslant M$ and

$$|x^{m+1}/(m+1)! + \ldots + x^n/n!|$$
$$\leqslant M^{m+1}/(m+1)! + \ldots + M^n/n! < \varepsilon$$

whenever $n > m > N$. Thus Corollary 8.2.1 implies that $\sum_{n=0}^{\infty} x^n/n!$ converges uniformly on S.

We conclude this section with a test for uniform convergence for series of functions, called the *Weierstrass M-Test*.

Theorem 8.2.2. (*Weierstrass M-Test*) *Let* $\{f_n\}$ *be a series of functions with common domain D, and let $S \subseteq D$. Let $\{M_n\}$ be a sequence of positive real numbers such that*

$$|f_n(x)| \leqslant M_n$$

for every $x \in S$ and each positive integer n. If ΣM_n converges, then Σf_n converges uniformly on S.

Proof. Let $\varepsilon > 0$. Since ΣM_n converges, Theorem 7.1.1 implies that there exists an N such that

$$M_{m+1} + \ldots + M_n = |M_{m+1} + \ldots + M_n| < \varepsilon$$

when $n > m > N$. Then

$$|f_{m+1}(x) + \dots + f_n(x)| \leqslant M_{m+1} + \dots + M_n < \varepsilon$$

for $n > m > N$. Thus Corollary 8.2.1 implies that the series Σf_n converges uniformly on S.

Problem Set 8.2

1. Discuss the uniform convergence of the sequence of functions defined in Exercise 2, Problem Set 8.1.
2. Discuss the uniform convergence of the sequence of functions defined in Exercise 3, Problem Set 8.1.
3. Discuss the uniform convergence of the sequence of functions defined in Exercise 4, Problem Set 8.1.
4. Discuss the uniform convergence of the sequence of functions defined in Exercise 5, Problem Set 8.1.
5. Discuss the uniform convergence of the sequence of functions defined in Exercise 6, Problem Set 8.1.
6. Discuss the uniform convergence of the type of sequence of functions defined in Exercise 7, Problem Set 8.1.
7. Let $\{f_n\}$ and $\{g_n\}$ be sequences of functions that converge uniformly on the set S. What can be said about the uniform convergence of the sequence $\{f_n + g_n\}$ on S?
8. Discuss the convergence and uniform convergence of the sequence of functions $\{e^{-nx^2}\}$ with common domain R.
9. Show that the sequence of functions $\{x/n\}$ with common domain R converges uniformly on any bounded subset of R, but does *not* converge uniformly on R.
10. Prove the Negation of Uniform Convergence.
11. Prove Theorem 8.2.1. [*Hint*: By Theorem 3.4.3, for $x \in S$, the sequence $\{f_n(x)\}$ converges; let

$$f(x) = \lim_n f_n(x),$$

for each $x \in S$. For $\varepsilon > 0$ there is an N such that

$$|f_n(x) - f_m(x)| < \varepsilon/2$$

for $n > m > N$ and all $x \in S$. Let $y \in S$, and let n be a fixed integer greater than N. There is an N_1 such that

$$|f_m(y) - f(y)| < \varepsilon/2$$

for $m > N_1$. For $m > \max(N, N_1)$,

$$|f_n(y) - f(y)| \leqslant |f_n(y) - f_m(y)| + |f_m(y) - f(y)|$$
$$< \varepsilon/2 + \varepsilon/2 = \varepsilon.]$$

12. Discuss the convergence and uniform convergence of the series of functions
$\sum\limits_{n=1}^{\infty} 1/n^x$ with domain $D = \{x: 1 < x < +\infty\}$.

13. Let Σf_n be a series of functions with domain D, and let g be a function which is bounded on D. If the series Σf_n converges uniformly on D, then show that the series $\Sigma g f_n$ converges uniformly on D.

14. Let $\{f_n\}$ be a sequence of continuous functions with common domain D. Let C be a dense subset of D. If $\{f_n\}$ converges uniformly on C, then show that $\{f_n\}$ converges uniformly on D.

15. Let $f: R \to R$ and f be uniformly continuous on R. For each positive integer n and each real number x, define the function f_n by

$$f_n(x) = f(x + 1/n).$$

Show that the sequence $\{f_n\}$ converges uniformly to f on R.

16. Let $\{f_n\}$ be a sequence of functions that converges uniformly on each of the sets S_i, where $i = 1, 2, ..., k$. Show that $\{f_n\}$ converges uniformly on $\bigcup\limits_{i=1}^{k} S_i$.

17. Let Σf_n and Σg_n be two series of functions with domain D, and let $|f_n(x)| \leqslant g_n(x)$ for each positive integer n and each $x \in S \subseteq D$. If Σg_n converges uniformly on S, then show that Σf_n converges uniformly on S.

18. Let $\{f_n\}$ be a sequence of functions with common domain D. If the sequence $\{f_n\}$ converges uniformly on $S \subseteq D$, then show that any subsequence of $\{f_n\}$ also converges uniformly on S.

19. Give an example to show that the converse of Exercise 18 is not true; that is, give a sequence of functions which does not converge uniformly on a set S but which has a subsequence that does converge uniformly on the set S.

20. Let $\{f_n\}$ be a sequence of functions with common domain D which converges uniformly on $S \subseteq D$. Show that the sequence $\{f_n\}$ converges uniformly on any subset $T \subseteq S$.

21. Give an example to show that the converse of Exercise 20 is not true; that is, give a sequence of functions whose common domain contains the set S and converges uniformly on $T \subset S$, but which does not converge uniformly on S.

22. Let $\{f_n\}$ be a sequence of functions with common domain $[a, b]$. If $\{f_n\}$ converges for a and b and $\{f_n\}$ converges uniformly on (a, b), then show that $\{f_n\}$ converges uniformly on $[a, b]$.

23. Give an example of two sequences of functions $\{f_n\}$ and $\{g_n\}$ such that the sequences $\{f_n\}$ and $\{g_n\}$ both converge uniformly on a set S, but the sequence of functions $\{f_n g_n\}$ does not converge uniformly on S.

24. Let $\{f_n\}$ and $\{g_n\}$ be two sequences of functions with common domain D. If the sequences $\{f_n\}$ and $\{g_n\}$ are both uniformly convergent on $S \subseteq D$ and are both uniformly bounded on S (see Exercise 7 of Problem Set 8.1), then show that $\{f_n g_n\}$ is uniformly convergent on S. What can be said if only one of the sequences is uniformly bounded on S?

25. Let $\{f_n\}$ be a sequence of functions with common domain D. Let $f_n(x) \neq 0$ for each positive integer n and for each $x \in S \subseteq D$. If there are positive numbers N and $r < 1$ such that

$$\left|\frac{f_{n+1}(x)}{f_n(x)}\right| < r$$

for each $n > N$ and $x \in S$, then show that the series Σf_n converges uniformly on S.

26. Let $\{f_n\}$ be a sequence of functions with common domain D. If the sequence $\{f_n\}$ converges uniformly on $S \subseteq D$, and if for each positive integer n the function f_n is bounded on S, then show that the function

$$f(x) = \lim_n f_n(x),$$

for each $x \in S$, is bounded on S.

27. Give an example of a sequence of continuous functions $\{f_n\}$ with common domain R which converges pointwise on R, but which does not converge uniformly on any interval of R.

28. Let D be any infinite subset of R. Give an example of a sequence of functions $\{f_n\}$ with common domain D which converges pointwise on D, but which does not converge uniformly on any infinite subset of D.

8.3. Continuous Functions

It was pointed out in Section 8.2 that there are pointwise convergent sequences of continuous functions whose limit functions are not continuous. That is, convergence is not strong enough to preserve continuity. Theorem 8.3.1 points out that uniform convergence is a strong enough type of convergence to preserve continuity.

Theorem 8.3.1. *Let $\{f_n\}$ be a sequence of continuous functions with common domain D, and let $S \subseteq D$. If the sequence $\{f_n\}$ converges uniformly on S, then the limit function f (where $f(x) = \lim_n f_n(x)$ for each $x \in S$) is continuous on S.*

Proof. Let $\varepsilon > 0$ and $x_0 \in S$. Since $\{f_n\}$ converges uniformly on S, there is an N such that $x \in S$ and $n > N$ imply that

$$|f(x) - f_n(x)| < \varepsilon/3.$$

Let $n_0 > N$ be a fixed positive integer. Since f_{n_0} is continuous on S, there is a $\delta > 0$ such that $x \in S$ and $|x - x_0| < \delta$ imply that

$$|f_{n_0}(x) - f_{n_0}(x_0)| < \varepsilon/3.$$

Assume that $x \in S$ and that $|x - x_0| < \delta$, then

$$
\begin{aligned}
|f(x) - f(x_0)| &= |f(x) - f_{n_0}(x) + f_{n_0}(x) - f_{n_0}(x_0) \\
&\quad + f_{n_0}(x_0) - f(x_0)| \\
&\leqslant |f(x) - f_{n_0}(x)| + |f_{n_0}(x) - f_{n_0}(x_0)| \\
&\quad + |f_{n_0}(x_0) - f(x_0)| \\
&< \varepsilon/3 + \varepsilon/3 + \varepsilon/3 = \varepsilon.
\end{aligned}
$$

Hence $x \in S$ and $|x - x_0| < \delta$ imply that

$$|f(x) - f(x_0)| < \varepsilon,$$

and so f is continuous at $x_0 \in S$. Since x_0 was any point of S, f is continuous on S.

We now state the analogous result for series.

Corollary 8.3.1. *Let Σf_n be a series of continuous functions with domain D, and let $S \subseteq D$. If the series Σf_n converges uniformly on S, then the limit function is continuous on S.*

Another property that is preserved by uniform convergence is the property of being bounded, that is, if $\{f_n\}$ is a sequence of functions, each being bounded, and the sequence $\{f_n\}$ converges uniformly, then the limit function is bounded.

Theorem 8.3.2. *Let $\{f_n\}$ be a sequence of functions with common domain D and let $S \subseteq D$. If for each n the function f_n is bounded on S and the sequence $\{f_n\}$ converges uniformly to f on S, then f is bounded on S.*

Proof. Since f_n is bounded on S for each n, there is an M_n such that $|f_n(x)| \leqslant M_n$ for each $x \in S$. Let $\varepsilon = 1$. Since $\{f_n\}$ converges uniformly to f on S, there is an N such that $x \in S$ and $n > N$ imply that

$$|f(x) - f_n(x)| < \varepsilon = 1.$$

Let n_0 be a fixed integer such that $n_0 > N$. Then

$$-1 < f(x) - f_{n_0}(x) < 1$$

or

$$-1 + f_{n_0}(x) < f(x) < 1 + f_{n_0}(x).$$

Since $|f_{n_0}(x)| \leqslant M_{n_0}$ for every $x \in S$, we have

$$
\begin{aligned}
-(1 + M_{n_0}) \leqslant -1 + f_{n_0}(x) &< f(x) \\
&< 1 + f_{n_0}(x) \leqslant 1 + M_{n_0}
\end{aligned}
$$

or

$$|f(x)| \leqslant 1 + M_{n_0}.$$

Thus, f is bounded on S.

Theorem 8.3.1 assures us that uniform convergence of a sequence of continuous functions preserves continuity. But Examples 8.2.4 and 8.2.5 imply that uniform convergence is not a necessary condition for continuity to be preserved. Theorem 8.3.3 will point out that, under certain conditions, convergence implies uniform convergence.

Definition 8.3.1. *Let* $\{f_n\}$ *be a sequence of functions with common domain* D. *The sequence* $\{f_n\}$ *is said to be **decreasing** on* $S \subseteq D$ *if and only if*

$$f_{n+1}(x) \leqslant f_n(x)$$

for every positive integer n *and every* $x \in S$. *The sequence* $\{f_n\}$ *is said to be **increasing** on* $S \subseteq D$ *if and only if*

$$f_{n+1}(x) \geqslant f_n(x)$$

for every positive integer n *and every* $x \in S$.

Theorem 8.3.3. *Let* $\{f_n\}$ *be a sequence of continuous functions with common domain* D. *If the sequence* $\{f_n\}$ *is increasing (or decreasing) on the compact set* $S \subseteq D$ *and* $\{f_n\}$ *converges pointwise to a continuous function* $f: S \to R$, *then* $\{f_n\}$ *converges uniformly to* f *on* S.

Proof. We prove the theorem for an increasing sequence of functions; the case where the sequence is decreasing is handled in a similar manner.

Let $g_n = f - f_n$ for each positive integer n. The sequence $\{g_n\}$ is a decreasing sequence of continuous functions which converge pointwise to the zero function. We need to show that the sequence $\{g_n\}$ converges uniformly to the zero function.

Let $\varepsilon > 0$. For each $x \in S$ there is an $N(x)$ such that $g_{N(x)}(x) < \varepsilon/2$. Since the function $g_{N(x)}$ is continuous on S, there is a $\delta_x > 0$ such that, for $y \in S$ and $|y - x| < \delta_x$, we have

$$-\varepsilon/2 < g_{N(x)}(y) - g_{N(x)}(x) < \varepsilon/2$$

or

$$0 \leqslant g_{N(x)}(y) < g_{N(x)}(x) + \varepsilon/2 < \varepsilon.$$

Let $I_x = \{y : |y - x| < \delta_x\}$. Then the family of open sets

$$I = \{I_x : x \in S\}$$

is an open cover of the compact set S. Theorem 3.5.7 implies that there is a finite subset of I which covers S, say, the sets $I_{x_1}, I_{x_2}, \ldots, I_{x_k}$. Let

$$N = \max [N(x_1), N(x_2), \ldots, N(x_k)].$$

Since $\{g_n\}$ is a decreasing sequence on S, it follows that $g_N(x) < \varepsilon$ for each $x \in S$. Also, since $\{g_n\}$ is a decreasing sequence on S, $g_n(x) \leqslant g_N(x) < \varepsilon$ for each $n > N$ and each $x \in S$. Since $g_n(x) \geqslant 0$ for each positive integer n and each $x \in S$, the sequence $\{g_n\}$ converges uniformly to the zero function on S.

Problem Set 8.3

1. Let $\{f_n\}$ be a sequence of functions with common domain D which converges pointwise to the function $f: S \to R$ on $S \subseteq D$. The sequence $\{f_n\}$ is said to **converge uniformly** at $a \in S$ if and only if to each $\varepsilon > 0$ there are real positive numbers N and δ such that $n > N$ and $x \in S$ with $|x - a| < \delta$ imply that

$$|f_n(x) - f(x)| < \varepsilon.$$

(a) If $\{f_n\}$ converges uniformly on S, then show that $\{f_n\}$ converges uniformly at a for each $a \in S$.

(b) If S is compact and $\{f_n\}$ converges uniformly at a for each $a \in S$, then show that $\{f_n\}$ converges uniformly on S.

2. Let $\{f_n\}$ be a sequence of functions with common domain D which converges pointwise to the function $f: S \to R$ on $S \subseteq D$. Let $a \in S$, and let infinitely many of the terms of the sequence $\{f_n\}$ be continuous at a. If $\{f_n\}$ converges uniformly at a, then show that the limit function f is continuous at a.

3. Let $\{f_n\}$ be a sequence of functions with common domain D which converges pointwise to the function $f: S \to R$ on $S \subseteq D$. Show that the sequence $\{f_n\}$ converges uniformly at $a \in S$ if and only if to each $\varepsilon > 0$ there are real numbers N and δ such that $n > m > N$ and $x \in S$ with $|x - a| < \delta$ imply that

$$|f_n(x) - f_m(x)| < \varepsilon.$$

4. Let

$$f_n(x) = \begin{cases} 1 - nx & \text{if } 0 \leqslant x \leqslant 1/n, \\ 0 & \text{if } 1/n < x \leqslant 1, \end{cases}$$

for each positive integer n. Show that the sequence $\{f_n\}$ converges pointwise on $[0, 1]$ but does not converge uniformly on $[0, 1]$. Show that the sequence $\{f_n\}$ converges uniformly on each closed interval $[\varepsilon, 1]$, where $0 < \varepsilon < 1$.

5. Let $\{f_n\}$ be a sequence of lower-semicontinuous functions with common domain D. If $\{f_n\}$ converges uniformly on $S \subseteq D$, then show that the limit function $f: S \to D$ is lower-semicontinuous on S.

6. A series

$$\sum_{n=0}^{\infty} a_n x^n$$

is called a **power series**. Show that if a power series converges at $a \neq 0$ and if $0 < b < |a|$, then the power series converges uniformly on the interval $[-b, b]$.

7. Let $\{f_n\}$ be a sequence of continuous functions with common domain $[a, b]$ which converges pointwise on $[a, b]$. Show there exists at least one point $c \in (a, b)$ such that $\{f_n\}$ converges uniformly at c.

8. Construct a sequence of functions $\{f_n\}$ with common domain R such that the function f_n is discontinuous for each n, but the sequence $\{f_n\}$ converges uniformly to a continuous function on R.

9. Let Σf_n be a series of functions with domain D whose sequence of partial sums $\{s_n\}$ is uniformly bounded on $S \subseteq D$. Let $\{g_n\}$ be a sequence of nonnegative functions with common domain D such that the sequence $\{g_n\}$ converges uniformly to the zero function on S. Show that the series of functions $\Sigma f_n g_n$ converges uniformly on S. [This result is called the **Abel test for uniform convergence**.] (Cf. Theorem 7.5.3.)

10. Let $\{f_n\}$ be a sequence of nonnegative functions with common domain D such that the sequence $\{f_n\}$ converges uniformly to the zero function on $S \subseteq D$. If the sequence $\{f_n\}$ is decreasing on S, then show that the series of functions $\Sigma (-1)^n f_n$ is uniformly convergent on S. [This result is called the **alternating test for uniform convergence**.] (Cf. Exercise 9.)

11. Show that the series of functions $\Sigma (-1)^n/(n + x)$ converges uniformly for all real $x \geq 0$. (Cf. Exercise 10.) Show that this series of functions does not converge absolutely for any real x.

8.4. Equicontinuous Sequences of Functions

The sequence of functions $\{f_n\}$ with common domain D is said to be **uniformly bounded** on $S \subseteq D$ if and only if there is an M such that, for each positive integer n and each $x \in S$, $|f_n(x)| \leq M$. The sequence $\{x^n\}$ is uniformly bounded on the set $[0, 1]$, but neither the sequence $\{x^n\}$ nor any subsequence of $\{x^n\}$ converge uniformly on $[0, 1]$.

It is often useful in analysis to know under what conditions a sequence of functions that converges pointwise on a set S has a subsequence that converges uniformly on S. We now give such a condition.

Definition 8.4.1. *Let $\{f_n\}$ be a sequence of functions with common domain D. The sequence $\{f_n\}$ is said to be **equicontinuous** on $S \subseteq D$ if and only if to each $\varepsilon > 0$ there is a $\delta = \delta(\varepsilon) > 0$ such that $x, y \in S$ and $|x - y| < \delta$ imply that*

$$|f_n(x) - f_n(y)| < \varepsilon$$

for each positive integer n.

From the definition of an equicontinuous sequence of functions $\{f_n\}$ it follows that the function f_n is uniformly continuous on S, for each positive integer n. Moreover, the uniform continuity must be very uniform in the sense that, given an $\varepsilon > 0$, one $\delta > 0$ must hold for all positive integers n. Hence we see that the condition of equicontinuity is a very strong property. We now give an example of an equicontinuous sequence of functions.

Example 8.4.1. For each positive integer n, let f_n be the function defined by

$$f_n(x) = x^n/n$$

for each $x \in [-1, 1]$. Let $\varepsilon > 0$ and let $\delta = \varepsilon$. Then, for any positive integer n and $x, y \in [-1, 1]$ with $|x - y| < \delta$, we have

$$|x^n/n - y^n/n| = |x - y||x^{n-1} + x^{n-2}y + \ldots + y^{n-2} + y^{n-1}|/n$$
$$\leqslant |x - y| < \delta = \varepsilon.$$

Thus the sequence $\{x^n/n\}$ is equicontinuous on $[-1, 1]$.

In Example 8.2.2 it was proved that the sequence of functions $\{x^n/n\}$ with common domain $[-1, 1]$ converges uniformly on $[-1, 1]$. It is easily seen that this sequence of functions is uniformly bounded, and now we see that this sequence is equicontinuous. One might ask if there is any relationship between these three concepts. We will show that if a sequence of continuous functions converges uniformly on a compact set, then the sequence is uniformly bounded and equicontinuous. As a partial converse, it will be proved that every uniformly bounded equicontinuous sequence of functions on a compact set has a uniformly convergent subsequence.

Theorem 8.4.1. *Let $\{f_n\}$ be a sequence of continuous functions with common domain D which converge uniformly on the compact set $S \subseteq D$. Then the sequence $\{f_n\}$ is uniformly bounded and equicontinuous on S.*

Proof. We first show that the sequence $\{f_n\}$ is uniformly bounded on S. For each positive integer n, the function f_n is continuous on the compact set S and Corollary 4.5.1 implies that f_n is bounded on S, say, by M_n. That is, for each positive integer n, $|f_n(x)| \leqslant M_n$ for each $x \in S$. Let $\varepsilon = 1$. Since the sequence $\{f_n\}$ converges uniformly on S, Theorem 8.2.1 implies that there exists an N such that for $n > N$ and $x \in S$ we have

$$|f_n(x) - f_N(x)| < 1.$$

We can rewrite this inequality as

$$-1 + f_N(x) < f_n(x) < 1 + f_N(x),$$

and so
$$-(1 + |f_N(x)|) < f_n(x) < 1 + |f_N(x)|,$$
that is,
$$|f_n(x)| < 1 + |f_N(x)|.$$

Define $M = \max(M_1, M_2, ..., M_N) + 1$. Then for each positive integer n and for each $x \in S$ we have

$$|f_n(x)| \leqslant M.$$

Thus the sequence $\{f_n\}$ is uniformly bounded on S.

We now show that the sequence $\{f_n\}$ is equicontinuous on S. Let $\varepsilon > 0$ be given. Since the sequence $\{f_n\}$ converges uniformly on S, Theorem 8.2.1 implies that there is an N such that, for $n > N$ and $x \in S$,
$$|f_n(x) - f_N(x)| < \varepsilon/3.$$

Since each function f_n, for $n = 1, 2, ...$, is continuous on the compact set S, it follows from Theorem 4.6.3 that f_n is uniformly continuous on S, for $n = 1, 2, ... $. Hence, for each positive integer n, there is a $\delta_n > 0$ such that $x, y \in S$ and $|x - y| < \delta_n$ imply that

$$|f_n(x) - f_n(y)| < \varepsilon/3.$$

Let $\delta = \min(\delta_1, \delta_2, ..., \delta_N)$. Since $\delta_k > 0$ for $k = 1, 2, ..., N$, it follows that $\delta > 0$.

If n is a positive integer such that $n \leqslant N$, then $\delta_n \geqslant \delta$ and, for $x, y \in S$ with $|x - y| < \delta$, we have

$$|f_n(x) - f_n(y)| < \varepsilon/3 < \varepsilon.$$

For $n > N$, let $x, y \in S$ and $|x - y| < \delta$, then

$$
\begin{aligned}
|f_n(x) - f_n(y)| \\
= |f_n(x) - f_N(x) + f_N(x) - f_N(y) + f_N(y) - f_n(y)| \\
\leqslant |f_n(x) - f_N(x)| + |f_N(x) - f_N(y)| + |f_N(y) - f_n(y)| \\
< \varepsilon/3 + \varepsilon/3 + \varepsilon/3 = \varepsilon.
\end{aligned}
$$

Thus, the sequence $\{f_n\}$ is equicontinuous on the set S.

This theorem permits us easily to construct equicontinuous sequences of functions. We need only look for sequences of continuous functions that converge uniformly on a compact set.

Theorem 8.4.2. *Let $\{f_n\}$ be a sequence of functions with common domain D. If the sequence $\{f_n\}$ is uniformly bounded and equicontinuous* *on the compact subset S of D, then there exists a subsequence of $\{f_n\}$ which converges uniformly on S.*

Proof. Since $\{f_n\}$ is uniformly bounded on S, there is an M such that $|f_n(x)| \leqslant M$ for each positive integer n and for each $x \in S$. Since S is a compact set, there exists a countable set $A = \{a_j : j = 1, 2, ...\}$ of points of S which is dense in S. Consider the sequence of constants $\{f_n(a_1)\}$. This is a bounded sequence, and it follows from Theorem 3.4.4 that the sequence $\{f_n(a_1)\}$ has a convergent subsequence. We denote this subsequence of $\{f_n(a_1)\}$ by $\{f_{1,n}(a_1)\}$. Then the sequence of functions $\{f_{1,n}\}$ is a subsequence of the sequence of functions $\{f_n\}$, and $\{f_{1,n}\}$ converges at a_1. Consider the sequence of constants $\{f_{1,n}(a_2)\}$. This is a bounded sequence, and Theorem 3.4.4 implies that the sequence $\{f_{1,n}(a_2)\}$ has a convergent subsequence. This subsequence of $\{f_{1,n}(a_2)\}$ is denoted by $\{f_{2,n}(a_2)\}$. Then the sequence of functions $\{f_{2,n}\}$ is a subsequence of the sequence of functions $\{f_{1,n}\}$ (as well as a subsequence of the sequence of functions $\{f_n\}$) and $\{f_{2,n}\}$ converges at a_1 and a_2. Consider the sequence of constants $\{f_{2,n}(a_3)\}$. This is a bounded sequence and it follows from Theorem 3.4.4 that the sequence $\{f_{2,n}(a_3)\}$ has a convergent subsequence. This subsequence of $\{f_{2,n}(a_3)\}$ is denoted by $\{f_{3,n}(a_3)\}$. Then the sequence of functions $\{f_{3,n}\}$ is a subsequence of the sequence of functions $\{f_{2,n}\}$ (as well as a subsequence of the sequences of functions $\{f_{1,n}\}$ and $\{f_n\}$) and $\{f_{3,n}\}$ converges at a_1, a_2, and a_3. By induction, we obtain a sequence of sequences of functions $\{\{f_{m,n}\}\}$, which we can represent by the following array,

$$f_{1,1}, f_{1,2}, ..., f_{1,n}, \cdots$$

$$f_{2,1}, f_{2,2}, ..., f_{2,n}, \cdots$$

$$\cdots$$

$$f_{m,1}, f_{m,2}, ..., f_{m,n}, \cdots$$

$$\cdots,$$

with the following properties:

(1) each sequence of functions $\{f_{m,n}\}$ is a subsequence of the preceeding one, $\{f_{m-1,n}\}$, and a subsequence of $\{f_n\}$;

(2) each sequence of functions $\{f_{m,n}\}$ converges at a_j, where $j = 1, 2, ..., m$.

We need a subsequence of the sequence of functions $\{f_n\}$ which converges at each point of A. We construct such a sequence by taking the *diagonal sequence* $\{f_{m,m}\}$.

By the construction of the sequence of functions $\{f_{m,m}\}$, it is a subsequence of the sequence of functions $\{f_n\}$. We show that $\{f_{m,m}\}$ converges at each point of A. Let $a_k \in A$. Then the sequence

$$f_{k,k}, f_{k+1,k+1}, \cdots$$

is a subsequence of the sequence $\{f_{k,n}\}$, and so the sequence $\{f_{m,m}\}$ converges at a_k. That is, the sequence of functions $\{f_{m,m}\}$ converges at each point of A.

For convenience of notation, let $g_n = f_{n,n}$ for each positive integer n. Then $\{g_n\}$ is a subsequence of the sequence of functions $\{f_n\}$, and $\{g_n\}$ converges at each point of A. The construction of this subsequence $\{g_n\}$ of $\{f_n\}$ required only the property that the sequence $\{f_n\}$ be uniformly bounded. Since $\{g_n\}$ is a subsequence of the sequence of functions $\{f_n\}$ which is equicontinuous on S, the sequence $\{g_n\}$ is equicontinuous on S. (Cf. Exercise 1 of Problem Set 8.4.)

We now prove that the sequence $\{g_n\}$ converges uniformly on S. Let $\varepsilon > 0$. Since the sequence $\{g_n\}$ is equicontinuous on S, there is a $\delta > 0$ such that $x, y \in S$ and $|x - y| < \delta$ imply that

$$|g_n(x) - g_n(y)| < \varepsilon/3$$

for each positive integer n. Since the set S is compact, there exist points $b_1, b_2, ..., b_k$ of A such that the sets $I_{b_j} = \{x : |x - b_j| < \delta\}$, $j = 1, 2, ..., k$, cover S (that is, $S \subseteq \bigcup_{j=1}^{k} I_{b_j}$). (Cf. Exercise 27 of Problem Set 3.5) Since the sequence of functions $\{g_n\}$ converges at each point of A, for each $j = 1, 2, ..., k$ there is an N_j such that $n > m > N_j$ implies that

$$|g_n(b_j) - g_m(b_j)| < \varepsilon/3.$$

Let $N = \max(N_1, N_2, ..., N_k)$ and let $x \in S$. Then there is a j, where $1 \leqslant j \leqslant k$, such that

$$|x - b_j| < \delta.$$

Thus, for $n > m > N$,

$$|g_n(x) - g_m(x)|$$
$$= |g_n(x) - g_n(b_j) + g_n(b_j) - g_m(b_j) + g_m(b_j) - g_m(x)|$$
$$\leqslant |g_n(x) - g_n(b_j)| + |g_n(b_j) - g_m(b_j)| + |g_m(b_j) - g_m(x)|$$
$$< \varepsilon/3 + \varepsilon/3 + \varepsilon/3 = \varepsilon.$$

Hence, $n > m > N$ and $x \in S$ imply that

$$|g_n(x) - g_m(x)| < \varepsilon,$$

and the sequence $\{g_n\}$ converges uniformly on S.

Problem Set 8.4

1. Let $\{f_n\}$ be a sequence of functions which are equicontinuous on a set S. Show that every subsequence of $\{f_n\}$ is equicontinuous on S.
2. Is the sequence of functions $\{\sin nx\}$ uniformly bounded on the interval $[0, 1]$?
3. Is the sequence of functions $\{\sin nx\}$ equicontinuous on the interval $[0, 1]$?
4. Is the sequence of functions $\{(\sin nx)/x\}$ uniformly bounded on $(0, 1)$.
5. Let $\{f_n\}$ be an increasing sequence of continuous functions with common domain D. If the sequence $\{f_n(x)\}$ is uniformly bounded on $S \subseteq D$, then show that the limit function $f: S \to R$ is lower semicontinuous on S.
6. Give an example of a sequence of continuous functions with common domain R which converges pointwise on R, but whose limit function is neither upper semicontinuous nor lower semicontinuous.
7. Let $\{f_n\}$ be a sequence of functions with common domain D. If $\{f_n\}$ is uniformly bounded and increasing on the compact set S, then show that there is a subsequence of $\{f_n\}$ which converges uniformly on S.
8. Let $\{f_n\}$ be a sequence of increasing functions with common domain D, and let S be a compact subset of D. Let $f: S \to R$ be a continuous function on S, and let $\{f_n\}$ converge to f on a dense subset of S. Show that $\{f_n\}$ converges pointwise to f on S and that the convergence is uniform on S.

8.5. Integration and Differentiation of Sequences and Series

Let $\{f_n\}$ be a sequence of functions with common domain D, and assume that the sequence $\{f_n\}$ converges pointwise to f on D, that is,

$$f(x) = \lim_n f_n(x)$$

for each $x \in D$. If $\zeta \in D$, is it true that

$$\lim_{x \to \zeta} f(x) \equiv \lim_{x \to \zeta} \left[\lim_n f_n(x) \right] = \lim_n \left[\lim_{x \to \zeta} f_n(x) \right]?$$

That is, can the order of taking limits be reversed? If each of the functions f_n is continuous at ζ and the limit function f is continuous at ζ, then we see that the answer is affirmative, since

$$\lim_{x \to \zeta} f(x) = f(\zeta) = \lim_n \left[\lim_{x \to \zeta} f_n(x) \right] = \lim_n f_n(\zeta).$$

$\{f_n\}$ is a sequence of continuous functions with common domain D that
nverge uniformly on D, then Theorem 8.3.1 implies that the limit function
continuous and, thus, the order of taking limits may be reversed.
The following example indicates that pointwise convergence of a sequence
continuous functions is not a strong enough condition to allow the order
taking limits to be interchanged.

Example 8.5.1. Let $f_n(x) = x^n$ for each positive integer n and each
$x \in [0, 1]$. Example 8.2.1 implies that

$$f(x) = \lim_n f_n(x) = \begin{cases} 0 & \text{if } x \in [0, 1), \\ 1 & \text{if } x = 1. \end{cases}$$

We see that

$$\lim_{x \to 1} f(x) = \lim_{x \to 1} \left[\lim_n f_n(x) \right] = 0,$$

while

$$\lim_n \left[\lim_{x \to 1} f_n(x) \right] = 1.$$

Thus

$$0 = \lim_{x \to 1} \left[\lim_n f_n(x) \right] \neq \lim_n \left[\lim_{x \to 1} f_n(x) \right] = 1.$$

The operations of differentiation and integration are defined in terms of
mits, thus it seems reasonable to ask when the statements

$$\frac{df(x)}{dx} \equiv \frac{d}{dx} \left[\lim_n f_n(x) \right] = \lim_n \frac{df_n(x)}{dx}$$

nd

$$\int_a^b f(x) dx \equiv \int_a^b \left[\lim_n f_n(x) \right] dx = \lim_n \int_a^b f_n(x) dx$$

re true. Again we would like to know under what conditions the order of
aking limits can be reversed.
As in the case of ordinary limits, the following example shows that point-
vise convergence of a sequence of integrable functions (in fact, continuous
unctions) is not a strong enough condition to allow the order of limits and
ntegration to be interchanged.

Example 8.5.2. Let

$$f_n(x) = \begin{cases} 4n^2x & \text{if } 0 \leqslant x \leqslant 1/2n, \\ 4n(1-nx) & \text{if } 1/2n < x < 1/n, \\ 0 & \text{if } 1/n \leqslant x \leqslant 1, \end{cases}$$

for each positive integer n and for each $x \in [0, 1]$. We can see that

$$f(x) = 0 = \lim_n f_n(x).$$

From the definition of the sequence $\{f_n\}$, it follows that

$$\int_0^1 f_n(x)dx = 1, \quad n = 1, 2, \ldots,$$

while

$$\int_0^1 f(x)dx = 0.$$

Hence,

$$0 = \int_0^1 f(x)dx = \int_0^1 \left[\lim_n f_n(x)\right]dx \neq \lim_n \int_0^1 f_n(x)dx = 1.$$

We now show that uniform convergence of a sequence of integrable functions guarantees that the limit function is integrable and that the order of integration and limit may be reversed.

Theorem 8.5.1. *Let $\{f_n\}$ be a sequence of functions with common domain $[a, b]$, and let f_n be integrable on $[a, b]$ for each positive integer n. If the sequence $\{f_n\}$ converges uniformly to the function f on $[a, b]$, then f is integrable on $[a, b]$ and*

$$\int_a^b f(x)dx \equiv \int_a^b \left[\lim_n f_n(x)\right]dx = \lim_n \int_a^b f_n(x)dx.$$

Proof. We first show that the function f is integrable on $[a, b]$. Since each function f_n is bounded on $[a, b]$, and since the sequence $\{f_n\}$ converges uniformly on $[a, b]$, it follows that the limit function f is bounded on $[a, b]$. Let $\varepsilon > 0$. Since the sequence $\{f_n\}$ converges uniformly to f on $[a, b]$, there is an N such that $n \geqslant N$ implies that

$$|f_n(x) - f(x)| < \varepsilon/3(b - a)$$

for each $x \in [a, b]$. From this inequality it follows that

$$f_N(x) - \varepsilon/3(b - a) < f(x) < f_N(x) + \varepsilon/3(b - a)$$

for each $x \in [a, b]$. Theorem 6.1.3 implies that there is a net $\mathfrak{n}$ of $[a, b]$ such that

$$u(f_N, \mathfrak{n}) - l(f_N, \mathfrak{n}) < \varepsilon/3.$$

These two inequalities imply that

$$l(f_N, \mathfrak{n}) - \varepsilon/3 < l(f, \mathfrak{n}) \leqslant u(f, \mathfrak{n}) < u(f_N, \mathfrak{n}) + \varepsilon/3.$$

Consequently

$$u(f, \mathfrak{n}) - l(f, \mathfrak{n}) < u(f_N, \mathfrak{n}) - l(f_N, \mathfrak{n}) + 2\varepsilon/3 < \varepsilon.$$

Therefore f is integrable on $[a, b]$.

We now show that the *limiting processes* may be interchanged. Let $\varepsilon > 0$ be given. Again there is an N such that $n \geqslant N$ implies that

$$|f_n(x) - f(x)| < \varepsilon/(b - a)$$

for each $x \in [a, b]$. For each positive integer $n \geqslant N$, we have

$$\left| \int_a^b f_n(x)dx - \int_a^b f(x)dx \right| = \left| \int_a^b [f_n(x) - f(x)]dx \right|$$

$$\leqslant \int_a^b |f_n(x) - f(x)|dx$$

$$< (b - a)\varepsilon/(b - a) = \varepsilon.$$

Thus, the sequence $\int_a^b f_n(x)dx$ converges to the integral $\int_a^b f(x)dx$ and we may write

$$\lim_n \int_a^b f_n(x)dx = \int_a^b f(x)dx.$$

Two important special cases of Theorem 8.5.1 are given in the following two corollaries.

Corollary 8.5.1a. *Let $\{f_n\}$ be a sequence of continuous functions with common domain $[a, b]$. If the sequence $\{f_n\}$ converges uniformly to f on $[a, b]$, then the function f is integrable (continuous) on $[a, b]$ and*

$$\int_a^b f(x)dx \equiv \int_a^b \left[\lim_n f_n(x) \right] dx = \lim_n \int_a^b f_n(x)dx.$$

Corollary 8.5.1b. *Let $\{f_n\}$ be a sequence of increasing (decreasing) functions with common domain $[a, b]$. If the sequence $\{f_n\}$ converges* *uniformly to the function f on* $[a, b]$, *then the function f is integrable on* $[a, b]$, *and*

$$\int_a^b f(x)dx \equiv \int_a^b \left[\lim_n f_n(x)\right]dx = \lim_n \int_a^b f_n(x)dx.$$

These three results can now be extended to series of functions. We state the first of these and leave the proof as Exercise 10 of Problem Set 8.5.

Theorem 8.5.2. *Let* $\{f_n\}$ *be a sequence of functions with common domain* $[a, b]$, *and let* f_n *be integrable on* $[a, b]$ *for each positive integer n. If the series of functions* Σf_n *converges uniformly to the function f on* $[a, b]$, *then f is integrable on* $[a, b]$ *and*

$$\int_a^b f(x)dx \equiv \int_a^b [\Sigma f_n(x)]dx = \Sigma \int_a^b f_n(x)dx.$$

We now present an example to illustrate Theorem 8.5.2.

Example 8.5.3. Let $f_n(x) = x^n$, where $n = 0, 1, 2, \ldots$, for each $x \in [-1, 1]$. Then the series Σf_n converges for each $x \in (-1, 1)$, and

$$f(x) \equiv 1/(1 - x) = \sum_{n=0}^{\infty} x^n$$

for each $x \in (-1, 1)$. Let r be any real number such that $0 < r < 1$. Then, for $|x| \leq r$,

$$|1/(1 - x) - (1 + x + \ldots + x^{n-1})| = |x^n/(1 - x)| \leq r^n/(1 - r)$$

and the series

$$\sum_{n=0}^{\infty} x^n$$

converges uniformly to $1/(1 - x)$ on $[-r, r]$. Let y be any real number with $|y| \leq r$. Then Theorem 8.5.2 implies that

$$-\ln(1 - y) = \int_0^y dx/(1 - x) = \int_0^y \left[\sum_{n=0}^{\infty} x^n\right]dx$$

$$= \sum_{n=0}^{\infty} \int_0^y x^n dx = \sum_{n=0}^{\infty} y^{n+1}/(n + 1)$$

$$= \sum_{n=1}^{\infty} y^n/n$$

for $|y| \leqslant r$. But, for any $|y| < 1$, there is an $r < 1$ such that $|y| \leqslant r < 1$, namely, $r = |y|$; and so for any y with $|y| < 1$ we have

$$-\ln(1 - y) = \sum_{n=1}^{\infty} y^n/n.$$

We now turn to the question of differentiating a sequence of functions term by term, that is, when does

$$\frac{d}{dx}\left[\lim_n f_n(x)\right] = \lim_n \frac{df_n(x)}{dx}?$$

For integration, uniform convergence of a sequence of integrable functions $\{f_n\}$ was sufficient to ensure that

$$\int_a^b \left[\lim_n f_n(x)\right] dx = \lim_n \int_a^b f_n(x)dx.$$

The following example proves that uniform convergence is not sufficient for differentiation.

Example 8.5.4. For each positive integer n, let

$$f_n(x) = xe^{-nx^2}$$

for each $x \in [-1, 1]$. Then the sequence $\{f_n\}$ converges uniformly to the zero function on $[-1, 1]$, that is,

$$\lim_n xe^{-nx^2} = 0 \equiv f(x)$$

for each $x \in [-1, 1]$. Thus

$$\frac{df(x)}{df} = \frac{d}{dx}\left[\lim_n xe^{-nx^2}\right] = 0 = f'(x).$$

Checking the derivative of xe^{-nx^2}, we see that

$$\frac{df_n(x)}{dx} = \frac{d(xe^{-nx^2})}{dx} = e^{-nx^2}[1 - 2nx^2].$$

It follows that

$$\lim_n \frac{df_n(0)}{dx} = \lim_n [1] = 1,$$

while

$$f'(0) = 0.$$

Therefore

$$0 = \frac{df(0)}{dx} = \frac{d}{dx}\left[\lim_n f_n(0)\right] \neq \lim_n \frac{df_n(0)}{dx} = 1.$$

Example 8.5.4 shows that uniform convergence is not enough to guarantee that

$$\frac{d}{dx}\lim_n f_n(x) = \lim_n \frac{df_n(x)}{dx}.$$

The following theorem gives a sufficient condition.

Theorem 8.5.3. *Let $\{f_n\}$ be a sequence of functions with common domain $[a, b]$, and for each positive integer n let f_n be differentiable on $[a, b]$ and f_n' be continuous on $[a, b]$. If the sequence $\{f_n\}$ converges at one point $y \in [a, b]$ and the sequence $\{f_n'\}$ converges uniformly on $[a, b]$, then the sequence $\{f_n\}$ converges uniformly on $[a, b]$, the limit function is differentiable on $[a, b]$, and*

$$\frac{d}{dx}\left[\lim_n f_n(x)\right] = \lim_n \frac{df_n(x)}{dx}$$

for each $x \in [a, b]$.

Proof. We shall first use the Cauchy Criterion (Theorem 8.2.1) to show that the sequence $\{f_n\}$ converges uniformly on $[a, b]$.

Let $\varepsilon > 0$ be given. Since the sequence $\{f_n\}$ converges at y, the Cauchy Criterion (Theorem 3.4.3) implies that there is an N_1 such that

$$|f_n(y) - f_m(y)| < \varepsilon/(b - a + 1)$$

when $n > m > N_1$. Also, since the sequence $\{f_n'\}$ converges uniformly on $[a, b]$, the Cauchy Criterion for Uniform Convergence (Theorem 8.2.1) implies that there is an N_2 such that $n > m > N_2$ implies that

$$|f_n'(\zeta) - f_m'(\zeta)| < \varepsilon/(b - a + 1)$$

for each $\zeta \in [a, b]$. Let $N = \max(N_1, N_2)$. The Fundamental Theorem of Integral Calculus (Theorem 6.4.5) implies that we may write

$$f_n(x) = \int_y^x f_n'(\zeta)d\zeta + f_n(y)$$

for each positive integer n and for each $x \in [a, b]$. Thus,

$$f_n(x) - f_m(x) = \int_y^x [f_n'(\zeta) - f_m'(\zeta)]d\zeta + [f_n(y) - f_m(y)],$$

and it follows that

$$|f_n(x) - f_m(x)| \leqslant \int_y^x |f_n'(\zeta) - f_m'(\zeta)| d\zeta + |f_n(y) - f_m(y)|.$$

Letting $n > m > N$ and using the estimates given above for $|f_n(y) - f_m(y)|$ and $|f_n'(\zeta) - f_m'(\zeta)|$, we have

$$|f_n(x) - f_m(x)| < (|x - y| \varepsilon + \varepsilon)/(b - a + 1)$$
$$\leqslant \varepsilon(b - a + 1)/(b - a + 1) = \varepsilon.$$

Therefore, the Cauchy Criterion for Uniform Convergence implies that the sequence $\{f_n\}$ converges uniformly on $[a, b]$.

Let

$$f(x) \equiv \lim_n f_n(x)$$

and

$$g(x) \equiv \lim_n f_n'(x)$$

for each $x \in [a, b]$. We need to show that $f'(x) = g(x)$ for each $x \in [a, b]$. Using the Fundamental Theorem of Integral Calculus (Theorem 6.4.5) we write

$$f_n(x) - f_n(y) = \int_y^x f_n'(\zeta) d\zeta.$$

Taking limits on both sides of this equation we have

$$f(x) - f(y) = \lim_n \int_y^x f_n'(\zeta) d\zeta,$$

and Theorem 8.5.1 implies that

$$\lim_n \int_y^x f_n'(\zeta) d\zeta = \int_y^x \left[\lim_n f_n'(\zeta) \right] d\zeta$$

$$= \int_y^x g(\zeta) d\zeta.$$

Thus we have

$$f(x) - f(y) = \int_y^x g(\zeta) d\zeta$$

for each $x \in [a, b]$. Since g is the limit of a sequence of continuous functions that converge uniformly on $[a, b]$, the function g is continuous on $[a, b]$. Theorem 6.4.4 now implies that

$$f'(x) = g(x)$$

for each $x \in [a, b]$. Thus we have

$$\frac{d}{dx}\left[\lim_n f_n(x)\right] \equiv \frac{df(x)}{dx} = g(x) = \lim_n \frac{df_n'(x)}{dx}.$$

Theorem 8.5.3 can be extended to series of functions. We state this result and leave the proof as Exercise 18 of Problem Set 8.5.

Theorem 8.5.4. *Let $\{f_n\}$ be a sequence of functions with common domain $[a, b]$, and for each positive integer n let f_n be differentiable on $[a, b]$ and f_n' be continuous on $[a, b]$. If the series Σf_n converges at one point $y \in [a, b]$ and the series $\Sigma f_n'$ converges uniformly on $[a, b]$, then the series Σf_n converges uniformly on $[a, b]$, the sum function is differentiable on $[a, b]$, and*

$$\frac{d}{dx}\left[\Sigma f_n(x)\right] = \Sigma \frac{df_n(x)}{dx}$$

for each $x \in [a, b]$.

Example 8.5.4. We use Theorem 8.5.4 to obtain the series expansion

$$1/(1 - x)^2 = 1 + 2x + 3x^2 + \cdots$$

for each $x \in (-1, 1)$. We begin with the geometric series

$$1/(1 - x) = 1 + x + x^2 + \cdots,$$

which is convergent for each $x \in (-1, 1)$, and which converges uniformly on $[-r, r]$ for any r, where $0 < r < 1$. We see that the series for $1/(1 - x)^2$ is derived from the series for $1/(1 - x)$ by differentiating term by term. In order to justify the procedure of Theorem 8.5.4, we must show that the series $1 + 2x + 3x^2 + \cdots$ converges uniformly on the interval $[-r, r]$, where r is any real number such that $0 < r < 1$. We use the Weierstrass M-Test (Theorem 8.2.2) to show the convergence is uniform.

Let $x \in [-r, r]$ where $0 < r < 1$. Then $|x| \leq r$ and

$$|(n + 1)x^n| \leq (n + 1)r^n$$

for each positive integer n. We let $M_n = (n + 1)r^n$ and show that

$$\Sigma M_n = \Sigma(n + 1)r^n$$

converges. The ratio

$$\frac{M_{n+1}}{M_n} = \frac{(n + 2)r^{n+1}}{(n + 1)r^n} = \frac{(n + 2)r}{(n + 1)}$$

and

$$\lim_{n} \frac{M_{n+1}}{M_n} = \lim_{n} \frac{(n+2)r}{(n+1)} = r.$$

Since $0 < r < 1$, the series

$$\Sigma M_n = \Sigma (n+1)r^n$$

converges, and the Weierstrass M-Test implies that the series

$$\sum_{n=0}^{\infty} (n+1)x^n$$

converges uniformly on $[-r, r]$. Thus

$$1/(1-x)^2 = 1 + 2x + 3x^2 + \dots$$

holds for the interval $[-r, r]$. Since r was an arbitrary real number such that $0 < r < 1$, the result also holds for the open interval $(-1, 1)$.

Problem Set 8.5

1. Let $f(x) = \displaystyle\sum_{n=1}^{\infty} (\cos nx)/n^2$. Show that

$$\int_0^{\pi/2} f(x)dx = \sum_{n=0}^{\infty} (-1)^n/(2n+1)^3.$$

2. Let $f(x) = \displaystyle\sum_{n=1}^{\infty} (\sin nx)/n^2$. Show that

$$\int_0^{\pi} f(x)dx = \sum_{n=1}^{\infty} 2/(2n-1)^3.$$

3. Let $f(x) = \displaystyle\sum_{n=1}^{\infty} (\ln nx)/n^2$. Show that

$$\int_1^2 f(x)dx = \sum_{n=1}^{\infty} (\ln 4n - 1)/n^2.$$

4. Let $f_n(x) = (\sin nx)/nx$. Show that

$$\lim_{n} \int_{\pi/2}^{\pi} f_n(x)dx = 0.$$

5. Let $f_n(x) = e^{-nx^2}$. Show that

$$\lim_n \int_1^2 f_n(x)dx = 0.$$

6. Let $f(x) = \sum_{n=1}^{\infty} (\cos nx)/n^4$. Find a series for $\int_0^x f(\zeta)d\zeta$.

7. Let $f(x) = \sum_{n=1}^{\infty} (\sin nx)/n^4$. Find a series for $\int_{\pi/2}^x f(\zeta)d\zeta$.

8. Let $f_n(x) = nx/(nx + 1)$.
   (a) Find the pointwise limit of $\{f_n\}$ on $[0, 1]$. Is the convergence uniform?

   (b) Find $\lim_n \int_0^1 f_n(x)dx$ and $\int_0^1 \left[\lim_n f_n(x)\right]dx$.

   Explain your answer.

9. Let $f_n(x) = nxe^{-nx^2}$.
   (a) Find the pointwise limit of $\{f_n\}$ on $[0, 1]$. Is the convergence uniform?

   (b) Find $\lim_n \int_0^1 f_n(x)dx$ and $\int_0^1 \left[\lim_n f_n(x)\right]dx$.

   Explain your answer.

10. Probe Theorem 8.5.2.

11. Let $f(x) = \sum_{n=1}^{\infty} (\sin nx)/n^2$ for $x \in [0, 2\pi]$. For what values of x can $f'(x)$ be written as the series obtained by differentiating the series for $f(x)$ term by term?

12. Let $f(x) = \sum_{n=1}^{\infty} (-1)^{n+1}x^n/n$ for $x \in [0, 1]$. Where can $f'(x)$ be written as the series obtained by differentiating the series for $f(x)$ term by term?

13. Let $f(x) = \sum_{n=1}^{\infty} [x/(1-x)]^n/n$. For what values of x does

$$f'(x) = \sum_{n=0}^{\infty} [x/(1-x)]^n/(1-x)^2?$$

14. Let $f(x) = \sum_{n=1}^{\infty} n^{-x}$. For what values of x does $f'(x) = -\sum_{n=1}^{\infty} n^{-x} \ln n$?

15. Let Σna_n converge absolutely and let

$$f(x) = \sum_{n=1}^{\infty} a_n \cos nx, \quad g(x) = \sum_{n=1}^{\infty} a_n \sin nx.$$

Show that

$$f'(x) = - \sum_{n=1}^{\infty} na_n \sin nx, \quad g'(x) = \sum_{n=1}^{\infty} na_n \cos nx$$

for every real x.

16. Show that $\dfrac{d}{dx}\left[\sum_{n=1}^{\infty} x^n/n(n+1) \right] = \sum_{n=0}^{\infty} x^n/(n+2)$ for $|x| < 1$.

17. Show that $\dfrac{d}{dx}\left[\sum_{n=1}^{\infty} 1/n^3(1 + nx^2) \right] = -2x \sum_{n=1}^{\infty} 1/n^2(1 + nx^2)^2$ for each real x.

18. Prove Theorem 8.5.4.

19. Let $f_n(x) = e^{-n^2x^2}/n$. Show that

$$\lim_n f_n'(x) = \frac{d}{dx}\left[\lim_n f_n(x) \right]$$

for each real number x. Show that the convergence of the sequence of derivatives is not uniform in any interval containing the origin. Show that the sequence $\{f_n\}$ converges uniformly on R.

20. Let $f_n(x) = (\sin nx)/n$. Show that the sequence $\{f_n\}$ converges uniformly on R, but that the sequence $\{f_n'\}$ converges only for integral multiples of 2π.

21. Let $\{f_n\}$ be a sequence of positive increasing functions with common domain $[a, b]$. If $\Sigma f_n(b)$ converges, show that

$$\int_a^b [\Sigma f_n(x)]dx = \Sigma \int_a^b f_n(x)dx.$$

22. Let $\{f_n\}$ be a sequence of continuous functions with common domain $[a, b]$ which converge uniformly on $[a, c]$ for each c in $[a, b)$. If the sequence $\{f_n\}$ is uniformly bounded on $[a, b]$, show that

$$\lim_n \int_a^b f_n(x)dx = \int_a^b \left[\lim_n f_n(x) \right] dx.$$

23. Let ϕ and f_0 be continuous functions with domain $[a, b]$. Inductively, define the sequence $\{f_n\}$ by

$$f_n(x) = \alpha + \int_a^x \phi(\zeta)f_{n-1}(\zeta)d\zeta$$

for each $x \in [a, b]$ and for each $n \geq 1$. Show that the sequence of functions $\{f_n\}$ converges uniformly on $[a, b]$ to a function f which is a solution of the equation $f'(x) = \phi(x)f(x)$, $f(a) = \alpha$. Show that

$$f(x) = \alpha \exp \int_a^x \phi(\zeta)d\zeta.$$

[*Note*: $\exp x = e^x$.]

24. Let $\{a_n\}$ be a sequence of real numbers. Let

$$f(x) \equiv \sum_{n=0}^{\infty} a_n x^n.$$

If this series converges for $x = x_0 > 0$, show that the derivative of f is given by the series

$$f'(x) = \sum_{n=1}^{\infty} n a_n x^{n-1}$$

for each $x \in (-x_0, x_0)$

Chapter Nine

POWER SERIES

9.1. The Interval of Convergence

In the previous chapter we discussed the concept of series of functions. We now turn our attention to one of the most important classes of series of functions—the *power series*. These are series whose terms are powers of a function multiplied by constants.

Definition 9.1.1. *A series of the form*

$$\sum a_n x^n = \sum_{n=0}^{\infty} a_n x^n = a_0 + a_1 x + a_2 x^2 + \dots$$

*is called a **power series** in x. A series of the form*

$$\sum a_n (x - a)^n = \sum_{n=0}^{\infty} a_n (x - a)^n$$

$$= a_0 + a_1 (x - a) + a_2 (x - a)^2 + \dots$$

*is called a **power series** in $(x - a)$. More generally, if f is a function of x, then a series of the form*

$$\sum a_n [f(x)]^n = \sum_{n=0}^{\infty} a_n [f(x)]^n$$

$$= a_0 + a_1 f(x) + a_2 [f(x)]^2 + \dots$$

*is called a **power series** in $f(x)$.*

In this chapter we shall be mainly interested in power series in x and power series in $(x - a)$. If we make the change of a variable $\zeta = (x - a)$ in a power series in $(x - a)$,

$$a_0 + a_1(x - a) + a_2(x - a)^2 + ...,$$

we reduce it to a power series in ζ,

$$a_0 + a_1\zeta + a_2\zeta^2 + ...$$

Thus the behavior of the power series in $(x - a)$ at the point x is exactly the same as the behavior of the corresponding power series in ζ at $\zeta = x - a$. Thus we shall only study power series in which $a = 0$, that is, we shall consider only power series in x.

Our first question involving power series is the following: For what values of x does the power series $\Sigma a_n x^n$ converge? For the value $x = 0$ we see that the power series $\Sigma a_n x^n$ reduces to the constant a_0. Thus every power series $\Sigma a_n x^n$ converges for at least one real number, namely, $x = 0$. Before pursuing this question further, we shall look at some examples of power series.

> **Example 9.1.1.** The geometric series $\Sigma a x^n$, where $a \neq 0$, is an example of a power series that converges for $|x| < 1$ and that diverges for $|x| \geqslant 1$. If $r \neq 0$, then the power series $\Sigma a r^n x^n$ is a geometric series for each value of x, and thus it converges for $|x| < 1/|r|$ and diverges for $|x| \geqslant 1/|r|$. In particular, the series $\Sigma x^n/2^n$ converges for $|x| < 2$ and diverges for $|x| \geqslant 2$.

> **Example 9.1.2.** As a second example, consider the power series $\Sigma (n!)x^n$. For each nonzero real number x,
>
> $$\lim_n |(n!)x^n| = +\infty,$$
>
> and thus the limit of the nth term is not zero. Corollary 7.1.1 implies that this power series does not converge if $x \neq 0$. This is an example of a power series which diverges for all nonzero values of x.

> **Example 9.1.3.** Lastly, we consider a power series that converges for all real x. We consider the power series $\Sigma x^n/n!$. Since
>
> $$\lim_n |[x^{n+1}/(n + 1)!]/[x^n/n!]| = \lim_n |x|/(n + 1) = 0,$$
>
> the Ratio Test (Theorem 7.2.2) implies that the power series converges for each real x.

We shall use the following lemma to show that the set of real numbers for which a power series converges is an interval.

> **Lemma 9.1.1.** If the power series $\Sigma a_n x^n$ converges for $x = x_0$, then it converges absolutely for all y with $|y| < |x_0|$.

Proof. Assume that the power series $\Sigma\, a_n x^n$ converges at $x = x_0$ and that $|y| < |x_0|$. Since the series $\Sigma\, a_n(x_0)^n$ converges, there is a positive real number M such that $|a_n(x_0)^n| \leqslant M$ for each nonnegative integer n. We now write

$$|a_n(y)^n| = |a_n(x_0)^n| \cdot |y/x_0|^n$$
$$\leqslant M \cdot |y/x_0|^n.$$

Since $|y| < |x_0|$, the geometric series

$$\sum_{n=0}^{\infty} M |y/x_0|^n$$

converges and the Comparison Test (Theorem 7.2.1) implies that the series $\Sigma\, a_n y^n$ converges absolutely.

It follows from Lemma 9.1.1 that if the series $\Sigma\, a_n x^n$ diverges at $x = x_0$, and if $|y| > |x_0|$, then the series $\Sigma\, a_n x^n$ also diverges at $x = y$. Using these facts we are now able to prove the following theorem.

Theorem 9.1.1. *Let* $\Sigma\, a_n\, x^n$ *be a power series. Then one of the following holds:*

(a) *the power series converges only at* $x = 0$;

(b) *the power series converges absolutely for all real* x; *or*

(c) *there is a positive real number* r *such that the power series converges absolutely for all* x *with* $|x| < r$ *and diverges for all* x *with* $|x| > r$.

Proof. Assume that neither (a) nor (b) holds; that is, assume that there is a real number $x_c \neq 0$ such that the power series converges for $x = x_c$, and there exists a real number x_d such that the power series diverges for $x = x_d$.

Let S denote the set of all positive real numbers y such that the power series converges at $x = y$. The set S is nonempty since $x_c \in S$. If S did not have an upper bound, then we could find a real number $x_1 \in S$ such that $|x_d| < x_1$. But then Lemma 9.1.1 implies that the power series also converges at $x = x_d$. This is a contradiction to our assumption that the power series diverges at $x = x_d$. Therefore the set S has an upper bound.

Let $r = \sup S$. We assert that the power series converges absolutely for any x with $|x| < r$ and diverges for any x with $|x| > r$. The proof of this is a consequence of Lemma 9.1.1.

Let x be any real number such that $|x| < r$. Since r is the least upper bound of the set S, there is an $x_0 \in S$ such that $x < x_0 < r$. Thus the power series converges at x_0 and Lemma 9.1.1 implies that the power series converges absolutely at x.

Let x be any real number with $|x| > r$. Assume that the power series converges at x. Then for any real number y with $r < y < x$, Lemma 9.1.1 implies that the power series converges absolutely at y, and thus $y \in S$. But this contradicts the fact that r is the least upper bound of the set S. Therefore the series diverges at x. This completes the proof.

Definition 9.1.2. *The real number* r *in Theorem 9.1.1 is called the* **radius of convergence** *of the power series. If the power series converges for all real numbers, then the radius of convergence is said to be* $+\infty$, *and we write* $r = +\infty$. *Let* S *be the set of real numbers* x *for which the power series converges. Then* S *is called the* **interval of convergence** *of the power series.*

In case (a) of Theorem 9.1.1 we say that the power series has radius of convergence 0, while in case (b) we say that its radius of convergence is infinite and write $r = +\infty$. In Example 9.1.1 the radius of convergence of the geometric series is $r = 1$, while $r = 0$ in Example 9.1.2 and $r = +\infty$ in Example 9.1.3.

Theorem 9.1.1 yields no information about the convergence of the power series at the endpoints of the interval of convergence. The following examples illustrate that anything can happen at the endpoints.

Example 9.1.4. The geometric series Σx^n converges for $|x| < 1$ and diverges for $|x| \geqslant 1$. Thus the radius of convergence is 1, and the interval of convergence is $(-1, 1)$. This is an example of a power series that diverges at each endpoint of the interval of convergence.

Example 9.1.5. The power series $\Sigma x^n/n$ becomes the harmonic series when 1 is substituted for x. Thus the power series diverges for $x = 1$. If -1 is substituted for x, then we have the alternating harmonic series, which converges. Therefore the radius of convergence of this power series is 1, and its interval of convergence is $[-1, 1)$.

It is easily seen that the power series $\Sigma (-1)^n x^n/n$ has radius of convergence 1, and it diverges at the left endpoint and converges at the right endpoint. Thus the interval of convergence of this power series is $(-1, 1]$.

Example 9.1.6. The power series $\Sigma x^n/n^2$ converges absolutely for $|x| \leqslant 1$ and diverges for $|x| > 1$. Thus its radius of convergence is 1, and the interval of convergence of this power series is $[-1, 1]$.

In Example 9.1.5, the power series converges at one endpoint, and diverges at the other endpoint, of the interval of convergence. The convergence in such a case is not absolute convergence. It is easily seen that if a power series converges absolutely at one endpoint of the interval of convergence then it also converges absolutely at the other endpoint. This is the case in Example 9.1.6.

In the study of analysis it is important to be able to find the radius of convergence exactly. It is relatively easy to find bounds for the radius of convergence. If the power series converges at x_1 and diverges at x_2, then it follows that

$$|x_1| \leqslant r \leqslant |x_2|$$

In some cases such bounds may be sufficient, but usually we need to know the radius of convergence precisely.

In many practical cases, the radius of convergence of a power series may be found by using the Ratio Test (Corollary 7.2.2). In order to apply the Ratio Test, the limit of the ratio of successive coefficients must exist. We now state this result formally.

Theorem 9.1.2. *The radius of convergence of the power series* $\Sigma a_n x^n$ *is given by*

$$r = \lim_n |a_n/a_{n+1}|$$

if this limit exists or is $+\infty$.

Proof. Assume that

$$\lim_n |a_{n+1}/a_n| = q.$$

(q is a nonnegative real number or $+\infty$.) Then

$$\lim_n |a_{n+1}x^{n+1}/a_n x^n| = |x| \lim_n |a_{n+1}/a_n|$$

$$= |x|q.$$

Thus the power series $\Sigma a_n x^n$ converges if $|x|q < 1$ and diverges if $|x|q > 1$. From this we may conclude that $r = 1/q$ if $q \neq 0$. If $q = 0$, then the limit is zero for each real number x and the power series converges for all real values of x. Thus, if $q = 0$, then $r = +\infty$. If $q = +\infty$, then the limit is $+\infty$ for each nonzero value of x and so $r = 0$.

We illustrate this theorem with the following examples.

Example 9.1.7. Consider the power series $\Sigma x^n/n^p$, where p is any real number. Then

$$\lim_n [n^p/(n + 1)^p] = 1$$

and the radius of convergence is 1.

Example 9.1.8. Let

$$a_n = \begin{cases} (2/3)^n & \text{if } n \text{ is even,} \\ (3/4)^n & \text{if } n \text{ is odd,} \end{cases}$$

and find the radius of convergence of the power series $\Sigma\, a_n x^n$. It is easily seen that

$$\lim_n |a_n/a_{n+1}|$$

does not exist, and so Theorem 9.1.2 fails.

In Example 9.1.8 the test given in Theorem 9.1.2 fails. We now present a test that always produces the precise value for the radius of convergence. The main problem with applying this test is that the actual calculations are difficult to carry out. The test is called the *Cauchy–Hadamard Test* and is named for A. L. Cauchy and the French mathematician Jacques Hadamard (1865–1963).

Theorem 9.1.3. (*Cauchy–Hadamard Test*). *The radius of convergence of the power series $\Sigma\, a_n x^n$ is given by*

$$r = \begin{cases} 0 & \text{if } \lim_n \sup |a_n|^{1/n} = +\infty, \\ +\infty & \text{if } \lim_n \sup |a_n|^{1/n} = 0, \text{ and} \\ 1/\lim_n \sup |a_n|^{1/n} & \text{otherwise.} \end{cases}$$

Proof. The proof of this theorem depends on Exercise 17 of Problem Set 7.2. Let

$$q = \lim_n \sup |a_n|^{1/n}.$$

Then

$$\lim_n \sup |a_n x^n|^{1/n} = |x| q.$$

If $0 < q < +\infty$, then

$$\lim_n \sup |a_n x^n|^{1/n} < 1$$

whenever $|x| < 1/q$. Thus the power series converges absolutely for $|x| < 1/q$. Also

$$\lim_n \sup |a_n x^n|^{1/n} > 1$$

whenever $|x| > 1/q$, and the power series diverges for $|x| > 1/q$. Therefore we have the radius of convergence for $r = 1/q$.

If $q = 0$, then

$$\lim_n \sup |a_n x^n|^{1/n} = 0$$

for each real x, and the series converges absolutely for each value of x. Thus the radius of convergence is $r = +\infty$.

If $q = +\infty$, then, for each nonzero x,

$$\lim_n \sup |a_n x^n|^{1/n} = +\infty$$

and the power series diverges. Hence the radius of convergence is $r = 0$.

We illustrate this theorem with the following examples.

Example 9.1.9. Find the radius of convergence of the power series $\Sigma \, 2^n x^n$. We see that
$$q = \lim_n \sup (2^n)^{1/n} = 2.$$
Thus the radius of convergence $r = 1/2$.

Example 9.1.10. Let
$$a_n = \begin{cases} (2/3)^n & \text{if } n \text{ is even,} \\ (3/4)^n & \text{if } n \text{ is odd,} \end{cases}$$
and find the radius of convergence of the power series $\Sigma \, a_n x^n$. (Cf. Example 9.1.8.) It is a routine matter to verify that
$$q = \lim_n \sup |a_n|^{1/n} = 3/4.$$
Hence the radius of convergence is $r = 4/3$.

If
$$\lim_n |a_n|^{1/n}$$
exists or is $+\infty$, then we have the following corollary to Theorem 9.1.3.

Corollary 9.1.3. *If* $\lim_n |a_n|^{1/n}$ *exists or equals* $+\infty$, *then the radius of convergence of the power series* $\Sigma \, a_n x^n$ *is given by*
$$r = \begin{cases} 0 & \text{if } \lim |a_n|^{1/n} = +\infty, \\ +\infty & \text{if } \lim_n |a_n|^{1/n} = 0, \text{ and} \\ 1/\lim_n |a_n|^{1/n} & \text{otherwise.} \end{cases}$$

We see that Corollary 9.1.3 would produce the radius of convergence in Example 9.1.9 but would fail in Example 9.1.10.

For many practical problems, it is necessary to know not only the radius of convergence of the power series, but also whether or not the power series converges at an endpoint of the interval of convergence. It is not possible to use the Ratio Test (Theorem 7.2.2) or the Root Test (Theorem 7.2.3 and Exercise 18 of Problem Set 7.2), since
$$r \cdot \left[\lim_n |a_{n+1}/a_n| \right] = 1,$$
if the limit exists, and
$$r \cdot [\lim_n \sup |a_n|^{1/n}] = 1.$$

Thus, to test the endpoints, the Comparison Test (Theorem 7.2.1) or some other test must be used, such as those given in Section 7.4.

Example 9.1.11. Find the radius of convergence of the power series

$$\sum_{n=1}^{\infty} [(2n)!/4^n(n!)^2]^3 x^n$$

and check the endpoints for convergence.

Let $a_n = [(2n)!/4^n(n!)^2]^3$. Then

$$\lim_n |a_{n+1}/a_n| = \lim_n [(2n + 2)(2n + 1)/4(n + 1)^2]^3 = 1.$$

Thus the radius of convergence is 1.

We use Raabe's Test (Theorem 7.4.3) to check the convergence of the power series at $x = 1$, since the Ratio Test fails in this case. For Raabe's Test, we calculate

$$n[(a_n/a_{n+1}) - 1] = (12n^3 + 18n^2 + 7n)/(2n + 1)^3.$$

Thus

$$\lim n[(a_n/a_{n+1}) - 1] = 3/2 > 1,$$

and the power series converges absolutely for $x = 1$.

Since the power series converges absolutely for $x = 1$, it also converges absolutely for $x = -1$. Hence the interval of convergence of this power series is $[-1, 1]$.

Problem Set 9.1

In Exercises 1–15, find the interval of convergence of each of the power series, and check the endpoints of the interval of convergence for convergence.

1. $\displaystyle\sum_{n=0}^{\infty} (-2)^n x^n/n!.$

2. $\displaystyle\sum_{n=1}^{\infty} x^n/2n.$

3. $\displaystyle\sum_{n=0}^{\infty} x^n/(2^n + 1)!.$

4. $\displaystyle\sum_{n=1}^{\infty} x^n/n(n + 1)(n + 2).$

5. $\displaystyle\sum_{n=1}^{\infty} (n!)^2 x^n/(2n)!.$

6. $\displaystyle\sum_{n=1}^{\infty} e^{2n} x^n.$

7. $\displaystyle\sum_{n=0}^{\infty} (3^n + 2n)x^n/(4^n + 5).$

8. $\displaystyle\sum_{n=0}^{\infty} (n!)^3 x^n/(3n)!.$

9. $\displaystyle\sum_{n=0}^{\infty} (n!)^k x^n(kn)!,\ k$ a positive integer.

10. $\displaystyle\sum_{n=0}^{\infty} (n!)^2 x^n/(3n)!.$

11. $\displaystyle\sum_{n=0}^{\infty} (n!)^{5/2} x^n(4n)!.$

12. $\displaystyle\sum_{n=0}^{\infty} (n + p)! x^n/n!(n + q)!,\ p$ and q positive integers.

13. $\displaystyle\sum_{n=0}^{\infty} a_n x^n$, where $a_n = 4^n$ if n is odd and $a_n = 3^n$ if n is even.

14. $\displaystyle\sum_{n=1}^{\infty} x^n/(\sqrt{n})^n.$

15. $\displaystyle\sum_{n=1}^{\infty} x^n/n^{\sqrt{n}}.$

16. Use Lemma 9.1.1 to show that if a power series diverges at $x = x_0$ and $|y| > |x_0|$, then the power series also diverges at $x = y$.

A series of the form

$$1 + \frac{\alpha \cdot \beta}{1 \cdot \gamma} x + \frac{\alpha \cdot (\alpha + 1) \cdot \beta \cdot (\beta + 1)}{1 \cdot 2 \cdot \gamma \cdot (\gamma + 1)} x^2 + \cdots,$$

where $\alpha, \beta, \gamma \notin \{0, -1, -2, \ldots\}$, is called a **hypergeometric series**. In Exercises 17–22 prove the following results for hypergeometric series. (Cf. Exercise 7 of Problem 7.4.)

17. The series converges absolutely if $|x| < 1$.

18. The series diverges if $|x| > 1$.

19. The series converges absolutely if $x = 1$ and $\gamma - \alpha - \beta > 0$.

20. The series diverges if $x = 1$ and $\gamma - \alpha - \beta \leqslant 0$.

21. The series converges absolutely if $x = -1$ and $\gamma - \alpha - \beta > 0$.

22. The series converges conditionally if $x = -1$ and $-1 < \gamma - \alpha - \beta \leqslant 0$.

For Exercises 23–27, let $\Sigma\, a_n x^n$ and $\Sigma\, b_n x^n$ be two power series with radii of convergence r and r', respectively.

23. What can be said about the radius of convergence of $\Sigma\, (a_n + b_n)x^n$?

24. What can be said about the radius of convergence of $\Sigma\, (a_n - b_n)x^n$?

25. What can be said about the radius of convergence of $\Sigma\, a_n b_n x^n$?

26. What can be said about the radius of convergence of $\Sigma\, (a_n/b_n)x^n$?

27. If $|a_n| \leqslant |b_n|$ for all sufficiently large values of n, show that $r \geqslant r'$.

For Exercises 28–34, let $\Sigma\, a_n x^n$ be a power series with radius of convergence r.

28. If the sequence $\{a_n\}$ is bounded, then show that $r \geqslant 1$.

29. If the sequence $\{a_n\}$ is bounded away from zero, that is, there is an $\varepsilon > 0$ and there is a positive integer N such that $|a_n| \geqslant \varepsilon$ when $n \geqslant N$, then show that $r \leqslant 1$.

30. If $\lim_n a_n = 0$ and $\Sigma\, |a_n| = +\infty$, then show that $r = 1$.

31. Prove that the radius of convergence of the power series $\Sigma\, a_n x^{2n}$ is $r^{1/2}$.

32. If p is a positive integer, then show that the radius of convergence of the power series $\Sigma\, a_n x^{pn}$ is $r^{1/p}$.

33. If p is a positive real number, then show that the radius of convergence of the power series $\Sigma\, a_n p^n x^n$ is r/p.

34. If $0 < r < +\infty$, then show that the radius of convergence of the power series $\Sigma\, a_n x^{n^2}$ is 1.

For Exercises 35–37, use Exercise 34 to determine $\lim_n a_n$.

35. Let $a_n = p$, p a positive constant.

36. Let $a_n = n$.

37. Let $a_n = \dfrac{1 \cdot 3 \dots (2n - 1)}{2 \cdot 4 \dots 2n} = \dfrac{(2n)!}{4^n (n!)^2}$.

38. If $\Sigma\, a_n x^n$ converges for all real values of x, then show that $\lim_n |a_n|^{1/n} = 0$.

39. Let $e = \lim_n (1 + 1/n)^n$. Show that $\lim_n [n/(n!)^{1/n}] = e$. [*Hint*: Determine the radius of convergence of the power series $\Sigma\, n^n x^n/n!$ by Theorem 9.1.2 and by the Cauchy–Hadamard Test.]

In Exercises 40–42, find the interval of convergence of each of the following power series and check the endpoints for convergence.

40. $\displaystyle\sum_{n=1}^{\infty} [(x - 1)/x]^n/n$.

41. $\displaystyle\sum_{n=0}^{\infty} 1/(2n+1)(2x+1)^n$.

42. $\displaystyle\sum_{n=0}^{\infty} [(x-1)/(x+1)]^n/(2n+1)$.

43. Find the values of x for which the series

$$\sin x - \frac{\sin^3 x}{3} + \frac{\sin^5 x}{5} - \frac{\sin^7 x}{7} + \dots$$

converges, and specify the type of convergence.

44. Find the value of x for which the power series

$$1 + xe^{-x} + 2x^2 e^{-2x} + ex^3 3^{-3x} + 4x^4 e^{-4x} + \dots$$

converges.

45. If $\Sigma\, a_n$ is convergent but not absolutely convergent, then prove that $\Sigma\, a_n x^n$ has radius of convergence 1.

46. Develop a theory for power series of the form $\Sigma\, a_n(x-a)^n$ similar to the theory in Section 9.1 and Problem Set 9.1.

9.2. The Function Defined by a Power Series

Let $\Sigma\, a_n x^n$ be a power series with radius of convergence $r > 0$. Theorem 9.1.1 implies that the power series converges absolutely for each x in $(-r, r)$. We are able to define a function $f: (-r, r) \to R$ such that

$$f(x) = \sum_{n=0}^{\infty} a_n x^n$$

for each $x \in (-r, r)$. In this section we propose to investigate the properties of functions defined by power series. We begin by proving that a power series converges uniformly on any compact subset of the interval of convergence.

Theorem 9.2.1. *Let $\Sigma\, a_n x^n$ be a power series with radius of convergence $r > 0$, and let K be any compact subset of $(-r, r)$. Then the power series converges uniformly on K.*

Proof. The set K is a compact subset of the open interval $(-r, r)$, and thus there is a real number ρ, with $0 < \rho < r$, such that $|x| \leqslant \rho$ for each $x \in K$. Hence $|a_n x^n| \leqslant |a_n \rho^n|$ for each $x \in K$. Since $\rho \in (-r, r)$, Theorem 9.1.1 implies that

$$\sum_{n=0}^{\infty} |a_n \rho^n|$$

converges. The Weierstrass M-Test (Theorem 8.2.2) now implies that the power series converges uniformly on K.

This theorem leads to the following important result.

Corollary 9.2.1. *Let $\Sigma\, a_n x^n$ be a power series with radius of convergence r, and let f be the function defined by this power series. Then f is continuous on $(-r, r)$.*

Proof. Let $x \in (-r, r)$. Then x belongs to the closed interval $[(x - r)/2, (x + r)/2]$, and Theorem 9.2.1 implies that the power series converges uniformly to f on this closed interval. Since the terms of the power series are continuous functions, Corollary 8.3.1 implies that f is continuous on the closed interval containing x. Since x was an arbitrary point of $(-r, r)$, the sum function f is continuous on $(-r, r)$.

Corollary 9.2.1 implies that the function f defined by a power series with radius of convergence $r > 0$ is continuous on the interval $(-r, r)$. We shall show that the function f has much stronger properties on the interval $(-r, r)$; for example, the function f is infinitely differentiable on the interval $(-r, r)$. We begin by showing that a power series and the series obtained by term-by-term differentiation have the same radius of convergence.

Theorem 9.2.2. *Let $\Sigma\, a_n x^n$ be a power series with radius of convergence $r > 0$. Then the power series $\Sigma\, na_n x^{n-1}$ has radius of convergence r.*

Proof. Let r and r' be the respective radii of convergence of the power series $\Sigma\, a_n x^n$ and $\Sigma\, na_n x^{n-1}$. Since the power series $\Sigma\, na_n x^n$ has the same radius of convergence as the power series $\Sigma\, na_n x^{n-1}$, and since $|na_n| \geqslant |a_n|$ for each positive integer n, it follows that $r' \leqslant r$.

In order to show that $r \leqslant r'$, let x_1 be any real number, with $0 < |x_1| < r$. We shall show that the power series $\Sigma\, na_n x^{n-1}$ converges for $x = x_1$. Select a real number x_0 such that $|x_1| < |x_0| < r$. Since the radius of convergence of the power series $\Sigma\, a_n x^n$ is r, this power series converges at $x = x_0$. Thus there is a real number M such that

$$|a_n x_0{}^n| \leqslant M$$

for each positive integer n. Since $0 < |x_1| < |x_0|$, we see that

$$n|a_n x_1{}^{n-1}| = n|a_n x_0{}^n| \left|\frac{x_1}{x_0}\right|^n \cdot \frac{1}{|x_1|}$$

$$\leqslant nM \left|\frac{x_1}{x_0}\right|^n \cdot \frac{1}{|x_1|}$$

for each positive integer n. The Ratio Test (Theorem 7.2.2) implies that the series

$$\Sigma\, nM\left|\frac{x_1}{x_0}\right|^n \cdot \frac{1}{|x_1|}$$

converges. Since each term of the series

$$\Sigma\, n|a_n x_1^{n-1}|$$

is less than or equal to the corresponding term of the series

$$\Sigma\, nM\left|\frac{x_1}{x_0}\right|^n \cdot \frac{1}{|x_1|},$$

the Comparison Test (Theorem 7.2.1) implies that the series

$$\Sigma\, n|a_n x_1^{n-1}|$$

converges. Since x_1 was an arbitrary real number, with $0 < |x_1| < r$, it follows that the power series $\Sigma\, na_n x^{n-1}$ converges absolutely for any real x with $|x| < r$. Hence the radius of convergence $r' \geqslant r$.

We now have the two inequalities $r' \leqslant r$ and $r \leqslant r'$. Therefore, $r = r'$, which concludes the proof.

We now apply Theorem 8.5.4 and 9.2.2 to show that if $f(x) = \Sigma\, a_n x^n$, where the power series $\Sigma\, a_n x^n$ has radius of convergence $r > 0$, then $f'(x) = \Sigma\, na_n x^{n-1}$ for each $x \in (-r, r)$.

Theorem 9.2.3. *Let $\Sigma\, a_n x^n$ be a power series with radius of convergence $r > 0$, and let f denote its sum function on $(-r, r)$. Then f is differentiable on $(-r, r)$, and*

$$f'(x) = \Sigma\, na_n x^{n-1}$$

for each $x \in (-r, r)$.

Proof. Let x be any point of the interval $(-r, r)$, and let a be any real number such that $x \in [-a, a] \subseteq (-r, r)$. Then the power series $\Sigma\, a_n x^n$ and $\Sigma\, na_n x^{n-1}$ both converge uniformly on the closed interval $[-a, a]$. Applying Theorem 8.5.4, we have

$$f'(x) = \frac{df(x)}{dx} = \frac{d}{dx}\left[\Sigma\, a_n x^n\right]$$

$$= \Sigma\, \frac{d(a_n x^n)}{dx} = \Sigma\, na_n x^{n-1}.$$

Since x was an arbitrary point of $(-r, r)$, this equation holds for each $x \in (-r, r)$. This completes the proof.

Example 9.2.1. We begin with the geometric series $\Sigma\, x^n$ which has a radius of convergence of 1. Then for each $x \in (-1, 1)$, we have

$$f(x) = \frac{1}{1-x} = 1 + x + x^2 + \dots + x^n + \dots .$$

Theorem 9.2.1 implies that the function f is continuous on the open interval $(-1, 1)$. Since

$$\frac{d[1/(1-x)]}{dx} = \frac{1}{(1-x)^2},$$

Theorems 9.2.2 and 9.2.3 imply that

$$f'(x) = \frac{1}{(1-x)^2} = 1 + 2x + 3x^2 + \dots + (n+1)\, x^n + \dots$$

for all $x \in (-1, 1)$. Theorem 9.2.1 implies that f' is continuous on $(-1, 1)$. Repeating the above process, we have

$$\frac{d[1/(1-x)^2]}{dx} = \frac{2}{(1-x)^3},$$

and Theorems 9.2.2 and 9.2.3 imply that

$$f''(x) = \frac{2}{(1-x)^3} = 2 + 6x + 12x^2$$

$$+ \dots + (n+1)(n+2)x^n + \dots$$

for each $x \in (-1, 1)$. We can now write

$$\frac{1}{(1-x)^3} = 1 + 3x + 6x^2 + \dots$$

$$+ \frac{(n+1)(n+2)}{2}\, x^n + \dots$$

for each $x \in (-1, 1)$. In general,

$$f^{(k)}(x) = \frac{k!}{(1-x)^{k+1}} = k! + \frac{(k+1)!}{1!}\, x + \frac{(k+2)!}{2!}\, x^2$$

$$+ \dots + \frac{(k+n)!}{n!}\, x^n + \dots$$

for each $x \in (-1, 1)$. Thus, we have the following general formula, valid for all $x \in (-1, 1)$:

$$\frac{1}{(1-x)^{k+1}} = 1 + (k+1)x + \frac{(k+1)(k+2)}{2}x^2$$

$$+ \cdots + \frac{(k+n)!}{n!k!}x^n + \cdots .$$

Example 9.2.1 demonstrates an application of Theorems 9.2.1, 9.2.2, and 9.2.3. We also notice that in Example 9.2.1 the function defined by the geometric series has derivatives of all orders within the interval $(-1, 1)$. The following corollary indicates that this is true for all functions defined by power series.

Corollary 9.2.3a. *Let $\Sigma a_n x^n$ be a power series with radius of convergence $r > 0$, and let f denote its sum function on $(-r, r)$. Then f has derivatives of all orders, and, for each positive integer k,*

$$f^{(k)}(x) = \sum_{n=0}^{\infty} a_{n+k} \frac{(k+n)!}{n!k!} x^n$$

for each $x \in (-r, r)$.

The proof of this corollary is based on an induction argument and is left as Exercise 18 of Problem Set 9.2.

From Corollary 9.2.3a it follows that a function defined by power series with radius of convergence $r > 0$ is infinitely differentiable on the interval $(-r, r)$, and all the derivatives can be obtained by term-by-term differentiation of power series. It then follows that the coefficients of the power series can be expressed in terms of the derivatives of the function. This result is expressed in the following corollary.

Corollary 9.2.3b. *Let $\Sigma a_n x^n$ be a power series with radius of convergence $r > 0$, and let f denote its sum function on $(-r, r)$. Then*

$$a_k = \frac{f^{(k)}(0)}{k!}$$

for each positive integer k. (Here $f^{(0)} \equiv f$.)

Proof. Let $x = 0$ in the formula given in Corollary 9.2.3a.

It is now possible to state a uniqueness result for a function defined by power series.

Corollary 9.2.3c. *Let* $f(x) = \Sigma a_n x^n$ *and* $f(x) = \Sigma b_n x^n$, *where both power series converge on some interval* $(-r, r)$ *with* $r > 0$. *Then*

$$a_n = b_n$$

for each positive integer n.

Proof. From Corollary 9.2.3b, we have

$$a_n = \frac{f^{(n)}(0)}{n!} = b_n$$

for each positive integer *n*, which completes the proof.

Theorem 9.2.3 indicates that a power series with radius of convergence $r > 0$ can be differentiated term-by-term within the interval $(-r, r)$. The following theorem proves that a power series can also be integrated term-by-term within the interval $(-r, r)$.

Theorem 9.2.4. *Let* $\Sigma a_n x^n$ *be a power series with radius of convergence* $r > 0$, *and let f denote its sum function on* $(-r, r)$. *If* $a, b \in (-r, r)$, *then*

$$\int_a^b f(x)dx = \int_a^b (\Sigma a_n x^n)dx = \Sigma a_n \int_a^b x^n dx$$

$$= \Sigma a_n(b^{n+1} - a^{n+1})/(n + 1).$$

Proof. Without loss of generality, we may assume that $a < b$. Then the $\{a_n x^n\}$ is a sequence of continuous functions that converge uniformly on the closed interval $[a, b]$. As an application of Theorem 8.5.2, we have the desired result.

Example 9.2.2. Define the exponential function by the following formula:

$$e^x = \sum_{n=0}^{\infty} x^n/n!$$

for each real number *x*. (Cf. Example 9.1.3.) This power series converges for each real *x* and converges uniformly on any compact subset of the set of real numbers. Use the series for e^x to evaluate the series

$$\frac{1}{1! \cdot 3} + \frac{1}{2! \cdot 4} + \frac{1}{3! \cdot 5} + \cdots + \frac{1}{n! \cdot (n + 2)} + \cdots .$$

We multiply the power series for e^x by x to get

$$xe^x = \sum_{n=0}^{\infty} x^{n+1}/n! = x + \frac{x^2}{1!} + \frac{x^3}{2!} + \ldots + \frac{x^{n+1}}{n!} + \ldots .$$

Since the power series defining e^x converges uniformly on compact subsets of R, the power series for xe^x also converges uniformly on compact subsets of R. Thus Theorem 9.2.4 implies that

$$\int_0^1 xe^x dx = \int_0^1 \left(\sum_{n=0}^{\infty} x^{n+1}/n! \right) dx$$

$$= \sum_{n=0}^{\infty} (1/n!) \int_0^1 x^{n+1} dx$$

$$= \sum_{n=0}^{\infty} (1/n!) \cdot \left[\frac{x^{n+2}}{n+2} \right]_0^1$$

$$= \sum_{n=0}^{\infty} (1/n!) \cdot [1/(n+2)]$$

$$= 1/2 + \sum_{n=1}^{\infty} 1/n! \cdot (n+2).$$

But $\int_0^1 xe^x dx = 1$ (cf. Example 6.5.1), and so

$$\sum_{n=1}^{\infty} 1/n! \cdot (n+2) = 1/2.$$

Problem Set 9.2

In Exercises 1–4, find f' and give the interval of convergence.

1. $f(x) = \sum_{n=1}^{\infty} x^n/n^{(n+1)}$.

2. $f(x) = \sum_{n=1}^{\infty} x^n/n^2$.

3. $f(x) = \sum_{n=0}^{\infty} (n!)^2 x^n/(2n)!$.

4. $f(x) = \sum_{n=1}^{\infty} (n+1)x^n/n$.

In Exercises 5–7, let

$$f(x) = \sum_{n=0}^{\infty} (-1)^n x^{2n}/(2n)! \quad \text{and} \quad g(x) = \sum_{n=0}^{\infty} (-1)^n x^{2n+1}/(2n+1)!.$$

5. Find the radius of convergence for f and g.

6. Find f' and g', and show that $f' = -g$ and $g' = f$.

7. Find f'' and g'', and show that $f'' = -f$ and $g'' = -g$.

In Exercises 8–11, compute $\int_0^x f(\zeta)d\zeta$.

8. $f(x) = \sum_{n=1}^{\infty} x^n/(n+1)^n$.

9. $f(x) = \sum_{n=1}^{\infty} x^n/(n+1)^2$.

10. $f(x) = \sum_{n=0}^{\infty} (n!)^2 x^n/(2n)!$.

11. $f(x) = \sum_{n=1}^{\infty} (n+1)x^n/n$.

12. Let $f(x) = \sum_{n=0}^{\infty} x^n/n!$. Find the radius of convergence of f. Show that

$$f'(x) = f(x) \text{ and } f(0) = 1.$$

In Exercises 13–17, find the sum of the given infinite series.

13. $\sum_{n=0}^{\infty} (n+1)x^n$.

14. $\sum_{n=1}^{\infty} x^n/n(n+1)$.

15. $\sum_{n=3}^{\infty} x^n/n(n-2)$.

16. $\sum_{n=0}^{\infty} (-1)^n(n+1)x^{2n}$.

17. $\sum_{n=1}^{\infty} n^2 x^n$.

18. Prove Corollary 9.2.3a.

19. Prove that the function defined by a power series is determined by its values on any set that has zero as a limit point.

20. Prove that the power series obtained from a given power series by term-by-term integration has the same radius of convergence as the given series.

21. Let $f(x) = \sum_{n=0}^{\infty} a_n x^n$ have radius of convergence $r > 0$. If f is an even function, show that $a_n = 0$ for all odd n. (See Exercise 23 of Problem Set 6.3.)

22. Let $f(x) = \sum_{n=0}^{\infty} a_n x^n$ have radius of convergence $r > 0$. If f is an odd function, show that $a_n = 0$ for all even n. (See Exercise 25 of Problem Set 6.3.)

23. Let $f(x) = \sum_{n=0}^{\infty} a_n x^n$ have radius of convergence $r > 0$. Let $g(x) = \sum_{n=0}^{\infty} n^3 a_n x^n$.

Express the function g in terms of the function f.

24. Let $e^x = \sum_{n=0}^{\infty} x^n/n!$. Find a power series in x for $\dfrac{d}{dx}\left(\dfrac{e^x-1}{x}\right)$. Show that

$$\sum_{n=1}^{\infty} n/(n+1)! = 1.$$

25. Show that $\dfrac{3}{2\cdot 1!} - \dfrac{5}{2^2\cdot 2!} + \dfrac{7}{2^3\cdot 3!} - \ldots = 1$ by considering $\dfrac{d^2}{dx^2}(e^{-x^2})$.

(Cf. Exercise 24.)

26. Show that $\sum_{n=1}^{\infty}(-2)^{n+1}(n+2)/n! = 4$ by considering $\dfrac{d}{dx}(x^2 e^{-x})$. (Cf. Exercise 24.)

27. Develop a theory for power series of the form $\Sigma\, a_n(x-a)^n$ similar to the theory in Section 9.2 and Problem Set 9.2.

9.3. Taylor Series

Definition 9.3.1. *The function f is said to be **representable** by the series of functions Σf_n on the set S if and only if the series $\Sigma f_n(x)$ converges to $f(x)$ for each $x \in S$. The series of functions Σf_n is said to **represent** the function f on the set S if and only if f is representable by Σf_n on S.*

Example 9.3.1. Let

$$f(x) = \sum_{n=0}^{\infty} a_n(x - a)^n$$

for each $x \in (a - r, a + r)$, that is, the power series $\Sigma a_n(x - a)^n$ has radius of convergence r. Since the power series converges to $f(x)$ for each $x \in (a - r, a + r)$, the function f is representable by the power series $\Sigma a_n(x - a)^n$ on the interval $(a - r, a + r)$. Corollary 9.2.3c implies that the power series $\Sigma a_n(x - a)^n$ is the only power series in terms of $(x - a)$ that represents the function f on $(a - r, a + r)$, that is, the power series representation of a function is unique.

Assume that the function f is represented by the power series $\Sigma a_n(x - a)^n$ on the interval $(a - r, a + r)$. Then Corollary 9.2.3b implies that

$$a_n = f^{(n)}(a)/n!$$

for each positive integer n. Thus, if a function is represented by a power series then that power series must be the power series

$$\sum_{n=0}^{\infty} f^{(n)}(a)(x - a)^n/n!.$$

This leads to the following definition.

Definition 9.3.2. *If f is representable by the power series*

$$\sum_{n=0}^{\infty} f^{(n)}(a)(x - a)^n/n!$$

*on the interval $(a - r, a + r)$, where $r > 0$, then this series is called the **Taylor series** of the function f at a.* *

If $a = 0$ in the Taylor series of the function f, then the Taylor series $\Sigma a_n x^n$ is sometimes called the **Maclaurin series** ** of the function f. However, we shall always refer to the series as the Taylor series whether or not $a = 0$.

We again emphasize the fact that we are only considering power series (Taylor series) of the form $\Sigma a_n x^n$ ($\Sigma f^n(0)x^n/n!$). In order to derive similar results for power series (Taylor series) of the form $\Sigma a_n(x - a)^n$ ($\Sigma f^n(a)(x - a)^n/n!$), it is only necessary to make the substitution $\zeta = x - a$ in the power series $\Sigma a_n \zeta^n$ (Taylor series $\Sigma f^{(n)}(0)\zeta^n/n!$). This remark holds for all the results derived in this section.

---

* The term *Taylor series* is in honor of the English mathematician Brook Taylor (1685-1731), who in 1715 first gave an infinite series expansion for a function.

** The term *Maclaurin series* is in honor of the mathematician Colin Maclaurin (1698-1746). He was the leading British mathematician of his time.

From Corollary 9.2.3a it follows that any function defined by a power series with radius of convergence $r > 0$ has derivatives of all orders on the interval $(-r, r)$. Thus, if a function is to have a Taylor series, then the function must be infinitely differentiable on some interval. Being infinitely differentiable is a necessary condition for a function to have a Taylor series, but we now show that this condition is not sufficient. That is, there exist functions which have derivatives of all orders, but which are not representable by a power series.

Example 9.3.2. Let

$$f(x) = \begin{cases} e^{-1/x^2} & \text{if } x \neq 0, \\ 0 & \text{if } x = 0. \end{cases}$$

Then f is infinitely differentiable for all real x, and $f^{(n)}(0) = 0$ for each positive integer n. (Cf. Exercise 10 of Problem Set 5.4.) If f were to have a Taylor series $\Sigma f^{(n)}(0)x^n/n!$, then Corollary 9.2.3b would imply that it would be the zero Taylor series, that is, the series with all terms zero. But the zero series converges to the zero function and not to f. Hence f has no Taylor series.

The remainder of this section is devoted to developing a sufficient condition for an infinitely differentiable function to have a Taylor series.

Assume that f is a function that is n times differentiable on the interval $(-r, r)$, $r > 0$. For each positive integer n and each $x \in (-r, r)$, define the function R_n by the equation

$$R_n(x) = f(x) - \sum_{k=0}^{n} f^{(k)}(0)x^k/k!.$$

Then

$$f(x) = \sum_{k=0}^{n} f^{(k)}(0)x^k/k! + R_n(x)$$

for each $x \in (-r, r)$. This equation is known as **Taylor's formula with remainder** and may be written for any function that is n times differentiable on an open interval containing the origin. The **remainder** $R_n(x)$ is simply the difference between the function f and the **approximating polynomial**

$$\sum_{k=0}^{n} f^{(k)}(0)x^k/k!.$$

We shall use Taylor's formula with remainder to find when an infinitely differentiable function has a Taylor series.

If the function f is $(n+1)$-times differentiable on the interval $(-r, r)$, where $r > 0$, then the Extended Mean Value Theorem (Theorem 5.4.1)

implies that
$$R_n(x) = f^{(n+1)}(c)x^{n+1}/(n+1)!$$

for each $x \in (-r, r)$, where c is a real number between 0 and x. This is called the **Lagrange form of the remainder** and is named in honor of the French mathematician Joseph-Louis Lagrange (1736–1813), who first discovered an explicit form for the remainder. There are many other forms for the remainder; if $f^{(n+1)}$ is an integrable function, then one of these is called the **integral form of the remainder** and is given by the formula

$$R_n(x) = \frac{1}{n!}\int_0^x (x - \zeta)^n f^{(n+1)}(\zeta)d\zeta.$$

It is left as an exercise to verify the above equation (Exercise 17 of Problem Set 9.3).

Returning to infinitely differentiable functions, we see that

$$R_n(x) = f(x) - \sum_{k=0}^{n} f^{(k)}(0)x^k/k!,$$

for each positive integer n, and that the power series

$$\sum_{n=0}^{\infty} f^{(n)}(0)x^n/n!$$

converges to $f(x)$ for each $x \in (-r, r)$ if and only if the sequence of functions (remainders) converges to the zero function on $(-r, r)$, that is, the sequence $\{R_n(x)\}$ converges to zero for each $x \in (-r, r)$. This proves the following theorem.

Theorem 9.3.1. *The power series*

$$\sum_{n=0}^{\infty} f^{(n)}(0)x^n/n!$$

is the Taylor series of the function f on the interval $(-r, r)$, where $r > 0$, if and only if the sequence of remainders $\{R_n(x)\}$ converges to the zero function on $(-r, r)$.

Example 9.3.3. Let f be an infinitely differentiable function such that the sequence $\{f^{(n)}\}$ is uniformly bounded on the interval $(-r, r)$, where $r > 0$, that is,

$$|f^{(n)}(x)| \leqslant M$$

for each $x \in (-r, r)$ and for each positive integer n. Since

$$R_n(x) = f^{(n+1)}(c)x^{n+1}/n!$$

for each $x \in (-r, r)$, where c is between 0 and x (see the Lagrange form of the remainder), we have

$$|R_n(x)| \leqslant Mr^{(n+1)}/n!$$

for each $x \in (-r, r)$ and for each positive integer n. Thus, the sequence of remainders $\{R_n(x)\}$ converges to the zero function on $(-r, r)$ and the power series

$$\sum_{n=0}^{\infty} f^{(n)}(0)x^n/n!$$

is the Taylor series for the function f on the interval $(-r, r)$.

If a function has a Taylor series, then one important use of the formulas for the remainder in Taylor's formula with remainder is to give bounds on the closeness of the approximating polynomials to the function.

Example 9.3.4. Let $f: R \to R$ be the exponential function defined by $f(x) = e^x$ for each real number x. Since

$$f'(x) = \frac{de^x}{dx} = e^x$$

for each real number x, the Taylor's formula with remainder for f is given by

$$e^x = \sum_{k=0}^{n} x^k/k! + R_n(x),$$

where $R_n(x) = e^c x^{n+1}/(n+1)!$ with c between 0 and x. If $x < 0$, then $c < 0$ and

$$|R_n(x)| < |x|^{n+1}/(n+1)!$$

for each positive integer n, while if $x > 0$ then $e^c < e^x$ and

$$|R_n(x)| < e^x x^{n+1}/(n+1)!$$

for each positive integer n. In either case, for fixed x we see that

$$\lim_n |R_n(x)| = 0.$$

In fact, the sequence $\{R_n(x)\}$ converges uniformly to zero on compact sets of real numbers. Therefore the Taylor series of the exponential function is given by the formula

$$e^x = \sum_{k=0}^{\infty} x^k/k!$$

Problem Set 9.3

In Exercise 1–7, write out the Taylor's formula with remainder for the function f. Estimate the size of the remainder when $n = 5$.

1. $f(x) = \sin x$.
2. $f(x) = \cos (2x)$.
3. $f(x) = 1/(1 + x)$.
4. $f(x) = \ln (1 + x)$.
5. $f(x) = e^{-x}$.
6. $f(x) = e^{x^2}$.
7. $f(x) = x^5 + 4x^4 + 2x^3 + x + 1$.

In Exercises 8–14, write out the first four nonzero terms of the Taylor series for f.

8. $f(x) = \sin x$.
9. $f(x) = \cos (2x)$.
10. $f(x) = 1/(1 + x)$.
11. $f(x) = \ln (1 + x)$.
12. $f(x) = e^{-x}$.
13. $f(x) = e^{x^2}$.
14. $f(x) = x^5 + 4x^4 + 2x^3 + x + 1$.
15. Let f be an even function which has a Taylor series. Prove that only even powers of x occur in the Taylor series of f. (See Exercise 23 of Problem Set 6.3.)
16. Let f be an odd function which has a Taylor series. Prove that only odd powers of x occur in the Taylor series of f. (See Exercise 25 of Problem Set 6.3.)

For Exercises 17–19, let n be a fixed positive integer, let f have n continuous derivatives on $(-r, r)$, where $r > 0$, and let $f^{(n+1)}$ exist on $(-r, r)$. Let $R_n(x)$ be the remainder in Taylor's formula with remainder.

17. If $f^{(n+1)}$ is integrable on closed subintervals of $(-r, r)$, prove that

$$R_n(x) = \frac{1}{n!} \int_0^x (x - \zeta)^n f^{(n+1)}(\zeta)d\zeta,$$

for each $x \in (-r, r)$. [*Hint*: Use induction, and integrate by parts.]

18. Prove that

$$R_n(x) = \frac{f^{(n+1)}(c)}{n!} x(x - c)^n$$

for each $x \in (-r, r)$, where c is between 0 and x. [This is called the **Cauchy form of the remainder.**]

19. Prove that

$$R_n(x) = \frac{f^{(n+1)}(c)}{p \cdot n!} x^p (x - c)^{n-p+1}$$

for each $x \in (-r, r)$, where c is between 0 and x, and for each integer p with $0 \leqslant p \leqslant n$. [This is called the **Schlömilch form of the remainder.**]

20. Show that $|x|$ has no Taylor series.
21. Show that $\sqrt{x}$ has no Taylor series.
22. Show that $\ln x$ has no Taylor series.
23. Show that $\cot x$ has no Taylor series.
24. Let f be infinitely differentiable on the interval $(-r, r)$, where $r > 0$. Prove that a necessary and sufficient condition that f has a Taylor series is that there exist positive numbers σ, ρ, and M such that

$$|f^{(n)}(x)| \leqslant n! \, M/\sigma^n$$

for each positive integer n and for each $x \in (-\rho, \rho)$.

For Exercises 25–26, let $\Sigma \, a_n x^n$, with radius of convergence r_1, be the Taylor series for the function f, and let $\Sigma \, b_n x^n$, with radius of convergence r_2, be the Taylor series for the function g. Let $r = \min (r_1, r_2)$.

25. Let α and β be any two real numbers. Prove that the power series $\Sigma \, (\alpha a_n + \beta b_n) x^n$ has radius of convergence of at least r and that this series is the Taylor series of the function $\alpha f + \beta g$.
26. Let $\Sigma \, c_n$ be the Cauchy product of the series $\Sigma \, a_n$ and $\Sigma \, b_n$. (Cf. Definition 7.5.1.) Prove that the power series $\Sigma \, c_n x^n$ has radius of convergence of at least r and that this series is the Taylor series of the function fg. [*Hint*: Use Theorem 7.5.4.]
27. Find the Taylor series for $\cosh x$. [*Hint*: Write $\cosh x = (e^x + e^{-x})/2$ and use Exercise 25.]
28. Find the Taylor series for $\cos (x + \pi/4)$. [*Hint*: Write $\cos (x + \pi/4) = (\cos x)(\cos \pi/4) - (\sin x)(\sin \pi/4)$ and use Exercise 25.]
29. Find the Taylor series for $e^x \cos x$. [*Hint*: Use Exercise 26.]
30. Find the Taylor series for $e^x \ln (1 + x)$. [*Hint*: Use Exercise 26.]
31. If $a_0 \neq 0$ and the power series $\Sigma \, a_n x^n$ has radius of convergence $r > 0$, prove that there exists a unique power series $\Sigma \, b_n x^n$ with positive radius of convergence such that $(\Sigma \, a_n x^n)(\Sigma \, b_n x^n) = 1$. What is the radius of convergence of the power series $\Sigma \, b_n x^n$? [*Hint*: Use Exercise 26.]
32. Determine the inverse power series of $\cos x$. That is, find a power series $\Sigma \, a_n x^n$ such that $(\cos x) \Sigma \, a_n x^n = 1$. [*Hint*: Use Exercise 31.]
33. Using the fact that

$$\tan^{-1} x = \int_0^x d\zeta/(1 + \zeta^2),$$

find a Taylor series for $\tan^{-1} x$, and evaluate π to four decimal places.
34. Develop a theory for Taylor series of the form $\Sigma \, f^{(n)}(a)(x - a)^n/n!$ similar to the theory in Section 9.3 and Problem Set 9.3.
35. Find the Taylor series in $(x - a)$ for e^x. [*Hint*: Write $e^x = e^a e^{x-a}$, and represent e^{x-a} by a Taylor series.]
36. Find the Taylor series in $(x - a)$ for $\cos x$.
   [*Hint*: Write $\cos x = \cos (a + (x - a))$.]

9.4. Abel's Theorem and Tauber's Theorem

Let $\Sigma\, a_n x^n$ be a power series with radius of convergence 1, and let f denote its sum function. The power series may or may not converge at $x = 1$. If the power series does converge at $x = 1$, then the function f is defined at $x = 1$ by the equation $f(1) = \Sigma\, a_n$. The function f is continuous on the open interval $(-1, 1)$. Knowing that the power series converges at $x = 1$, can we conclude anything about the continuity of f on the interval $(-1, 1]$? This question is answered by the following theorem due to Niels Abel. *Abel's Theorem* shows that the power series converges uniformly on the interval $[0, 1]$, and thus the sum function is continuous on the interval $(-1, 1]$.

Theorem 9.4.1. (*Abel's Theorem*) *Let $\Sigma\, a_n x^n$ be a power series with radius of convergence 1. If $\Sigma\, a_n$ converges, then the power series $\Sigma\, a_n x^n$ converges uniformly on the closed interval $[0, 1]$.*

Proof. Let $\varepsilon > 0$. We need to show that there exists a positive number N such that $n > m > N$ implies that

$$\left| \sum_{k=m+1}^{n} a_k x^k \right| < \varepsilon$$

for all values of $x \in [0, 1]$.

Since the series $\Sigma\, a_n$ converges, there is a positive number N such that

$$\left| \sum_{k=m+1}^{n} a_k \right| < \varepsilon$$

whenever $n > m > N$. If $0 \leqslant x \leqslant 1$, then the sequence $\{x^n\}$ is a decreasing sequence of nonnegative real numbers, and, applying Abel's Lemma (Lemma 7.5.1) to the sequences $\{a_k\}_{k=m+1}^{+\infty}$ and $\{x^k\}_{k=m+1}^{+\infty}$, we have

$$\left| \sum_{k=m+1}^{n} a_k x^k \right| < \varepsilon\, x^{m+1} \leqslant \varepsilon$$

whenever $n > m > N$. Thus Corollary 8.2.1 implies that the power series $\Sigma\, a_n x^n$ converges uniformly on the closed interval $[0, 1]$.

Corollary 9.4.1a. *Let $\Sigma\, a_n x^n$ be a power series with radius of convergence 1, and let f be its sum function. If the power series $\Sigma\, a_n x^n$ converges at $x = -1$, then it converges uniformly on the closed interval $[-1, 0]$.*

The proof is left as Exercise 7 of Problem Set 7.4.

Corollary 9.4.1b. *Let $\Sigma a_n x^n$ be a power series with radius of convergence 1, and let f be its sum function. If Σa_n converges, then f is continuous on the closed interval $[0, 1]$ and*

$$\lim_{x \to 1^-} f(x)$$

exists and equals Σa_n.

Example 9.4.1. We consider the series

$$\ln(1 + x) = x - \frac{x^2}{2} + \frac{x^3}{3} - \frac{x^4}{4} + \ldots.$$

This series converges for $x = 1$ (since for $x = 1$ it is the alternating harmonic series) and Corollary 9.4.1b implies that

$$\ln 2 = \lim_{x \to 1^-} \ln(1 + x)$$

$$= 1 - \frac{1}{2} + \frac{1}{3} - \frac{1}{4} + \ldots.$$

Thus the sum of the alternating harmonic series is $\ln 2$.

Example 9.4.2. If, in the simple geometric series

$$\frac{1}{1 - x} = 1 + x + x^2 + x^3 + \ldots$$

with radius of convergence 1, we replace x by $-x^2$, we obtain the series

$$\frac{1}{1 + x^2} = 1 - x^2 + x^4 - x^6 + \ldots,$$

which also has radius of convergence 1. If $|x| < 1$ and we integrate 0 to x, we get

$$\tan^{-1} x = x - \frac{x^3}{3} + \frac{x^5}{5} - \frac{x^7}{7} + \ldots,$$

with radius of convergence 1. This series converges at $x = 1$, and it follows from Corollary 9.4.1b that

$$\tan^{-1} 1 = \frac{\pi}{4} = 1 - \frac{1}{3} + \frac{1}{5} - \frac{1}{7} + \ldots.$$

An important aspect of Corollary 9.4.1b is that it suggests a method of assigning limits to certain divergent series. If Σa_n is an infinite series such that

$$\lim_n \sup |a_n|^{1/n} = 1,$$

then the power series $\Sigma a_n x^n$ has radius of convergence 1, and it defines a function f on the interval $(-1, 1)$. If

$$\lim_{x \to 1^-} f(x)$$

exists and equals A (finite or infinite), then the series Σa_n is said to be **Abel summable** to A, and A is called its **Abel sum**. Corollary 9.4.1b asserts that if a series is convergent, then it is Abel summable to the sum of the series.

The converse of Corollary 9.4.1b is not true, as the following example indicates.

Example 9.4.3. For the series

$$\sum_{n=0}^{\infty} (-1)^n,$$

we form the power series

$$\sum_{n=0}^{\infty} (-1)^n x^n.$$

The power series is divergent for $x = 1$, since the limit of its nth term is not zero. But for each real x with $|x| < 1$ we have

$$\frac{1}{1 + x} = \sum_{n=0}^{\infty} (-1)^n x^n.$$

Since

$$\lim_{x \to 1^-} \frac{1}{1 + x} = \frac{1}{2},$$

the divergent series

$$\sum_{n=0}^{\infty} (-1)^n$$

is Abel summable to 1/2.

Example 9.4.3 demonstrates that a series which is Abel summable need not be convergent. If a series is Abel summable, there are other conditions that may be added to ensure that the series is also convergent. Results along this line are called **Tauberian theorems.** Such theorems are important because the added hypothesis allows one to go from a *weak form of convergence* to actual convergence of the series.

We shall prove the original Tauberian theorem. This result was formulated in 1897 by Alfred Tauber (1866–1947), a Viennese mathematician, and is called *Tauber's Theorem*. Example 9.4.3 asserts that the converse of Abel's Theorem is not true, but Tauber's Theorem does give a partial converse.

Theorem 9.4.2. (*Tauber's Theorem*) *Let*

$$\sum_{n=0}^{\infty} a_n x^n$$

be a power series with radius of convergence 1 *such that*

$$\lim_n na_n = 0,$$

and let f be its sum function. If

$$\lim_{x \to 1^-} f(x) = A \ (\textit{finite}),$$

then the series

$$\sum_{n=0}^{\infty} a_n$$

converges to A.

This theorem may also be stated as follows: *If Σa_n is Abel summable to A (finite) and $\lim_n na_n = 0$, then Σa_n converges to A.*

Proof. Let $\varepsilon > 0$ be given. We need to show that there is a positive number N such that

$$\left| \sum_{k=0}^{n} a_k - A \right| < \varepsilon$$

when $n > N$.

Since

$$\lim_n na_n = 0,$$

there is a positive number N_1 such that

$$|na_n| < \varepsilon/3$$

when $n > N_1$. Also, it follows that

$$\lim_n |na_n| = 0$$

(cf. Exercise 14 of Problem Set 3.1) and that

$$\lim_n \frac{1}{n} \sum_{k=1}^{n} |ka_k| = 0$$

(cf. Exercise 11 of Problem Set 3.2). Thus there is a positive number N_2 such that

$$\frac{1}{n} \sum_{k=1}^{n} |ka_k| < \varepsilon/3$$

when $n > N_2$. Finally, since

$$\lim_{x \to 1-} f(x) = A,$$

there is a positive number N_3 such that

$$|A - f(x)| < \varepsilon/3$$

when $1 - 1/N_3 < x < 1$. Let $N = \max(N_1, N_2, N_3)$. Then $n > N$ implies that

$$|na_n| < \varepsilon/3$$

and

$$\frac{1}{n} \sum_{k=1}^{n} |ka_k| < \varepsilon/3,$$

and $1 - 1/N < x < 1$ implies that

$$|A - f(x)| < \varepsilon/3.$$

For any positive integer n and for any $x \in (0, 1)$, we may write

$$A - \sum_{k=0}^{n} a_k = A - f(x) + f(x) - \sum_{k=0}^{n} a_k$$

$$= A - f(x) + \sum_{k=0}^{\infty} a_k x^k - \sum_{k=0}^{n} a_k$$

$$= A - f(x) - \sum_{k=1}^{n} a_k(1 - x^k) + \sum_{k=n+1}^{\infty} a_k x^k.$$

Hence

$$\left| A - \sum_{k=0}^{n} a_k \right| \leqslant |A - f(x)| + \sum_{k=0}^{n} |a_k|(1 - x^k) + \sum_{k=n+1}^{\infty} |a_k| x^k.$$

Let

$$S_1 = |A - f(x)|,$$

$$S_2 = \sum_{k=1}^{n} |a_k|(1 - x^k),$$

and

$$S_3 = \sum_{k=n+1}^{\infty} |a_k| x^k.$$

Let n be any fixed positive integer greater than N. Choose x such that

$$1 - 1/n < x < 1 - 1/(n + 1).$$

Then

$$1 - 1/N < 1 - 1/n < x < 1,$$

and so
$$S_1 = |A - f(x)| < \varepsilon/3.$$
Now
$$1 - x^k = (1 - x)(1 + x + \dots + x^{k-1}) \leqslant k(1 - x)$$
for any positive integer k. Since
$$1 - 1/n < x,$$
we have
$$1 - x < 1/n$$
and
$$1 - x^k \leqslant k(1 - x) < k/n.$$
It follows that
$$S_2 = \sum_{k=1}^{n} |a_k|(1 - x^k) < \frac{1}{n} \sum_{k=1}^{n} |ka_k| < \varepsilon/3.$$
Finally, since
$$|na_n| < \varepsilon/3,$$
we obtain
$$S_3 = \sum_{k=n+1}^{\infty} |a_k|x^k = \sum_{k=n+1}^{\infty} |ka_k|x^k/k < \frac{\varepsilon}{3} \sum_{k=n+1}^{\infty} x^k/k$$
$$< \frac{\varepsilon}{3(n+1)} \sum_{k=n+1}^{\infty} x^k < \frac{\varepsilon}{3(n+1)} \sum_{k=0}^{\infty} x^k = \frac{\varepsilon}{3(n+1)} \cdot \frac{1}{1-x}.$$
But
$$x < 1 - 1/(n+1),$$
and so
$$1 - x > 1/(n+1).$$
Thus
$$(n+1)(1 - x) > 1,$$
and
$$S_3 \leqslant \frac{\varepsilon}{3(n+1)(1-x)} < \frac{\varepsilon}{3}.$$
Therefore
$$\left| \sum_{k=0}^{n} a_k - A \right| \leqslant S_1 + S_2 + S_3 < \varepsilon$$
for $n > N$, which proves that the series
$$\sum_{n=0}^{\infty} a_n$$
converges to A.

Problem Set 9.4

For Exercises 1–4, show that each series is Abel summable.

1. $1 - 1/4 + 1/9 - 1/16 + \dots$.
2. $1 - 1/2 + 1/3 - 1/4 + \dots$.
3. $1 - 3 + 5 - 7 \dots$.
4. $1 + 2 + 3 + 4 + \dots$.
5. If Σa_n is Abel summable to A and Σb_n is Abel summable to B, then prove that $\Sigma (\alpha a_n + \beta b_n)$ is Abel summable to $\alpha A + \beta B$, where α and β are any real numbers.
6. If Σa_n is Abel summable to A, then prove that

$$0 + a_0 + 0 + a_1 + 0 + a_2 + \dots$$

is also Abel summable to A.
7. Prove Corollary 9.4.1a.
8. Let $r > 0$. If the series $\Sigma a_n r^n$ converges, then prove that the power series $\Sigma a_n x^n$ converges uniformly on the closed interval $[0, r]$.
9. Let $r > 0$. If the series $\Sigma a_n (-r)^n$ converges, then prove that the power series $\Sigma a_n x^n$ converges uniformly on the closed interval $[-r, 0]$.
10. Use Abel's Theorem to prove the following: If Σa_n converges to A, Σb_n converges to B, and their Cauchy product Σc_n is convergent, then Σc_n converges to AB.
11. Let $\Sigma a_n x^n$ be a power series with radius of convergence 1 and let f be its sum function. If $a_n \geqslant 0$ for each positive integer n and Σa_n diverges, then prove that f is unbounded.
12. Let $\Sigma a_n x^n$ be a power series with radius of convergence 1, and let f be its sum function. If the series

$$\sum_{n=0}^{\infty} a_n/(n + 1)$$

is convergent, then prove that

$$\int_0^1 f(x)dx = \sum_{n=0}^{\infty} a_n/(n + 1)$$

even though the series Σa_n may not converge.
13. Let $\Sigma b_n x^n$ be a power series with radius of convergence 1, and let $b_n > 0$ for each positive integer n. If $s = \lim_{n} (a_n/b_n)$, then prove that the radius of convergence of the power series $\Sigma a_n x^n$ is 1 and that

$$\lim_{x \to 1^-} \frac{\Sigma a_n x^n}{\Sigma b_n x^n} = s.$$

[*Hint*: First consider the case where $s = 0$. Use the fact that

$$\lim_{x \to 1-} [1/\Sigma b_n x^n] = 0.]$$

14. Apply Exercise 13 with $b_n = 1$ for each positive integer n to obtain Abel's Theorem.

15. Let $\{a_n\}$ be a sequence of real numbers, and let $a_0 = 0$. Let $s_n = a_1 + \ldots + a_n$, and let $\sigma_n = (s_1 + \ldots + s_n)/n$. Prove that if

$$s = \lim_n \sigma_n,$$

then

$$s = \lim_{x \to 1-} \sum_{n=0}^{\infty} a_n x^n.$$

[*Hint*: Apply Exercise 13 to

$$f(x) = \frac{1}{(1-x)^2} = \sum_{n=1}^{\infty} n x^{n-1}$$

and note that

$$\sum_{n=1}^{\infty} n\sigma_n x^n = f(x) \sum_{n=1}^{\infty} a_n x^n.]$$

Chapter Ten

IMPROPER INTEGRALS

10.1. Improper Integrals

In Chapter 6 the Riemann integral

$$\int_a^b f(x)\,dx$$

was defined and elementary properties of the integral were derived. In the definition of this integral it was required that the function f be bounded on the compact interval $[a, b]$. For many applications requiring integrals these conditions are not always satisfied. We now consider integrals having unbounded intervals of integration or unbounded integrands and extend the concept of integration. We begin with the following definition.

> **Definition 10.1.1.** *Let f be a function with domain $[a, b)$, where b is a real number or $+\infty$. If f is integrable on every closed interval $[a, c]$, where $a < c < b$, but is not integrable on the interval $[a, b]$, then the symbol*
>
> $$\int_a^b f(x)dx$$
>
> *is called the **improper integral** of f on $[a, b)$. If $b = +\infty$, then the improper integral is called an **improper integral of the first kind**; while if b is a real number, then the improper integral is called an **improper integral of the second kind**.*

The improper integral

$$\int_a^b f(x)dx$$

is said to be **convergent** *if and only if the limit*

$$\lim_{c \to b^-} \int_a^c f(x)dx$$

exists and is finite. If

$$\lim_{c \to b^-} \int_a^c f(x)dx = L,$$

where L is a real number, then the improper integral is said to **converge** *to L or have* **value** *L. If the limit does not exist or is not finite, then the improper integral is said to* **diverge** *or to be* **divergent**.

The following examples illustrate some improper integrals.

Example 10.1.1. Let $f(x) = x^p$, where p is a real number, for each $x \geqslant 1$. Then

$$\int_1^{+\infty} x^p dx$$

is an improper integral of the first kind.

If we consider the integral

$$\int_1^c x^p dx,$$

where c is any real number greater than 1 and $p \neq -1$, we see that

$$\int_1^c x^p dx = \frac{x^{p+1}}{p+1}\bigg]_1^c = \frac{c^{p+1}}{p+1} - \frac{1}{p+1}.$$

If $p > -1$, then

$$\lim_{c \to +\infty} \int_1^c x^p dx = +\infty,$$

and the improper integral

$$\int_1^{+\infty} x^p dx$$

diverges. If $p < -1$, then

$$\lim_{c \to +\infty} \int_1^c x^p dx = \frac{-1}{p+1},$$

and the improper integral converges to $-1/(p + 1)$.

For the case where $p = -1$, we have

$$\int_1^c x^{-1} dx = \ln x \Big]_1^c = \ln c.$$

Thus

$$\lim_{c \to +\infty} \int_1^c x^{-1} dx = +\infty,$$

and the improper integral diverges.

Example 10.1.2. Let $f(x) = (1 - x)^p$, where p is a negative real number, for each x such that $0 \leqslant x < 1$. Then

$$\int_0^1 (1 - x)^p dx$$

is an improper integral of the second kind.

If we consider the integral

$$\int_0^c (1 - x)^p dx,$$

where $0 \leqslant c < 1$ and $p \neq -1$, we see that

$$\int_0^c (1 - x)^p dx = \frac{-(1 - x)^{p+1}}{p+1} \Big]_0^c = \frac{1}{p+1} - \frac{(1 - c)^{p+1}}{p+1}.$$

If $p > -1$, then

$$\lim_{c \to 1-} \int_0^c (1 - x)^p dx = \frac{1}{p+1},$$

and the improper integral is convergent and has value $1/(p + 1)$.

If $p < -1$, then

$$\lim_{c \to 1-} \int_0^c (1 - x)^p dx = +\infty,$$

and the improper integral diverges.

If $p = -1$, then

$$\int_0^c (1 - x)^{-1} dx = -\ln(1 - c).$$

Hence

$$\lim_{c \to 1^-} \int_0^c (1 - x)^{-1} dx = +\infty,$$

and the improper integral diverges.

As we shall see, Definition 10.1.1 does not exhaust all the possibilities for defining improper integrals; Definitions 10.1.2 and 10.1.3 below define the other types.

Definition 10.1.2. *Let f be a function with domain* $(a, b]$*, where a is a real number or* $-\infty$*. If f is integrable on every closed interval* $[c, b]$*, where* $a < c < b$*, but is not integrable on the interval* $[a, b]$*, then the symbol*

$$\int_a^b f(x) dx$$

is called the **improper integral** *of f on* $(a, b]$*. If* $a = -\infty$*, then the improper integral is called an* **improper integral of the first kind**, *while if a is a real number, then the improper integral is called an* **improper integral of the second kind***. The improper integral,*

$$\int_a^b f(x) dx,$$

is said to be **convergent** *if and only if the limit*

$$\lim_{c \to a^+} \int_c^b f(x) dx$$

exists and is finite. If

$$\lim_{c \to a^+} \int_c^b f(x) dx = L,$$

where L is a real number, then the improper integral is said to **converge** *to L or have* **value** *L. If the limit does not exist or is not finite, then the improper integral is said to* **diverge** *or be* **divergent***.

Example 10.1.3. Let $f(x) = 1/(1 + x^2)$ for each $x \leqslant 0$. Then

$$\int_{-\infty}^{0} \frac{dx}{1 + x^2}$$

is an improper integral of the first kind.

For any $c < 0$, we have

$$\int_{c}^{0} \frac{dx}{1 + x^2} = \tan^{-1} x \Bigg]_{c}^{0} = -\tan^{-1} c.$$

Thus

$$\lim_{c \to -\infty} \int_{c}^{0} \frac{dx}{1 + x^2} = \frac{\pi}{2},$$

and the improper integral converges to $\pi/2$.

Example 10.1.4. Let $f(x) = \ln x$ for each x, with $0 < x \leqslant 1$, and let $f(0) = 0$. Then

$$\int_{0}^{1} \ln x \, dx$$

is an improper integral of the second kind. If c is any real number, with $0 < c < 1$, we have

$$\int_{c}^{1} \ln x \, dx = x \ln x - x \Bigg]_{c}^{1}$$

$$= -1 - c \ln c + c.$$

Thus

$$\lim_{c \to 0+} \int_{c}^{1} \ln x \, dx = -1,$$

and the improper integral is convergent and has value -1.

The last type of improper integrals that we shall consider involves improper integrals of both the first and the second kinds, and these are called *improper integrals of mixed kind.*

Definition 10.1.3. *Let a be a real number or $-\infty$, let b be a real number or $+\infty$, and let $a < c < b$. If the integral*

$$\int_{a}^{c} f(x) dx$$

is an improper integral of either first kind or second kind, and if the integral

$$\int_c^b f(x)dx$$

is an improper integral of either first kind or second kind, then the sum

$$\int_a^c f(x)dx + \int_c^b f(x)dx$$

is called an **improper integral of mixed kind** *and is denoted by*

$$\int_a^b f(x)dx.$$

If both of the improper integrals

$$\int_a^c f(x)dx \quad and \quad \int_c^b f(x)dx$$

are convergent, then the improper integral of mixed kind,

$$\int_a^b f(x)dx,$$

is said to be **convergent**. *In this case, the* **value** *of the improper integral of mixed kind is the sum*

$$\int_a^c f(x)dx + \int_c^b f(x)dx.$$

If at least one of the improper integrals

$$\int_a^c f(x)dx \quad and \quad \int_c^b f(x)dx$$

is divergent, then the improper integral of mixed kind is said to be **divergent**.

There are a number of ways that the improper integral of mixed kind,

$$\int_a^b f(x)dx,$$

may be improper. We consider these in the following cases.

CASE 1. If the values of a and b are both finite, then the function f is unbounded in neighborhoods of the points (i) c^- and c^+; (ii) c^- and b^-; (iii) a^+ and c^+; or (iv) a^+ and b^-. In (iv) the definition of the value of an improper integral of mixed kind is meaningful only if the value of

$$\int_a^b f(x)dx$$

is independent of the interior point c. It is left as an exercise to show that this is the case (Exercise 21 of Problem Set 10.1).

Example 10.1.5. The integral

$$\int_{-1}^{1} \frac{dx}{x^2}$$

is an improper integral of mixed kind since it can be written as

$$\int_{-1}^{1} \frac{dx}{x^2} = \int_{-1}^{0} \frac{dx}{x^2} + \int_{0}^{1} \frac{dx}{x^2},$$

where the integrals on the right are improper integrals of the second kind.

If we consider the improper integral

$$\int_{-1}^{0} \frac{dx}{x^2},$$

we see that

$$\int_{-1}^{t} \frac{dx}{x^2} = -\frac{1}{x} \Big]_{-1}^{t} = -\frac{1}{t} - 1,$$

for each t such that $-1 < t < 0$. Thus

$$\lim_{t \to 0-} \int_{-1}^{t} \frac{dx}{x^2} = +\infty,$$

and the improper integral

$$\int_{-1}^{0} \frac{dx}{x^2}$$

is divergent. Therefore, the improper integral

$$\int_{-1}^{1} \frac{dx}{x^2}$$

is also divergent.

Example 10.1.6. The integral

$$\int_{-1}^{1} \frac{dx}{(1 - x^2)^{1/2}}$$

is an improper integral of mixed kind since it can be written as

$$\int_{-1}^{1} \frac{dx}{(1-x^2)^{1/2}} = \int_{-1}^{c} \frac{dx}{(1-x^2)^{1/2}} + \int_{c}^{1} \frac{dx}{(1-x^2)^{1/2}},$$

where c is any real number such that $-1 < c < 1$ and the integrals on the right are improper integrals of the second kind.

If we first consider the improper integral

$$\int_{c}^{1} \frac{dx}{(1-x^2)^{1/2}},$$

we see that

$$\int_{c}^{t} \frac{dx}{(1-x^2)^{1/2}} = \sin^{-1}x \Big]_{c}^{t} = \sin^{-1}t - \sin^{-1}c$$

for each t such that $c < t < 1$. Hence,

$$\lim_{t \to 1-} \int_{c}^{t} \frac{dx}{(1-x^2)^{1/2}} = \frac{\pi}{2} - \sin^{-1}c,$$

and the improper integral

$$\int_{c}^{1} \frac{dx}{(1-x^2)^{1/2}}$$

converges to $\pi/2 - \sin^{-1}c$. In a similar manner, we see that

$$\int_{-1}^{c} \frac{dx}{(1-x^2)^{1/2}} = \lim_{s \to -1+} \left[\sin^{-1}c - \sin^{-1}s \right] = \sin^{-1}c + \frac{\pi}{2}.$$

Thus the improper integral

$$\int_{-1}^{c} \frac{dx}{(1-x^2)^{1/2}}$$

converges to $\sin^{-1}c + \pi/2$. Since both of the improper integrals

$$\int_{-1}^{c} \frac{dx}{(1-x^2)^{1/2}} \quad \text{and} \quad \int_{c}^{1} \frac{dx}{(1-x^2)^{1/2}}$$

converge, the improper integral

$$\int_{-1}^{1} \frac{dx}{(1-x^2)^{1/2}}$$

converges and has value π.

We see that the value of the improper integral

$$\int_{-1}^{1} \frac{dx}{(1 - x^2)^{1/2}}$$

is independent of the real number c.

CASE 2. If $a = -\infty$ and b is a real number, then the function f is unbounded in neighborhoods of the points (i) c^+ or (ii) b^-. In (ii) the definition of the value of an improper integral of mixed kind is meaningful only if the value of

$$\int_{-\infty}^{b} f(x)dx$$

is independent of the interior point c. It is left as an exercise to show that this is the case (Exercise 22 of Problem Set 10.1).

Example 10.1.7. The integral

$$\int_{-\infty}^{0} \frac{dx}{x^2}$$

is an improper integral of mixed kind since it can be written as

$$\int_{-\infty}^{0} \frac{dx}{x^2} = \int_{-\infty}^{c} \frac{dx}{x^2} + \int_{c}^{0} \frac{dx}{x^2},$$

where c is a negative number and the first integral on the right is an improper integral of the first kind and the second integral on the right is an improper integral of the second kind.

If we consider the improper integral

$$\int_{c}^{0} \frac{dx}{x^2},$$

we see that

$$\int_{c}^{t} \frac{dx}{x^2} = -\frac{1}{x}\bigg]_{c}^{t} = -\frac{1}{t} + \frac{1}{c},$$

for each t such that $c < t < 0$. Hence

$$\lim_{t \to 0^-} \int_{c}^{t} \frac{dx}{x^2} = +\infty,$$

and the improper integral

$$\int_{c}^{0} \frac{dx}{x^2}$$

is divergent. Therefore, the improper integral

$$\int_{-\infty}^{0} \frac{dx}{x^2}$$

is also divergent.

Since, for any negative c, the improper integral

$$\int_{c}^{0} \frac{dx}{x^2}$$

is divergent, we see that the divergence of the improper integral

$$\int_{-\infty}^{0} \frac{dx}{x^2}$$

is independent of the value of c.

CASE 3. If $b = +\infty$ and a is a real number, then the function f is unbounded in neighborhoods of the points (i) c^- or (ii) a^+. In (ii) the definition of the value of an improper integral of mixed kind is meaningful only if the value of

$$\int_{a}^{+\infty} f(x)dx$$

is independent of the interior point c. It is left as an exercise to show that this is the case (Exercise 23 of Problem Set 10.1).

Example 10.1.8. The integral

$$\int_{1}^{+\infty} \frac{dx}{(x-1)}$$

is an integral of mixed kind since it can be written as

$$\int_{1}^{+\infty} \frac{dx}{(x-1)} = \int_{1}^{c} \frac{dx}{(x-1)} + \int_{c}^{+\infty} \frac{dx}{(x-1)},$$

where c is any real number greater than 1 and the first integral on the right is an improper integral of the second kind and the second integral on the right is an improper integral of the first kind.

If we consider the improper integral

$$\int_{1}^{c} \frac{dx}{(x-1)},$$

we see that

$$\int_{s}^{c} \frac{dx}{(x-1)} = \ln(x-1)\Big]_{s}^{c} = \ln(c-1) - \ln(s-1),$$

for each s such that $1 < s < c$. Thus

$$\lim_{s \to 1+} \int_s^c \frac{dx}{(x-1)} = +\infty,$$

and the improper integral

$$\int_1^c \frac{dx}{(x-1)}$$

is divergent. Therefore, the improper integral

$$\int_1^{+\infty} \frac{dx}{(x-1)}$$

is also divergent.

We again see that the divergence of the improper integral

$$\int_1^{+\infty} \frac{dx}{(x-1)}$$

is independent of the value of c.

CASE 4. If $a = -\infty$ and $b = +\infty$, then the improper integral of mixed kind

$$\int_{-\infty}^{+\infty} f(x)dx$$

may be written as

$$\int_{-\infty}^{+\infty} f(x)dx = \int_{-\infty}^c f(x)dx + \int_c^{+\infty} f(x)dx,$$

where c is any real number. For the definition of the value of such an improper integral of mixed kind to be meaningful, the value on the right must be independent of the real number c. It is left as an exercise to show that this is the case (Exercise 24 of Problem Set 10.1).

Example 10.1.9. The integral

$$\int_{-\infty}^{+\infty} xe^{-x^2}dx$$

is an improper integral of mixed kind since it can be written as

$$\int_{-\infty}^{+\infty} xe^{-x^2}dx = \int_{-\infty}^0 xe^{-x^2}dx + \int_0^{+\infty} xe^{-x^2}dx,$$

where the integrals on the right-hand side are improper integrals of the first kind.

If we first consider the improper integral

$$\int_0^{+\infty} xe^{-x^2}dx,$$

we see that

$$\int_0^t xe^{-x^2}dx = \frac{-e^{-x^2}}{2}\Big]_0^t = \frac{1}{2} - \frac{e^{-t^2}}{2}$$

for each real positive t. Thus

$$\lim_{t\to+\infty}\int_0^t xe^{-x^2}dx = 1/2,$$

and the improper integral

$$\int_0^{+\infty} xe^{-x^2}dx$$

converges to 1/2. In a similar manner, we see that

$$\lim_{s\to-\infty}\int_s^0 xe^{-x^2}dx = \lim_{s\to-\infty}\left[-\frac{1}{2}+\frac{e^{-s^2}}{2}\right] = -\frac{1}{2}.$$

Hence the improper integral

$$\int_{-\infty}^0 xe^{-x^2}dx$$

converges to the value $-1/2$. Since both of the improper integrals

$$\int_{-\infty}^0 xe^{-x^2}dx \quad \text{and} \quad \int_0^{+\infty} xe^{-x^2}dx$$

converge, the improper integral

$$\int_{-\infty}^{+\infty} xe^{-x^2}dx$$

converges and has value 0.

We shall call any integral that is the sum of the improper integrals an **improper integral of mixed kind** and define convergence and the value of such improper integrals in the usual way. The following example illustrates such an improper integral of mixed kind.

Example 10.1.10. Investigate the improper integral

$$\int_0^{+\infty} \frac{dx}{1-x},$$

for convergence.

Since we may write

$$\int_0^{+\infty} \frac{dx}{1-x} = \int_0^1 \frac{dx}{1-x} + \int_1^2 \frac{dx}{1-x} + \int_2^{+\infty} \frac{dx}{1-x},$$

this is an improper integral of mixed kind. Example 10.1.2 implies that the improper integral

$$\int_0^1 \frac{dx}{1-x}$$

diverges, and thus the improper integral

$$\int_0^{+\infty} \frac{dx}{1-x}$$

diverges.

Problem Set 10.1

In Exercises 1–20, classify the given improper integral and check it for convergence. If the improper integral converges, find its value.

1. $\int_{-\infty}^{+\infty} \frac{dx}{1+x^2}.$

2. $\int_0^1 \csc x \, dx.$

3. $\int_{-\infty}^{+\infty} (1-x^2)dx.$

4. $\int_0^1 \frac{dx}{(1-x)^{1/2}}.$

5. $\int_{-\infty}^{+\infty} e^{-x}dx.$

6. $\int_{-\infty}^{+\infty} x \, e^{-x}dx.$

7. $\displaystyle\int_{-\infty}^{+\infty} x^2 e^{-x}\,dx.$

8. $\displaystyle\int_{0}^{+\infty} \sin x\,dx.$

9. $\displaystyle\int_{0}^{+\infty} \frac{dx}{1+x}.$

10. $\displaystyle\int_{0}^{+\infty} \frac{dx}{(1+x)^2}.$

11. $\displaystyle\int_{0}^{+\infty} \frac{dx}{(1+x)^p},$ where p is a real number.

12. $\displaystyle\int_{-\infty}^{+\infty} x \cos x^2\,dx.$

13. $\displaystyle\int_{-1}^{0} \frac{dx}{(1-x^2)^{1/2}}.$

14. $\displaystyle\int_{-\infty}^{+\infty} \frac{x\,dx}{1+x^2}.$

15. $\displaystyle\int_{-\infty}^{+\infty} \frac{x\,dx}{(1+x^2)^{1/2}}.$

16. $\displaystyle\int_{-\infty}^{+\infty} \frac{x\,dx}{(1+x^2)^p},$ where p is a real number.

17. $\displaystyle\int_{3}^{+\infty} \frac{dx}{x \ln x}.$

18. $\displaystyle\int_{0}^{+\infty} x \ln x\,dx.$

19. $\displaystyle\int_{1}^{+\infty} \frac{dx}{x(\ln x)^p},$ where p is a real number.

20. $\displaystyle\int_{1}^{+\infty} \frac{e^{-\sqrt{x}}\,dx}{\sqrt{x}}.$

21. Prove the independence of c in the improper integral of mixed kind

$$\int_a^b f(x)dx = \int_a^c f(x)dx + \int_c^b f(x)dx,$$

where a and b are real numbers, $a < c < b$, and the integrals

$$\int_a^c f(x)dx \quad \text{and} \quad \int_c^b f(x)dx$$

are improper integrals of the second kind with f unbounded in neighborhoods of a^+ and b^-.

22. Prove the independence of c in the improper integral of mixed kind

$$\int_{-\infty}^b f(x)dx = \int_{-\infty}^c f(x)dx + \int_c^b f(x)dx,$$

where b is a real number, $-\infty < c < b$, the integral

$$\int_c^b f(x)dx$$

is an improper integral of second kind with f unbounded in neighborhoods of b^-, and the integral

$$\int_{-\infty}^c f(x)dx$$

is an improper integral of the first kind.

23. Prove the independence of c in the improper integral of mixed kind

$$\int_a^{+\infty} f(x)dx = \int_a^c f(x)dx + \int_c^{+\infty} f(x)dx,$$

where a is a real number, $a < c < +\infty$, the integral

$$\int_a^c f(x)dx$$

is an improper integral of second kind with f unbounded in neighborhoods of a^+, and the integral

$$\int_c^{+\infty} f(x)dx$$

is an improper integral of first kind.

24. Prove the independence of c in the improper integral of mixed kind

$$\int_{-\infty}^{+\infty} f(x)dx = \int_{-\infty}^c f(x)dx + \int_c^{+\infty} f(x)dx,$$

where c is a real number, and the integrals

$$\int_{-\infty}^{c} f(x)dx \quad \text{and} \quad \int_{c}^{+\infty} f(x)dx$$

are improper integrals of first kind.

25. Illustrate Exercises 21–24 with examples different than those given in the text.

26. Prove that

$$\int_{0}^{+\infty} \frac{\sin^2 x}{x^2}\, dx = \int_{0}^{+\infty} \frac{\sin x}{x}\, dx.$$

27. Give an example to show that if $\int_{a}^{+\infty} f(x)dx$ converges, it is not necessary that

$$\lim_{x \to +\infty} f(x) = 0.$$

28. If $\int_{a}^{+\infty} f(x)dx$ converges and c is a constant, then prove that

$$\lim_{y \to +\infty} \int_{y}^{y+c} f(x)dx = 0.$$

29. If $\int_{a}^{+\infty} f(x)dx$ converges and $\lim_{x \to +\infty} f(x) = L$, then prove that $L = 0$.

30. Prove the following analogue to Exercise 13 of Problem Set 7.1. If $f(x) \geqslant 0$ for $1 \leqslant x < +\infty$, if f is decreasing on $[1, +\infty)$, and if

$$\int_{1}^{+\infty} f(x)dx$$

converges, then $\lim_{x \to +\infty} xf(x) = 0$. [*Hint*: Use Exercise 13 of Problem Set 7.1.]

31. Give an example to show that Exercise 30 is no longer true if the hypothesis that f is decreasing on $[1, +\infty)$ is dropped.

32. Prove the following analogue to Theorem 7.5.3. Let f be continuous on $[1, +\infty)$, and let

$$F(x) = \int_{1}^{x} f(t)dt$$

for each $x \in [1, +\infty)$. Let g be a function with domain $[1, +\infty)$ such that g' is continuous on $[1, +\infty)$, $g'(x) \leqslant 0$ for $x \in [1, +\infty)$, and $\lim_{x \to +\infty} g(x) = 0$.

If $f(x)$ is bounded on $[1, +\infty)$, then

$$\int_1^{+\infty} f(x)g(x)dx$$

converges. [*Hint*: Use integration by parts.]

33. Use Exercise 32 to show that

$$\int_3^{+\infty} \frac{\sin x}{\ln x} dx$$

converges.

34. Let f be a bounded function on the interval $[a, b]$, and let f be integrable on $[a, c]$ for each $a < c < b$. Prove that the f is Riemann integrable on $[a, b]$.

35. If the improper integrals

$$\int_a^b f(x)dx \quad \text{and} \quad \int_a^b g(x)dx$$

both converge, then prove that, for any real numbers α and β, the improper integral

$$\int_a^b (\alpha f + \beta g)(x)dx$$

converges and

$$\int_a^b (\alpha f + \beta g)(x)dx = \alpha \int_a^b f(x)dx + \beta \int_a^b f(x)dx.$$

10.2. Convergence of Improper Integrals

Since the convergence of improper integrals of mixed kind is defined in terms of the convergence of improper integrals of the first and second kinds, the study of convergence of improper integrals may be limited to the investigation of convergence of improper integrals of the first and second kinds. Let f be integrable on every interval $[c, b]$, where a is a real number or $a = -\infty$ and $a < c < b$. Then the improper integral

$$\int_a^b f(x)dx$$

can be written as

$$\int_a^b f(x)dx = \lim_{c \to a^+} \int_c^b f(x)dx = \lim_{c \to -a^-} \int_{-b}^c f(-u)du = \int_{-b}^{-a} f(-u)du,$$

if we make the change of variable $x = -u$.

Thus the convergence of the improper integral

$$\int_a^b f(x)dx,$$

which is *improper* at its lower limit of integration, depends on the convergence of the improper integral

$$\int_{-b}^{-a} f(-u)du,$$

which is *improper* at its upper limit of integration. Therefore, we shall restrict our attention to the study of convergence of improper integrals which are *improper* at their upper limits of integration. (Cf. Definition 10.1.1.)

From this point on the study of improper integrals is very similar to the theory of infinite series, and hence many of the details shall be omitted. We first prove a *Cauchy Criterion for Improper Integrals* similar to the Cauchy Criterion for Infinite Series.

Theorem 10.2.1. (*Cauchy Criterion*). *The improper integral*

$$\int_a^b f(x)dx$$

converges if and only if to each $\varepsilon > 0$ there exists a real number $d \in (a, b)$ such that for every $x, y \in (d,b)$ we have

$$\left| \int_y^x f(\zeta)d\zeta \right| < \varepsilon.$$

Proof. Define the function $F: [a, b) \to R$ such that

$$F(x) = \int_a^x f(\zeta)d\zeta.$$

Then we can write

$$\int_y^x f(\zeta)d\zeta = \int_a^x f(\zeta)d\zeta - \int_a^y f(\zeta)d\zeta = F(x) - F(y)$$

for any real numbers $x, y \in [a, b)$.

Suppose the Cauchy Criterion is satisfied, that is, to each $\varepsilon > 0$ there exists $d \in (a, b)$ such that, for every $x, y \in (d, b)$,

$$|F(x) - F(y)| = \left| \int_y^x f(\zeta)d\zeta \right| < \varepsilon.$$

It follows from Theorem 4.2.7 (Cauchy Criterion for Functions) that

$$\lim_{c \to b-} F(c) = \lim_{c \to b-} \int_a^c f(\zeta)d\zeta$$

exists and is finite, that is, the improper integral is convergent.
Conversely, if the improper integral converges and

$$\lim_{c \to b-} \int_a^c f(\zeta)d\zeta = \lim_{c \to b-} F(c) = L,$$

then for each $\varepsilon = 0$ there is a $d \in [a, b)$ such that
$$|F(x) - L| < \varepsilon/2$$
for any $x \in (d, b)$. Thus, for $x, y \in (d, b)$, we have

$$\left| \int_y^x f(\zeta)d\zeta \right| = |F(x) - F(y)|$$

$$\begin{aligned}
&= |F(x) - L + L - F(y)| \\
&\leqslant |F(x) - L| + |F(y) - L| \\
&< \varepsilon/2 + \varepsilon/2 = \varepsilon,
\end{aligned}$$

and the Cauchy Criterion is satisfied.

We continue by defining *absolute* and *conditional convergence of improper integrals.*

Definition 10.2.1. *The improper integral*

$$\int_a^b f(x)dx$$

*is said to be **absolutely convergent** if and only if the improper integral*

$$\int_a^b |f(x)|dx$$

*is convergent. An improper integral is said to be **conditionally convergent** if and only if it is convergent, but not absolutely convergent.*

Example 10.2.1. Let $f(x) = x^p$, where p is a real number less than -1, for each $x \geqslant 1$. Then

$$\int_1^{+\infty} x^p dx$$

is absolutely convergent since

$$|x^p| = x^p.$$

(Cf. Example 10.1.1.)

We now apply the Cauchy Criterion to show that an improper integral that converges absolutely also converges.

Theorem 10.2.2. *If the improper integral*

$$\int_a^b f(x)dx$$

converges absolutely, and if f is integrable on compact subintervals of (a, b), *then it converges.*

Proof. Since the improper integral

$$\int_a^b f(x)dx$$

converges absolutely, Theorem 10.2.1 implies that to each $\varepsilon > 0$ there is a $d \in (a, b)$ such that

$$\left| \int_y^x |f(\zeta)|d\zeta \right| < \varepsilon$$

whenever $x, y \in (d, b)$. But since

$$\left| \int_y^x f(\zeta)d\zeta \right| \leqslant \int_y^x |f(\zeta)|d\zeta,$$

it follows that

$$\left| \int_y^x f(\zeta)d\zeta \right| < \varepsilon$$

whenever $x, y \in (d, b)$ with $y < x$. Thus the improper integral

$$\int_a^b f(x)dx$$

satisfies the Cauchy Criterion, and so it is convergent.

As a second application of the Cauchy criterion, we prove the *Comparison Test* for improper integrals.

Theorem 10.2.3. *(Comparison Test) Let* $0 \leqslant f(x) \leqslant g(x)$ *for all* $x \in [a, b)$. *If the improper integral*

$$\int_a^b g(x)dx$$

converges, then the improper integral

$$\int_a^b f(x)dx$$

converges and

$$0 \le \int_a^b f(x)dx \le \int_a^b g(x)dx.$$

Proof. Since the improper integral

$$\int_a^b g(x)dx$$

converges, Theorem 10.2.1 implies that to each $\varepsilon > 0$ there is a $d \in (a, b)$ such that

$$\left| \int_y^x g(\zeta)d\zeta \right| < \varepsilon$$

whenever $x, y \in (d, b)$. Since $0 \le f(\zeta) \le g(\zeta)$ for each $\zeta \in [a, b)$, we have

$$\left| \int_y^x f(\zeta)d\zeta \right| = \int_y^x f(\zeta)d\zeta$$

$$\le \int_y^x g(\zeta)d\zeta$$

$$= \left| \int_y^x g(\zeta)d\zeta \right| < \varepsilon,$$

whenever $x, y \in (d, b)$ with $y < x$. Thus the improper integral

$$\int_a^b f(\zeta)d\zeta$$

satisfies the Cauchy criterion and is convergent. Since $0 \le f(x) \le g(x)$ for each $x \in [a, b)$,

$$0 \le \int_a^c f(x)dx \le \int_a^c g(x)dx,$$

for every $c \in (a, b)$. Thus

$$0 \le \lim_{c \to b-} \int_a^c f(x)dx \le \lim_{c \to b-} \int_a^c g(x)dx,$$

and so

$$0 \le \int_a^b f(x)dx \le \int_a^b g(x)dx.$$

Example 10.2.2. Let $f(x) = 1/(1 + x^3)^{1/2}$ for each $x \geqslant 1$. Let $g(x) = x^{-3/2}$ for each $x \geqslant 1$. Then

$$\int_1^{+\infty} x^{-3/2} dx$$

converges. Also,

$$0 \leqslant \frac{1}{(1 + x^3)^{1/2}} \leqslant \frac{1}{x^{3/2}}$$

for each $x \geqslant 1$. It follows from the Comparison Test that the improper integral

$$\int_1^{+\infty} \frac{dx}{(1 + x^3)^{1/2}}$$

converges and

$$0 \leqslant \int_1^{+\infty} \frac{dx}{(1 + x^3)^{1/2}} \leqslant \int_1^{+\infty} x^{-3/2} dx = 2.$$

To avoid the troublesome details of working with the inequalities of the Comparison Test, it is often more convenient to use the following theorem.

Theorem 10.2.4 (*Limit Comparison Test*) *Let f and g be nonnegative on the interval* $[a, b)$, *and assume that the limit*

$$\lim_{x \to b-} \frac{f(x)}{g(x)}$$

exists, is finite, and is nonzero. Then the improper integral

$$\int_a^b f(x) dx$$

converges if and only if the improper integral

$$\int_a^b g(x) dx$$

converges. If the limit is zero, and if the improper integral

$$\int_a^b g(x) dx$$

converges, then the improper integral

$$\int_a^b f(x) dx$$

converges. If the limit is $+\infty$, *and if the improper integral*

$$\int_a^b g(x)dx$$

diverges, then the improper integral

$$\int_a^b f(x)dx$$

diverges.

The proof is left as Exercise 14 of Problem Set 10.2.

Example 10.2.3. Check the improper integral

$$\int_2^{+\infty} \frac{dx}{\sqrt{x(x^2-1)}}$$

for convergence. We compare the integrand with the function

$$\frac{1}{x^{3/2}}.$$

Since the improper integral

$$\int_2^{+\infty} x^{-3/2}dx$$

converges and

$$\lim_{x \to +\infty} \left(\frac{1}{\sqrt{x(x^2-1)}} \bigg/ \frac{1}{x^{3/2}} \right) = 1,$$

the Limit Comparison Test implies that the improper integral

$$\int_2^{+\infty} \frac{dx}{\sqrt{x(x^2-1)}}$$

converges.

Example 10.2.4. We show that the improper integral

$$\int_1^{+\infty} x^\alpha e^{-x}dx$$

converges for each real number α. Since the improper integral

$$\int_1^{+\infty} x^{-2}dx$$

converges,

$$\frac{x^\alpha e^{-x}}{x^{-2}} = \frac{x^{\alpha+2}}{e^x},$$

and

$$\lim_{x \to +\infty} \frac{x^{\alpha+2}}{e^x} = 0,$$

the Limit Comparison Test implies that the improper integral

$$\int_1^{+\infty} x^\alpha e^{-x} dx$$

converges.

Example 10.2.5. Check the improper integral

$$\int_0^1 \frac{dx}{(x-1)\ln(1-x)}$$

for convergence. We compare the integrand with the function

$$\frac{1}{(1-x)}.$$

Since

$$\int_0^1 \frac{dx}{1-x}$$

diverges,

$$\frac{1/(1-x)}{1/(x-1)\ln(1-x)} = -\ln(1-x),$$

and

$$\lim_{x \to 1^-} [-\ln(1-x)] = +\infty,$$

the Limit Comparison Test implies that the improper integral

$$\int_0^1 \frac{dx}{(x-1)\ln(1-x)}$$

diverges.

We conclude this section with a test for improper integrals analogous to the Dirichlet Test for infinite series; it is a corollary to the following theorem.

Theorem 10.2.5. *Let f, g, and g' be functions continuous on the interval $[a, b)$, where b is a real number greater than a, or $b = +\infty$ If.*

(i) $\displaystyle\lim_{x\to b-} g(x) = 0$,

(ii) *the improper integral*

$$\int_a^b g'(x)dx$$

converges absolutely, and
(iii) *the function*

$$F(x) = \int_a^x f(\zeta)d\zeta$$

is bounded on $[a, b)$,
then the improper integral

$$\int_a^b f(x)g(x)dx$$

converges.

Proof. Let the function F be bounded by M on $[a, b)$, and let $\varepsilon = 0$. For any real numbers $\alpha, \beta \in [a, b)$, we use integration by parts (Theorem 6.5.1) to write

$$\int_\alpha^\beta f(x)g(x)dx = \int_\alpha^\beta g(x)F'(x)dx$$

$$= g(x)F(x)\Big]_\alpha^\beta - \int_\alpha^\beta F(x)g'(x)dx$$

$$= g(\beta)F(\beta) - g(\alpha)F(\alpha) - \int_\alpha^\beta F(x)g'(x)dx.$$

Since $|F(x)g'(x)| \leqslant M|g'(x)|$ for each $x \in [a, b)$, the improper integral

$$\int_a^b F(x)g'(x)dx$$

converges absolutely (cf. Theorem 10.2.3), and the Cauchy Criterion (Theorem 10.2.1) implies that there is a $d_0 \in (a, b)$ such that

$$\left| \int_\alpha^\beta F(x)g'(x)dx \right| < \varepsilon/3$$

when $\alpha, \beta \in (d_0, b)$. Since

$$\lim_{x \to b^-} g(x) = 0$$

and F is bounded by M on $[a, b)$, there is a real number $d \in (d_0, b)$ such that

$$|F(x)g(x)| < \varepsilon/3$$

whenever $x \in (d, b)$. Thus for $\alpha, \beta \in (d, b)$,

$$\left| \int_\alpha^\beta f(x)g(x)dx \right| \leq |g(\beta)F(\beta)| + |g(\alpha)F(\alpha)| + \left| \int_\alpha^\beta F(x)g'(x)dx \right|$$

$$< \varepsilon/3 + \varepsilon/3 + \varepsilon/3 = \varepsilon.$$

Hence it follows from Theorem 10.2.1 (Cauchy Criterion) that the improper integral

$$\int_a^b f(x)g(x)dx$$

converges.

If the function g in Theorem 10.2.5 decreases to zero as $x \to b^-$, then g' is negative near b and

$$\int_a^c |g'(x)|dx = -\int_a^c g'(x)dx = -g(x)\Big]_a^c = g(a) - g(c).$$

Hence

$$\lim_{c \to b^-} \int_a^c |g'(x)|dx = g(a)$$

and the improper integral

$$\int_a^b g'(x)dx$$

is absolutely convergent. This and Theorem 10.2.5 lead to the following corollary.

Corollary 10.2.5. (*Dirichlet Test*). *Let f, g, and g' be functions continuous on the interval $[a, b)$, where b is a real number greater than a or $b = +\infty$. If g decreases to zero as $x \to b^-$ and the function*

$$F(x) = \int_a^x f(\zeta)d\zeta$$

is bounded on $[a, b)$, then the improper integral

$$\int_a^b f(x)g(x)dx$$

converges.

Example 10.2.6. If g has a continuous first derivative on the interval $[c, +\infty)$, and if g decreases to zero as $x \to +\infty$, then the improper integrals

$$\int_c^{+\infty} g(x) \sin x \, dx \quad \text{and} \quad \int_c^{+\infty} g(x) \cos x \, dx$$

both converge. In particular, the following four integrals converge:

$$\int_1^{+\infty} \frac{\sin x}{x} \, dx, \qquad \int_1^{+\infty} \frac{\cos x}{x} \, dx,$$

$$\int_2^{+\infty} \frac{\sin x}{\ln x} \, dx, \qquad \int_2^{+\infty} \frac{\cos x}{\ln x} \, dx.$$

It is not difficult to see that if $g : [c, +\infty) \to R$ and g decreases to zero as $x \to +\infty$, then the improper integrals

$$\int_c^{+\infty} g(x) \sin x \, dx \quad \text{and} \quad \int_c^{+\infty} g(x) \cos x \, dx$$

both converge. The proof is left as Exercise 29 of Problem Set 10.2.

Problem Set 10.2

In Exercises 1–10, check the improper integrals for convergence.

1. $\displaystyle\int_0^{+\infty} \sin x^3 dx.$

2. $\displaystyle\int_0^{+\infty} \cos x^\alpha dx, \ \alpha \geqslant 1.$

3. $\displaystyle\int_1^{+\infty} e^{-t} dt.$

4. $\displaystyle\int_1^{+\infty} \frac{e^{1/x}}{1 + x^2} \, dx.$

5. $\displaystyle\int_1^{+\infty} \frac{\sin(1/x)}{x} \, dx.$

6. $\displaystyle\int_0^1 \ln(1 - x) dx.$

7. $\displaystyle\int_0^1 \frac{x^\alpha}{\sqrt{1 - x}} \, dx.$

8. $\displaystyle\int_{\pi/2}^{\pi} \frac{x}{\sin x} \, dx.$

9. $\displaystyle\int_{1/2}^1 \frac{\ln(1 - x)}{x^2} \, dx.$

10. $\displaystyle\int_0^{\pi/2} \frac{dx}{\sqrt{1 - \sin^2 x}}.$

11. For what values of α does the improper integral

$$\int_1^{+\infty} \frac{dx}{1 + x^\alpha}$$

converge?

12. For what values of α does the improper integral

$$\int_0^1 \frac{dx}{1 - x^\alpha}$$

converge?

13. Let b be a real number greater than a or $b = +\infty$. Let f be a nonnegative function on the interval $[a, b)$, and let f be integrable on every interval $[a, c]$ where $a < c < b$. Show that the improper integral

$$\int_a^b f(x)dx$$

is convergent if and only if the set

$$\left\{ \int_a^x f(\zeta)d\zeta : x \in [a, b) \right\}$$

is bounded.

14. Prove Theorem 10.2.4.

15. Show that the last four integrals in Example 10.2.6 are not absolutely convergent.

16. Let f be a nonnegative function on the interval $[a, +\infty)$, and let f be integrable on each closed interval $[a, b]$ where $a < b < +\infty$. If there is an $r < 1$ and an $M \geqslant a$ such that $f(x) \leqslant r^x$ for all $x \geqslant M$, show that the improper integral

$$\int_a^{+\infty} f(x)dx$$

converges.

17. Let f be a positive continuous function on the interval $[a, +\infty)$. If

$$\lim_{x \to +\infty} \frac{f(x + 1)}{f(x)} = r < 1,$$

show that the improper integral

$$\int_a^{+\infty} f(x)dx$$

converges.

18. Let f be defined on the interval $[a, b]$, where a and b are either finite or infinite. Let $c \in [a, b]$. If the integral

$$\int_a^b f(x)dx$$

does not exist, but

$$\lim_{\varepsilon \to 0+} \left[\int_a^{c-\varepsilon} f(x)dx + \int_{c+\varepsilon}^b f(x)dx \right]$$

does exist, then this value is called the **Cauchy principal value** of the integral $\int_a^b f(x)dx$ and is denoted by

$$\text{C.P.V.} \int_a^b f(x)dx.$$

Compute

$$\text{C.P.V.} \int_{-1}^{+1} \frac{dx}{x}.$$

For Exercises 19–21, use the definition of Exercise 18 to compute the Cauchy principal value for the given integrals.

19. $\text{C.P.V.} \displaystyle\int_{-1}^{+1} \frac{dx}{x^2}.$ **20.** $\text{C.P.V.} \displaystyle\int_{-1}^{+1} \frac{dx}{x^\alpha}, \quad \alpha > 1.$

21. $\text{C.P.V.} \displaystyle\int_{-\infty}^{+\infty} \frac{(1 + x)dx}{1 + x^2}.$

22. Let f be defined on the interval $[a, c)$ and $(c, b]$. If the improper integral

$$\int_a^b f(x)dx$$

exists, then show that

$$\text{C.P.V.} \int_a^b f(x)dx$$

also exists and that both integrals have the same value. Give an example to show that the converse is not true.

In Exercises 23–27, let f and g be defined on the interval $[a, b)$, where b is a real number greater than a or $b = +\infty$, and let

$$\int_a^b f(x)dx \quad \text{and} \quad \int_a^b g(x)dx$$

exist as proper integrals or a convergent improper integrals.

23. Show that

$$\int_a^b f(x)g(x)dx$$

exists as a proper integral or as a convergent improper integral.

24. Show that

$$\int_a^b [f(x) + g(x)]^2 dx$$

exists as a proper integral or as a convergent improper integral.

25. Show that

$$\left| \int_a^b f(x)g(x)dx \right| \leq \frac{1}{2} \int_a^b f^2(x)dx + \frac{1}{2} \int_a^b g^2(x)dx.$$

26. Show that

$$\left(\int_a^b f(x)g(x)dx \right)^2 \leq \left(\int_a^b f^2(x)dx \right) \left(\int_a^b g^2(x)dx \right).$$

[This inequality is called the **Schwarz (or Cauchy) inequality**.]

27. Show that

$$\left(\int_a^b [f(x) + g(x)]^2 dx \right)^{1/2} \leq \left(\int_a^b f^2(x)dx \right)^{1/2} + \left(\int_a^b g^2(x)dx \right)^{1/2}.$$

[This inequality is called the **Minkowski inequality**.]

28. Let a and b be real numbers with $a < b$. If the improper integral

$$\int_a^b f(x)dx$$

converges absolutely, then prove that

$$\int_a^b |f(x)|^{1/2} dx$$

converges and that

$$\int_a^b |f(x)|^{1/2} dx \leq (b - a)^{1/2} \left(\int_a^b |f(x)|dx \right)^{1/2}.$$

29. Let $g: [c, +\infty) \to R$. If g decreases to zero as $x \to +\infty$, show that the improper integrals

$$\int_c^{+\infty} g(x) \sin x \, dx \quad \text{and} \quad \int_c^{+\infty} g(x) \cos x \, dx$$

both converge. [*Hint*: Compare with a suitable alternating series.]

Chapter Eleven

STIELTJES
INTEGRATION

11.1. Functions of Bounded Variation

In this chapter we will present important generalizations of the Riemann integral, called the *Stieltjes integrals*. Associated with Stieltjes integrals is the special class of functions, the *functions of bounded variation*. We begin by defining and investigating some of their special properties.

Definition 11.1.1. *Let* $I = [a, b]$, *where* a *and* b *are real numbers with* $a < b$, *and let* $f: I \to R$. *For the net*

$$\mathfrak{n} = \{a = x_0 < x_1 < ... < x_{n-1} < x_n = b\}$$

of I, *the number*

$$v(f, \mathfrak{n}) = \sum_{k=1}^{n} |f(x_k) - f(x_{k-1})|$$

*is called the **variation** of* f *over the net* $\mathfrak{n}$. *If* $\overline{N}$ *denotes the set of all nets of* I, *and if there is a real number* M *such that*

$$v(f, \mathfrak{n}) \leqslant M$$

for each $\mathfrak{n} \in \overline{N}$, *then the function* f *is said to be of **bounded variation** on* I. *The real number*

$$V(f) = V(f; [a, b]) = \sup_{\mathfrak{n} \in \overline{N}} v(f, \mathfrak{n})$$

*is called the **variation** or **total variation** of* f *on* $[a, b]$.

357

We demonstrate these concepts with the following examples.

Example 11.1.1. Let f be any monotone function on $[a, b]$. Then for any net $\mathfrak{n} = \{a = x_0 < x_1 < \ldots < x_n = b\}$ of $[a, b]$, the variation of f over the net $\mathfrak{n}$ is

$$v(f, \mathfrak{n}) = \sum_{k=1}^{n} |f(x_k) - f(x_{k-1})| = |f(b) - f(a)|.$$

Thus this function is of bounded variation, and its variation is $|f(b) - f(a)|$.

Example 11.1.2. Let $f: [a, b] \to R$ be continuous on $[a, b]$ and be differentiable on (a, b), and assume that $|f'(x)| \leq M$ for each $x \in (a, b)$. Then, for any $x, y \in [a, b]$, the Mean Value Theorem (Theorem 5.3.4) implies that

$$|f(x) - f(y)| = |f'(\xi)||x - y| \leq M|x - y|,$$

where ξ is a real number between x and y. Hence, for any net $\mathfrak{n} = \{a = x_0 < x_1 < \ldots < x_n = b\}$ of $[a, b]$, we have

$$v(f, \mathfrak{n}) = \sum_{k=1}^{n} |f(x_k) - f(x_{k-1})|$$

$$= \sum_{k=1}^{n} |f'(\xi_k)||x_k - x_{k-1}|$$

$$\leq \sum_{k=1}^{n} M|x_k - x_{k-1}|$$

$$= M(b - a),$$

where $\xi_k \in [x_{k-1}, x_k]$, $k = 1, 2, \ldots, n$. Thus $V(f) \leq M(b - a)$, and f is of bounded variation on $[a, b]$. That is, any function which is continuous on $[a, b]$ and has a bounded derivative on (a, b) is of bounded variation on $[a, b]$.

Example 11.1.3. Let $f: [0, 1] \to R$ be defined by

$$f(x) = \begin{cases} x \cos(\pi/2x) & \text{if } 0 < x \leq 1, \\ 0 & \text{if } x = 0. \end{cases}$$

Since f is continuous on $[0, 1]$, f is uniformly continuous on $[0, 1]$ but is not of bounded variation on $[0, 1]$. To show this, we define the net

$$\mathfrak{n}_n = \left\{ 0, \frac{1}{2n}, \frac{1}{2n-1}, \ldots, \frac{1}{3}, \frac{1}{2}, 1 \right\}.$$

Then the variation of f over the net $\mathfrak{n}_n$ is

$$v(f, \mathfrak{n}_n) = \frac{1}{n} + \frac{1}{n-1} + \dots + \frac{1}{2} + 1,$$

which is the nth partial sum of the harmonic series. Since the harmonic series diverges to $+\infty$,

$$\lim_n v(f, \mathfrak{n}_n) = +\infty,$$

and f is not of bounded variation.

We now present some elementary properties of functions of bounded variation.

Theorem 11.1.1. *If f is of bounded variation on $[a, b]$, then f is bounded on $[a, b]$.*

The proof is left as Exercise 7 of Problem Set 11.1.

Theorem 11.1.2. *If $f: [a, b] \to R$ and $g: [a, b] \to R$, then*

$$V((f + g); [a, b]) \leqslant V(f; [a, b]) + V(g; [a, b]).$$

The proof is left as Exercise 9 of Problem Set 11.1.

It immediately follows from Theorem 11.1.2 that the sum of two functions of bounded variation is a function of bounded variation.

Corollary 11.1.2. *If f and g are of bounded variation on $[a, b]$, then $f + g$ is of bounded variation on $[a, b]$.*

Theorem 11.1.3. *If $f: [a, b] \to R$ and α is a real number, then*

$$V(\alpha f; [a, b]) = |\alpha| V(f; [a, b]).$$

The proof is left as Exercise 11 of Problem Set 11.1.

From Theorem 11.1.3 it follows that if f is of bounded variation and α is any real number, then the function αf is also of bounded variation.

Corollary 11.1.3a. *If f is of bounded variation on $[a, b]$ and α is a real number, then the function αf is also of bounded variation on $[a, b]$.*

Corollary 11.1.3b. *If f and g are of bounded variation on $[a, b]$ and α and β are real numbers, then the function $\alpha f + \beta g$ is also of bounded variation on $[a, b]$.*

Theorem 11.1.4. *If f and g are of bounded variation on $[a, b]$, then the function fg is of bounded variation on $[a, b]$.*

Proof. Since f and g are of bounded variation on $[a, b]$, Theorem 11.1.1 implies that there is a real number M such that $|f(x)| \leqslant M$ and $|g(x)| \leqslant M$ for each $x \in [a, b]$. Let

$$\mathfrak{n} = \{a = x_0 < x_1 < \ldots < x_n = b\}$$

be any net of the interval $[a, b]$. Then

$$\sum_{k=1}^{n} |(fg)(x_k) - (fg)(x_{k-1})| = \sum_{k=1}^{n} |f(x_k)g(x_k) - f(x_{k-1})g(x_{k-1})|$$

$$= \sum_{k=1}^{n} |f(x_k)g(x_k) - f(x_k)g(x_{k-1}) + f(x_k)g(x_{k-1}) - f(x_{k-1})g(x_{k-1})|$$

$$\leqslant \sum_{k=1}^{n} |f(x_k)||g(x_k) - g(x_{k-1})| + \sum_{k=1}^{n} |g(x_{k-1})||f(x_k) - f(x_{k-1})|$$

$$\leqslant MV(g; [a, b]) + MV(f; [a, b]).$$

Since f and g are of bounded variation, the right-hand quantity is finite, and thus the function fg is of bounded variation.

Thus the set of all functions of bounded variation forms an **algebra** *or* **vector space**, that is, a system which contains sums, products, and scalar multiples of its elements.

Theorem 11.1.5. *If f is of bounded variation on $[a, b]$ and $c \in (a, b)$, then f is of bounded variation on each of the intervals $[a, c]$ and $[c, b]$ and*

$$V(f; [a, b]) = V(f; [a, c]) + V(f; [c, b]).$$

The proof is left as Exercise 12 of Problem Set 11.1.

The definition of the variation of a function f over a net $\mathfrak{n}$ implies that if the net $\mathfrak{n}^*$ is a refinement of the net $\mathfrak{n}'$, then $v(f, \mathfrak{n}^*) \geqslant v(f, \mathfrak{n}')$. We make use of this fact in proving the following theorem.

Theorem 11.1.6. *If f is continuous and of bounded variation on $[a, b]$, and if $\{\mathfrak{n}_n\}$ is a sequence of nets of $[a, b]$ such that*

$$\lim_{n} \|\mathfrak{n}_n\| = 0,$$

then

$$V(f; [a, b]) = \lim_{n} v(f, \mathfrak{n}_n).$$

Proof. Let $\varepsilon > 0$. Since $V(f; [a, b])$ is the supremum of the set

$$\{v(f, \overline{\mathfrak{n}}): \overline{\mathfrak{n}} \text{ is a net of } [a, b]\},$$

there is a net

$$\mathfrak{n} = \{a = x_0 < x_1 < \ldots < x_m = b\}$$

of $[a, b]$ such that

$$v(f, \mathfrak{n}) > V(f; [a, b]) - \varepsilon/2.$$

Since f is continuous on $[a, b]$, it is uniformly continuous on $[a, b]$; and so there is a $\delta > 0$ such that

$$|f(x) - f(y)| < \varepsilon/4m$$

whenever $x, y \in [a, b]$ and $|x - y| < \delta$. Let $\mathfrak{n}'$ be any net of $[a, b]$ with $\|\mathfrak{n}'\| < \delta$, and let $\mathfrak{n}^* = \mathfrak{n} \cup \mathfrak{n}'$. Then, since $\mathfrak{n}^*$ is a refinement of each of the nets $\mathfrak{n}$ and $\mathfrak{n}'$, we know that

$$v(f, \mathfrak{n}^*) \geqslant v(f, \mathfrak{n}) \quad \text{and} \quad v(f, \mathfrak{n}^*) \geqslant v(f, \mathfrak{n}').$$

Thus we have

$$v(f, \mathfrak{n}^*) - v(f, \mathfrak{n}') \geqslant v(f, \mathfrak{n}) - v(f, \mathfrak{n}').$$

Since the net $\mathfrak{n}^*$ is a refinement of the net $\mathfrak{n}'$ constructed by adding the points of the net $\mathfrak{n}$, for each point of the net $\mathfrak{n}$ added to the net $\mathfrak{n}'$ at most $2\varepsilon/4m = \varepsilon/2m$ is added to the variation of f over the new net. Since there are at most $m - 1$ points of the net $\mathfrak{n}$ distinct from the points of the net $\mathfrak{n}'$, the maximum possible difference between $v(f, \mathfrak{n}^*)$ and $v(f, \mathfrak{n}')$ is $(m - 1)\varepsilon/2m$. Hence we obtain

$$v(f, \mathfrak{n}) - v(f, \mathfrak{n}') \leqslant v(f, \mathfrak{n}^*) - v(f, \mathfrak{n}') \leqslant (m - 1)\varepsilon/2m < \varepsilon/2.$$

Therefore, for any net $\mathfrak{n}'$ with $\|\mathfrak{n}'\| < \delta$, we have

$$v(f, \mathfrak{n}') + \varepsilon/2 \geqslant v(f, \mathfrak{n}) > V(f; [a, b]) - \varepsilon/2$$

or

$$V(f; [a, b]) \geqslant v(f, \mathfrak{n}') > V(f; [a, b]) - \varepsilon,$$

and the desired result follows.

If the function f' is integrable, then we are able to give a formula for the variation of f in terms of the integral of $|f'|$.

Theorem 11.1.7. *If f is differentiable on $[a, b]$ and $|f'|$ is integrable on $[a, b]$, then f is of bounded variation on $[a, b]$ and*

$$V(f; [a, b]) = \int_a^b |f'(x)| dx.$$

Proof. Let $\mathfrak{n} = \{a = x_0 < x_1 < \dots < x_n = b\}$ be any net of $[a, b]$. For each k, where $k = 1, 2, \dots, n$, the Mean Value Theorem implies that there is a $\xi_k \in [x_{k-1}, x_k]$ such that

$$f(x_k) - f(x_{k-1}) = f'(\xi_k)(x_k - x_{k-1}).$$

Then

$$v(f, \mathfrak{n}) = \sum_{k=1}^{n} |f(x_k) - f(x_{k-1})|$$

$$= \sum_{k=1}^{n} |f'(\xi_k)|(x_k - x_{k-1}).$$

By choosing any sequence of nets whose norms converge to zero, Theorem 11.1.6 implies that

$$V(f; [a, b]) = \int_a^b |f'(x)| dx.$$

We now look at the structure of functions of bounded variation. It follows directly from the definition that every increasing function defined on a closed interval is of bounded variation on that interval. We can also show that every function of bounded variation is the difference of two increasing functions. We begin with the following definition.

Definition 11.1.2. *Let f be of bounded variation on the closed interval $[a, b]$. Then the **total variation function** $v_f : [a, b] \to R$ is defined by the equation*

$$v_f(x) = V(f; [a, x]).$$

Theorem 11.1.8. *If f is of bounded variation on $[a, b]$, then*

(i) *v_f is an increasing function,*

(ii) *$v_f - f$ is an increasing function, and*

(iii) *f is continuous at $x_0 \in [a, b]$ if and only if v_f is continuous at x_0.*

The proof of this theorem is left as Exercises 14 and 15 of Problem Set 11.1.

Since $f = v_f - (v_f - f)$, Theorem 11.1.8 implies that any function of bounded variation can be written as the difference of two increasing functions. Since any increasing function is of bounded variation, Corollary 11.1.3b implies that the difference of two increasing functions is of bounded variation. These lead to the following important characterization of functions of bounded variation.

Theorem 11.1.9. *The function $f : [a, b] \to R$ is of bounded variation if and only if it may be represented as the difference of two increasing functions.*

The representation of a function of bounded variation into the difference of two increasing functions is not unique. For example, if f is of bounded variation on $[a, b]$ and g is any increasing function on $[a, b]$, then we can write

$$f = v_f + g - (v_f + g - f),$$

and here both $v_f + g$ and $v_f + g - f$ are increasing functions. However, the decomposition of f as $v_f - (v_f - f)$ is often the most desirable one.

Theorems 11.1.8 and 11.1.9 imply that if f is a continuous function of bounded variation, then f can be represented as the difference of two continuous increasing functions, since f is continuous if and only if v_f is continuous. We state this result in the form of a corollary.

Corollary 11.1.9. *If f is a continuous function of bounded variation on $[a, b]$, then f may be written as the difference of two continuous increasing functions on $[a, b]$. In particular,*

$$f = v_f - (v_f - f),$$

where v_f and $(v_f - f)$ are continuous and increasing on $[a, b]$.

Problem Set 11.1

In Exercises 1–6, find the variation of the given function over the given interval.

1. $f(x) = 2$ for $x \in [0, 1]$.
2. $f(x) = x$ for $x \in [0, 1]$.
3. $f(x) = x^k$ for $x \in [1, 2]$, k any real number.
4. $f(x) = e^x$ for $x \in [0, 1]$.
5. $f(x) = \cos x$ for $x \in [0, 2\pi]$.
6. $f(x) = \begin{cases} x^2 \sin(1/x) & \text{for } x \in (0, 1], \\ 0 & \text{for } x = 0. \end{cases}$

7. Prove Theorem 11.1.1.
8. If f is monotone on $[a, b]$, then show that f is of bounded variation on $[a, b]$ and $V(f; [a, b]) = |f(b) - f(a)|$.
9. Prove Theorem 11.1.2.
10. Give an example of two functions f and g of bounded variation on $[a, b]$ such that

$$V(f + g; [a, b]) \neq V(f; [a, b]) + V(g; [a, b]).$$

11. Prove Theorem 11.1.3.
12. Prove Theorem 11.1.5.

13. If f is integrable on $[a, b]$ and $F(x) = \int_a^x f(\zeta)d\zeta$ for each $x \in [a, b]$, then show

that F is of bounded variation.

14. Let f be of bounded variation on the interval $[a, b]$, and let

$$v_f(x) = V(f; [a, x])$$

for each $x \in [a, b]$. Show that the function v_f and $v_f - f$ are increasing functions·

15. Show that the function v_f defined in Exercise 14 is continuous at x_0 if and only if f is continuous at x_0. [*Hint*: If f is continuous at x_0, to prove that v_f is continuous at x_0 use nets of the form $\{x_0 = a_0 < a_1 < ... < a_n = x_0 + h\}$ of $[x_0, x_0 + h]$ to show that

$$V(f; [x_0, x_0 + h]) \leq \sup_{\overline{N}} \{|f(a_1) - f(x_0)| + V(f_1; [a_1, x_0 + h])\}.]$$

16. Give an example of a function f of bounded variation on a closed interval such that at some point x_0 of the closed interval f is not differentiable but v_f is differentiable. (Cf. Exercise 14.)

17. Let the function f be defined on the interval $[0, 1]$ by

$$f(x) = \begin{cases} 0 & \text{if } x \text{ is irrational,} \\ 1 & \text{if } x = 0, \\ 1/q & \text{if } x = p/q, \text{ where } p \text{ and } q \text{ are} \\ & \text{relative prime positive integers.} \end{cases}$$

Is the function f of bounded variation on $[0, 1]$?

18. Let the function f be defined on the interval $[0, 1\ 1]$ by

$$f(x) = \begin{cases} 0 & \text{if } x \text{ is irrational,} \\ 1 & \text{if } x = 0, \\ 1/q^2 & \text{if } x = p/q, \text{ where } p \text{ and } q \text{ are} \\ & \text{relative prime positive integers.} \end{cases}$$

Is the function f of bounded variation on $[0, 1]$?

19. Give an example of a function of bounded variation on $[a, b]$ which does *not* satisfy a Lipschitz condition on $[a, b]$. (Cf. Exercise 26 of Problem Set 5.3.)

20. Give an example of a function of bounded variation on $[a, b]$ which does *not* satisfy a Lipschitz condition on any subinterval of $[a, b]$. (Cf. Exercise 18.)

21. If f is a function with bounded derivative on $[a, b]$, then show that f satisfies a Lipschitz condition on $[a, b]$ and so is of bounded variation on $[a, b]$.

22. Prove the following theorem: If $\{f_n\}$ is a uniformly bounded sequence of functions with common domain $[a, b]$ and there exists a real number M such that $V(f_n; [a, b]) \leq M$ for each n, where $n = 1, 2, ...,$ then $\{f_n\}$ has a subsequence which converges everywhere on $[a, b]$ to a function f of bounded variation on $[a, b]$ and $V(f; [a, b]) \leq M$. [*Hint*: Use Theorem 11.1.9 and Exercise 8 of Problem Set 8.4.]

23. Prove that a function of bounded variation has only jump or removable discontinuities. [*Hint*: Cf. Theorem 11.1.9 and Exercise 14 of Problem Set 4.3.]

24. Prove that the set of points of discontinuity of a function of bounded variation is countable. [*Hint*: Cf. Theorem 11.1.9 and Exercise 15 of Problem Set 4.3.]

25. Prove that a function of bounded variation on $[a, b]$ is integrable on $[a, b]$. [*Hint*: Cf. Theorems 6.2.4 and 11.1.9.]

11.2. The Darboux-Stieltjes Integral

In the remainder of this chapter, we shall develop two new integrals, the *Darboux–Stieltjes integral* and the *Riemann–Stieltjes integral*. The Darboux–Stieltjes integral is defined in terms of *upper* and *lower sums*, while the Riemann–Stieltjes integral is defined in terms of the limit of *Riemann–Stieltjes sums*. We shall consider bounded functions on compact intervals, define the integrals of one function with respect to another, and derive the main properties of these two types of integrals. The integrals considered here are somewhat more general than the integral developed in Chapter 6. The added generality makes these integrals very useful in applications, especially in statistics. The theoretical machinery needed for rigorous treatment of these integrals is very similar to that required for the ordinary Riemann integral. We begin by defining the Darboux–Stieltjes integral.

If I is a compact interval and f is a bounded function on I, define the symbols $M(f, I)$ and $m(f, I)$ by

$$M(f, I) = \sup\{f(x) : x \in I\},$$

and

$$m(f, I) = \inf\{f(x) : x \in I\}.$$

Definition 11.2.1. *Let*

$$\mathfrak{n} = \{a = x_0 < x_1 < \dots < x_n = b\}$$

be a net of the compact interval $I = [a, b]$, *let* $I_k = [x_{k-1}, x_k]$, $k = 1, 2, \dots, n$, *let* f *be a bounded function on* I, *and let* g *be an increasing function on* I. *Then define*

$$u(f, g, \mathfrak{n}) = \sum_{k=1}^{n} M(f, I_k)[g(x_k) - g(x_{k-1})]$$

and

$$l(f, g, \mathfrak{n}) = \sum_{k=1}^{n} m(f, I_k)[g(x_k) - g(x_{k-1})].$$

The numbers $u(f, g, \mathfrak{n})$ *and* $l(f, g, \mathfrak{n})$ *are respectively called the **upper** and **lower sums** of* f *with respect to* g.

We see that if $g(x) = x$ for each $x \in I$, then the upper and lower sums of f with respect to g are exactly the upper and lower sums given in Definition 6.1.3.

Since the least upper bound of a nonempty set is always greater than or equal to the greatest lower bound of the set, and since g is an increasing function, that is, $g(x_k) - g(x_{k-1}) \geq 0$, where $k = 1, 2, ..., n$, it follows that

$$l(f, g, \mathfrak{n}) \leq u(f, g, \mathfrak{n}).$$

The concepts of upper and lower sums are illustrated by the following example.

Example 11.2.1. Let $I = [0, 1]$, and let $f : I \to R$ and $g : I \to R$ such that $f(x) = g(x) = x^2$. For each natural number n, define the net $\mathfrak{n}_n$ by

$$\mathfrak{n}_n = \{0 = 0/n, 1/n, 2/n, ..., n/n = 1\}.$$

Denote the subintervals of $\mathfrak{n}_n$ by

$$I_{n,k} = [(k-1)/n, k/n],$$

where $k = 1, 2, ..., n$. It is easily seen that

$$M(f, I_{n,k}) = k^2/n^2$$

and

$$m(f, I_{n,k}) = (k-1)^2/n^2,$$

where $k = 1, 2, ..., n$. Also,

$$g(k/n) - g((k-1)/n) = \frac{k^2}{n^2} - \frac{(k-1)^2}{n^2} = \frac{2k-1}{n^2}.$$

Thus,

$$u(f, g, \mathfrak{n}_n) = \sum_{k=1}^{n} M(f, I_{n,k})[g(k/n) - g((k-1)/n)]$$

$$= \sum_{k=1}^{n} (k^2/n^2)(2k-1)/n^2$$

$$= (3n^3 + 4n^2 - 1)/6n^3,$$

and

$$l(f, g, \mathfrak{n}_n) = \sum_{k=1}^{n} m(f, I_{n,k})[g(k/n) - g((k-1)/n)]$$

$$= \sum_{k=1}^{n} [(k-1)^2/n^2](2k-1)/n^2$$

$$= (3n^4 - 4n^3 - 6n^2 + n - 1)/6n^4.$$

It should be noted that, for any positive integers m and n,

$$l(f, g, \mathfrak{n}_m) < 1/2 < u(f, g, \mathfrak{n}_n).$$

From the definition of upper and lower sums of a bounded function f with respect to an increasing function g, we see that if $\mathfrak{n}^*$ is a refinement of the net $\mathfrak{n}$, then

$$u(f, g, \mathfrak{n}^*) \leqslant u(f, g, \mathfrak{n})$$

and

$$l(f, g, \mathfrak{n}^*) \geqslant l(f, g, \mathfrak{n}).$$

If we let $\mathfrak{n}_1$ and $\mathfrak{n}_2$ be any two nets of the compact interval I, and let $\mathfrak{n}$ be a common refinement of $\mathfrak{n}_1$ and $\mathfrak{n}_2$, then we have that

$$l(f, g, \mathfrak{n}_1) \leqslant l(f, g, \mathfrak{n}) \leqslant u(f, g, \mathfrak{n}) \leqslant u(f, g, \mathfrak{n}_2).$$

That is, any lower sum is less than or equal to any upper sum.

We are now able to define *upper and lower integrals of a bounded function with respect to an increasing function.*

Definition 11.2.2. *Let I be a compact interval, let f be a bounded function on I, let g be an increasing function on I, and let $\overline{N}$ be the set of all nets of the interval I. The **upper integral** of f with respect to g over I, denoted by $\overline{\int} f \, dg$, is defined by*

$$\overline{\int} f \, dg = \inf \{u(f, g, \mathfrak{n}) : \mathfrak{n} \in \overline{N}\}.$$

*The **lower integral** of f with respect to g over I, denoted by $\underline{\int} f \, dg$, is defined by*

$$\underline{\int} f \, dg = \sup \{l(f, g, \mathfrak{n}) : \mathfrak{n} \in \overline{N}\}.$$

It follows from the definition of upper and lower integrals that

$$\underline{\int} f \, dg \leqslant \overline{\int} f \, dg.$$

Definition 11.2.3. *Let $I = [a, b]$ be a closed interval, let f be a bounded function on I, and let g be an increasing function on I. The function f is said to be **Darboux–Stieltjes integrable** with respect to the function g over I if and only if the upper and lower integrals of f with respect to g over I are equal. If f is Darboux–Stieltjes integrable with respect to g over I, then the common value of the upper and lower integrals is called the **Darboux–Stieltjes integral** of f with respect to g over I. The Darboux–Stieltjes integral is denoted by*

$$\text{DS} \int_I f \, dg, \quad \text{DS} \int_a^b f \, dg, \quad \text{or} \quad \text{DS} \int_a^b f(x) \, dg(x).$$

*The numbers a and b are called the **lower** and **upper limits of integration**, respectively. The function f is called the **integrand** and g is called the **integrator**.*

The Darboux–Stieltjes integral is named in honor of the French mathematician Gaston Darboux (1842–1917) and the Dutch astronomer and mathematician Thomas Joannes Stieltjes (1856–1894).

Example 11.2.2. For the functions f and g and the interval I defined in Example 11.2.1, we see that

$$\overline{\int} f\, dg \leqslant u(f, g, \mathfrak{n}_n) = (3n^3 + 4n^2 - 1)/6n^3$$

for each positive integer n. Hence

$$\overline{\int} f\, dg \leqslant 1/2.$$

Likewise, we see that

$$\underline{\int} f\, dg \geqslant l(f, g, \mathfrak{n}_n) = (3n^4 - 4n^3 - 6n^2 + n - 1)/6n^4$$

for each positive integer n, and

$$\underline{\int} f\, dg \geqslant 1/2.$$

Since the lower integral is always less than or equal to the upper integral, we obtain

$$\underline{\int} f\, dg \leqslant \overline{\int} f\, dg \leqslant 1/2 \leqslant \underline{\int} f\, dg.$$

Therefore,

$$\underline{\int} f\, dg = \overline{\int} f\, dg = \mathrm{DS}\int_I f\, dg = \mathrm{DS}\int_0^1 f\, dg = 1/2,$$

and the function f is Darboux–Stieltjes integrable with respect to g over the interval I. Thus we may write

$$\mathrm{DS}\int_0^1 x^2\, dx^2 = 1/2.$$

Example 11.2.3. If $g(x) = x$, then the Darboux–Stieltjes integral reduces to the ordinary Riemann integral.

Many of the elementary properties of the Riemann integral are extendable to the Darboux–Stieltjes integral and are left as Exercises 12–20 of Problem Set 11.2. We now discuss some other results concerning the Darboux–Stieltjes integral.

Theorem 11.2.1. *The function f is Darboux–Stieltjes integrable with respect to the function g over the compact interval* $[a, b]$ *if and only if to each* $\varepsilon > 0$ *there is a net* $\mathfrak{n}$ *of* $[a, b]$ *such that*

$$u(f, g, \mathfrak{n}) - l(f, g, \mathfrak{n}) < \varepsilon.$$

This theorem is analogous to Theorem 6.1.3.

Proof. We first assume that to each $\varepsilon > 0$ there is a net $\mathfrak{n}$ of $[a, b]$ such that

$$u(f, g, \mathfrak{n}) - l(f, g, \mathfrak{n}) < \varepsilon.$$

Then for the net $\mathfrak{n}$ we have

$$\bar{\int} f \, dg \leqslant u(f, g, \mathfrak{n}) < l(f, g, \mathfrak{n}) + \varepsilon \leqslant \underline{\int} f \, dg + \varepsilon.$$

Thus

$$\bar{\int} f \, dg < \underline{\int} f \, dg + \varepsilon.$$

Since ε was an arbitrary positive number, we see that

$$\bar{\int} f \, dg \leqslant \underline{\int} f \, dg,$$

and thus

$$\bar{\int} f \, dg = \underline{\int} f \, dg.$$

Hence f is Darboux–Stieltjes integrable with respect to g over $[a, b]$.

Conversely, we assume that f is Darboux–Stieltjes integrable with respect to g over $[a, b]$ and let $\varepsilon > 0$ be given. From the definition of the upper and lower sums there are nets $\mathfrak{n}_1$ and $\mathfrak{n}_2$ of $[a, b]$ such that

$$\underline{\int} f \, dg < l(f, g, \mathfrak{n}_1) + \varepsilon/2$$

and

$$\bar{\int} f \, dg > u(f, g, \mathfrak{n}_2) - \varepsilon/2.$$

Let $\mathfrak{n}$ be a common refinement of the nets $\mathfrak{n}_1$ and $\mathfrak{n}_2$. Then

$$\underline{\int} f \, dg < l(f, g, \mathfrak{n}_1) + \varepsilon/2 \leqslant l(f, g, \mathfrak{n}) + \varepsilon/2$$

and

$$\bar{\int} f \, dg > u(f, g, \mathfrak{n}_2) - \varepsilon/2 \geqslant u(f, g, \mathfrak{n}) - \varepsilon/2.$$

Using the fact that the upper and lower integrals are equal, and combining the above inequalities, we obtain

$$u(f, g, \mathfrak{n}) - l(f, g, \mathfrak{n}) < \varepsilon.$$

Theorem 11.2.2. *Let g be an increasing function on the compact interval [a, b]. If f is continuous on [a, b], then f is Darboux–Stieltjes integrable with respect to g over [a, b].*

Proof. Let f be continuous, let g be increasing on [a, b], and let $\varepsilon > 0$ be given. Since f is continuous on [a, b], f is uniformly continuous on [a, b], and so there is a $\delta > 0$ such that

$$|f(x) - f(y)| < \varepsilon/[g(b) - g(a) + 1]$$

whenever $x, y \in [a, b]$ and $|x - y| < \delta$. Let

$$\mathfrak{n} = \{a = x_0 < x_1 < \ldots < x_n = b\}$$

be any net of [a, b] with norm less than δ, and let I_k, where $k = 1, 2, \ldots, n$, be the sub-intervals of the net $\mathfrak{n}$. Then

$$u(f, g, \mathfrak{n}) - l(f, g, \mathfrak{n}) = \sum_{k=1}^{n} M(f, I_k)[g(x_k) - g(x_{k-1})]$$

$$- \sum_{k=1}^{n} m(f, I_k)[g(x_k) - g(x_{k-1})]$$

$$= \sum_{k=1}^{n} [M(f, I_k) - m(f, I_k)][g(x_k) - g(x_{k-1})]$$

$$\leqslant \sum_{k=1}^{n} [\varepsilon/\{g(b) - g(a) + 1\}][g(x_k) - g(x_{k-1})]$$

$$= [\varepsilon/\{g(b) - g(a) + 1\}][g(b) - g(a)]$$

$$< \varepsilon.$$

Therefore, Theorem 11.2.1 implies that f is Darboux–Stieltjes integrable with respect to g over [a, b].

Conversely, if f is not continuous on [a, b], then there is an increasing function g such that f is not Darboux–Stieltjes integrable with respect to g over [a, b]. The following example illustrates such a function g.

Example 11.2.4. Let the functions f and g be defined as follows:

$$f(x) = g(x) = \begin{cases} 0 & \text{for } 0 \leqslant x \leqslant 1, \\ 1 & \text{for } 1 < x \leqslant 2. \end{cases}$$

For any net $\mathfrak{n}$ of [0, 2], there is a subinterval [c, d] which contains the point 1 and some point greater than 1. Then

$$M(f, [c, d])[g(d) - g(c)] = 1$$

and

$$m(f, [c, d])[g(d) - g(c)] = 0.$$

Thus it follows that

$$u(f, g, \mathfrak{n}) \geqslant l(f, g, \mathfrak{n}) + 1.$$

Since $\mathfrak{n}$ was an arbitrary net of $[0, 2]$, we have

$$\overline{\int} f \, dg \geqslant \underline{\int} f \, dg + 1,$$

which implies that f is not Darboux–Stieltjes integrable with respect to g.

With regard to this example, we note that, for any function f which is bounded on the compact interval $[a, b]$, there is always an increasing function g (namely, a constant function) such that f is Darboux–Stieltjes integrable with respect to g over $[a, b]$.

Theorem 11.2.3. *If f is a monotone function on $[a, b]$, and if g is a continuous increasing function on $[a, b]$, then f is Darboux–Stieltjes integrable with respect to g over $[a, b]$.*

Proof. We assume that f is an increasing function on $[a, b]$. The case where f is decreasing is handled in a similar manner. Let g be increasing and continuous on $[a, b]$, and let $\varepsilon > 0$ be given. It follows from the fact that f is increasing on $[a, b]$ that f is bounded on $[a, b]$, and there is a real positive number λ such that

$$\lambda[f(b) - f(a)] < \varepsilon.$$

Since g is continuous on $[a, b]$, g is uniformly continuous on $[a, b]$, and so there is a $\delta > 0$ such that

$$|g(x) - g(y)| < \lambda$$

whenever $x, y \in [a, b]$ and $|x - y| < \delta$. If

$$\mathfrak{n} = \{a = x_0 < x_1 < ... < x_n = b\}$$

is any net with norm less than δ, then, since g is increasing, we have

$$g(x_k) - g(x_{k-1}) < \lambda,$$

for $k = 1, 2, ..., n$. Since f is increasing we have

$$M(f, [x_{k-1}, x_k]) = f(x_k),$$
$$m(f, [x_{k-1}, x_k]) = f(x_{k-1}),$$

and we write

$$u(f, g, \mathfrak{n}) - l(f, g, \mathfrak{n})$$

$$= \sum_{k=1}^{n} [M(f, [x_{k-1}, x_k]) - m(f, [x_{k-1}, x_k])][g(x_k) - g(x_{k-1})]$$

$$= \sum_{k=1}^{n} [f(x_k) - f(x_{k-1})][g(x_k) - g(x_{k-1})]$$

$$\leqslant \sum_{k=1}^{n} [f(x_k) - f(x_{k-1})]\lambda$$

$$= \lambda[f(b) - f(a)] < \varepsilon.$$

Therefore, Theorem 11.2.1 implies that f is Darboux–Stieltjes integrable with respect to g over $[a, b]$.

Problem Set 11.2

For Exercises 1–7, evaluate the given Darboux–Stieltjes integral.

1. DS $\int_0^1 x \, d(x^2)$.

2. DS $\int_0^2 x^2 d(x^3)$.

3. DS $\int_0^2 x \, d([x])$, where $[x]$ denotes the bracket function.

4. DS $\int_0^2 x^2 \, d([x])$.

5. DS $\int_0^1 x \, dg(x)$, where $g(x) = \begin{cases} -1 & \text{if } x = 0, \\ x & \text{if } 0 < x < 1, \\ 2 & \text{if } x = 1. \end{cases}$

6. DS $\int_{-1}^1 x \, d(e^x)$.

7. DS $\int_0^1 x^2 \, d([x^2])$.

8. Prove directly that if g is increasing on $[a, b]$, then

$$\text{DS} \int_a^b dg(x)$$

exists and is equal to $g(b) - g(a)$.

9. Let f be Darboux–Stieltjes integrable with respect to every increasing function g over $[a, b]$. Prove that f is continuous.

10. Let $f(x) = x^2 - 2x$ for $0 \leqslant x \leqslant 2$, and let

$$g(x) = \begin{cases} x^2 & \text{if } 0 \leqslant x < 1, \\ x^3 + 1 & \text{if } 1 \leqslant x \leqslant 2. \end{cases}$$

Find $V(g; [0, 2])$ and $\text{DS} \int_0^2 f(x) dg(x)$.

11. Let $f(x) = x^2 - 2x$ for $0 \leqslant x \leqslant 2$, and let

$$g(x) = \begin{cases} x & \text{if } 0 \leqslant x < 1, \\ 2 & \text{if } x = 1, \\ 2+x & \text{if } 1 < x \leqslant 2. \end{cases}$$

Find $V(g; [0, 2])$ and $\text{DS} \int_0^2 f(x) dg(x)$.

For Exercises 12–20, assume that f_1 and f_2 are Darboux–Stieltjes integrable with respect to the increasing functions g_1 and g_2 over $[a, b]$.

12. If c is a constant, then prove that the function cf_1 is Darboux–Stieltjes integrable with respect to g_1 over $[a, b]$ and that

$$\text{DS} \int_a^b cf_1(x) dg_1(x) = c \left(\text{DS} \int_a^b f_1(x) dg_1(x) \right).$$

13. Prove that the function $f_1 + f_2$ is Darboux–Stieltjes integrable with respect to g_1 over $[a, b]$ and that

$$\text{DS} \int_a^b [f_1(x) + f_2(x)] dg_1(x) = \text{DS} \int_a^b f_1(x) dg_1(x) + \text{DS} \int_a^b f_2(x) dg_1(x).$$

14. If c_1 and c_2 are constants, then prove that the function $c_1 f_1 + c_2 f_2$ is Darboux–Stieltjes integrable with respect to g_1 over $[a, b]$ and that

$$\text{DS} \int_a^b [c_1 f_1(x) + c_2 f_2(x)] dg_1(x) =$$

$$c_1 \left(\text{DS} \int_a^b f_1(x) dg_1(x) \right) + c_2 \left(\text{DS} \int_a^b f_2(x) dg_1(x) \right).$$

15. If $m \leqslant f_1(x) \leqslant M$ for $x \in [a, b]$, then prove that

$$m[g_1(b) - g_1(a)] \leqslant \mathrm{DS} \int_a^b f_1(x)dg(x) \leqslant M[g_1(b) - g_1(a)].$$

16. If $c \geqslant 0$ is a constant, then prove that the function f_1 is Darboux–Stieltjes integrable with respect to cg_1 over $[a, b]$ and that

$$\mathrm{DS} \int_a^b f_1(x)d[cg_1(x)] = c\left(\mathrm{DS} \int_a^b f_1(x)dg_1(x)\right).$$

17. Prove that the function f_1 is Darboux–Stieltjes integrable with respect to $g_1 + g_2$ over $[a, b]$ and that

$$\mathrm{DS} \int_a^b f_1(x)d[g_1(x) + g_2(x)] = \mathrm{DS} \int_a^b f_1(x)dg_1(x) + \mathrm{DS} \int_a^b f_1(x)dg_2(x).$$

18. If $c_1 \geqslant 0$ and $c_2 \geqslant 0$ are constants, then prove that the function f_1 is Darboux–Stieltjes integrable with respect to the function $c_1g_1 + c_2g_2$ over $[a, b]$ and that

$$\mathrm{DS} \int_a^b f_1(x)d[c_1g_1(x) + c_2g_2(x)]$$
$$= c_1\left(\mathrm{DS} \int_a^b f_1(x)dg_1(x)\right) + c_2\left(\mathrm{DS} \int_a^c f_1(x)dg_2(x)\right).$$

19. Prove that the function $f_1 f_2$ is Darboux–Stieltjes integrable with respect to g_1 over $[a, b]$.

20. Prove that the function $|f_1|$ is Darboux–Stieltjes integrable with respect to g_1 over $[a, b]$ and that

$$\left|\mathrm{DS} \int_a^b f_1(x)dg_1(x)\right| \leqslant \mathrm{DS} \int_a^b |f_1(x)|dg_1(x).$$

21. Let f be bounded and g be increasing on the compact interval $[a, b]$ and assume $c \in [a, b]$. Prove that f is Darboux–Stieltjes integrable with respect to g over $[a, b]$ if and only if f is Darboux–Stieltjes integrable with respect to g over each of the intervals $[a, c]$ and $[c, b]$. If f is Darboux–Stieltjes integrable with respect to g over $[a, b]$, then show that

$$\mathrm{DS} \int_a^b f(x)dg(x) = \mathrm{DS} \int_a^c f(x)dg(x) + \mathrm{DS} \int_c^b f(x)dg(x).$$

11.3. The Riemann-Stieltjes Integral

In this section we define the *Riemann–Stieltjes integral* and develop some of its properties. We begin by defining *Riemann–Stieltjes sums*. This type of sum need not be an upper or lower sum, but it always lies between them.

Definition 11.3.1. *Let f be a bounded function and let g be an increasing function on the compact interval* $I = [a, b]$. *If*

$$\mathfrak{n} = \{a = x_0 < x_1 < \ldots < x_n = b\}$$

is any net of I with subintervals $I_k = [x_{k-1}, x_k]$, $k = 1, 2, \ldots, n$, *let* ξ_k *be any point in the subinterval* $I_k, k = 1, 2, \ldots, n$. *Then the sum*

$$R(f, g, \mathfrak{n}, \xi) = \sum_{k=1}^{n} f(\xi_n)[g(x_k) - g(x_{k-1})],$$

where $\xi = \{\xi_1, \xi_2, \ldots, \xi_n\}$, *will be called a* **Riemann–Stieltjes sum** *of f with respect to g on I.*

We present an example to illustrate the concept of a Riemann–Stieltjes sum.

Example 11.3.1. Let $I = [0, 1]$, and let $f: I \to R$ and $g: I \to R$ such that $f(x) = g(x) = x^2$. If $\mathfrak{n} = \{0, 1/3, 2/3, 1\}$, then the subintervals of $\mathfrak{n}$ are $I_1 = [0, 1/3]$, $I_2 = [1/3, 2/3]$, and $I_3 = [2/3, 1]$. Let $\xi_1 = 1/4$, $\xi_2 = 1/2$, and $\xi_3 = 3/4$. Then, for $\xi = \{\xi_1, \xi_2, \xi_3\} = \{1/4, 1/2, 3/4\}$, we have

$$R(f, g, \mathfrak{n}, \xi) = \sum_{k=1}^{3} f(\xi_k)[g(x_k) - g(x_{k-1})]$$

$$= (1/4)^2 \,[(1/3)^2 - 0^2]$$

$$+ (1/2)^2 \,[(2/3)^2 - (1/3)^2]$$

$$+ (3/4)^2 \,[1^2 - (2/3)^2]$$

$$= 29/72.$$

More generally, if for each natural number n we define the net $\mathfrak{n}_n$ by

$$\mathfrak{n}_n = \{0 = 0/n < 1/n < 2/n < \ldots < n/n = 1\},$$

then the subintervals of $\mathfrak{n}_n$ are

$$I_{n,k} = [(k-1)/n, k/n],$$

$k = 1, 2, \ldots, n$. For each k, where $k = 1, 2, \ldots, n$, let

$$\xi_{n,k} = (k - 1/2)/n.$$

Then $\xi_{n,k}$ is the midpoint of the interval $I_{n,k}$. For

$$\xi_n = \{\xi_{n,k} : k = 1, 2, \ldots, n\},$$

we have

$$R(f, g, \mathfrak{n}_n, \xi_n) = \sum_{k=1}^{n} f(\xi_{k,n})[g(k/n) - g((k-1)/n)]$$

$$= \sum_{k=1}^{n} [(k - 1/2)/n]^2 [(k/n)^2 - (k-1)/n^2]$$

$$= (1/n^4) \sum_{k=1}^{n} (k - 1/2)^2 [k^2 - (k-1)^2]$$

$$= (1/n^4) \sum_{k=1}^{n} (2k - 1)^3/4$$

$$= (1/2n^3)[n^3 - n/2].$$

From Example 11.2.1, we see that

$$u(f, g, \mathfrak{n}_n) = (3n^3 + 4n^2 - 1)/6n^3,$$

$$l(f, g, \mathfrak{n}_n) = (3n^4 - 4n^3 - 6n^2 + n - 1)/6n^4,$$

and so

$$l(f, g, \mathfrak{n}_n) < R(f, g, \mathfrak{n}_n, \xi_n) < u(f, g, \mathfrak{n}_n),$$

for each natural number n.

In Example 11.3.1 we observed that the Riemann–Stieltjes sum was between the upper and lower sums. This fact is now put in the form of a theorem, with the proof left as Exercise 12 of Problem Set 11.3.

Theorem 11.3.1. *Let f be bounded and g be increasing on the compact interval* $[a, b]$. *If* $\mathfrak{n}$ *is any net of* $[a, b]$, *then*

$$l(f, g, \mathfrak{n}) \leqslant R(f, g, \mathfrak{n}, \xi) \leqslant u(f, g, \mathfrak{n}).$$

We now define a special kind of limit associated with Riemann–Stieltjes sums.

Definition 11.3.2. *Let f be bounded and g be increasing on the compact interval I, and let* $\mathfrak{R}$ *be the set of all Riemann–Stieltjes sums of f with respect to g on I. Then the set* $\mathfrak{R}$ *is said to have* **limit** $\mathfrak{s}$ *if and only if to each* $\varepsilon > 0$ *there is a* $\delta = \delta(\varepsilon) > 0$ *such that*

$$|R(f, g, \mathfrak{n}, \xi) - \mathfrak{s}| < \varepsilon$$

whenever $\|\mathfrak{n}\| < \delta$. *This limit is denoted by*

$$\lim_{\|\mathfrak{n}\| \to 0} R(f, g, \mathfrak{n}, \xi) = \mathfrak{s}.$$

This limit is similar to the limit of a function in the sense that if it exists, then it is unique. However, it must be emphasized that the ε-condition must hold for each net $\mathfrak{n}$ with $\|\mathfrak{n}\| < \delta$ as well as for every choice of ξ.

We are now ready to define the Riemann–Stieltjes integral.

Definition 11.3.3. *Let f be bounded and g be increasing on the compact interval $I = [a, b]$. Then f is said to be **Riemann–Stieltjes integrable with respect to g over I** if and only if*

$$\lim_{\|\mathfrak{n}\| \to 0} R(f, g, \mathfrak{n}, \xi)$$

*exists and is finite. If this limit exists and is finite then its value is called the **Riemann–Stieltjes integral** of f with respect to g over I and is denoted by*

$$RS \int f \, dg, \quad RS \int_a^b f \, dg, \quad \text{or} \quad RS \int_a^b f(x) dg(x).$$

*The function f is called the **integrand**, and g is called the **integrator**.*

We note that if $g(x) = x$ on the interval $[a, b]$, then the Riemann–Stieltjes integral

$$RS \int_a^b f(x) dg(x),$$

as well as the Darboux–Stieltjes integral

$$DS \int_a^b f(x) dg(x),$$

reduce to the ordinary Riemann integral. Since the Riemann integral can be defined either in terms of upper and lower integrals or as the limit of Riemann sums (cf. Definition 6.1.5 and Theorem 6.2.3), one might suspect that the Darboux–Stieltjes and Riemann–Stieltjes integrals are equivalent. This is not the case, however, as the following example indicates.

Example 11.3.2. Let f and g be defined on the interval $[0, 2]$ by

$$f(x) = \begin{cases} 0 & \text{if } 0 \leqslant x < 1, \\ 1 & \text{if } 1 \leqslant x \leqslant 2, \end{cases}$$

and

$$g(x) = \begin{cases} 0 & \text{if } 0 \leqslant x \leqslant 1, \\ 1 & \text{if } 1 < x \leqslant 2. \end{cases}$$

Since g is constant in $[0, 1]$, f is Darboux–Stieltjes integrable with respect to g over $[0, 1]$ and

$$\text{DS} \int_0^1 f(x) dg(x) = 0.$$

The function f is constant on $[1, 2]$ and Theorem 11.2.2 implies that f is Darboux–Stieltjes integrable with respect to g over $[1, 2]$. Furthermore,

$$\text{DS} \int_1^2 f(x) dg(x) = 1.$$

Then f is Darboux–Stieltjes integrable with respect to g over $[0, 2]$ and

$$\text{DS} \int_0^2 f(x) dg(x) = 1.$$

(Cf. Exercise 21 of Problem Set 11.2.)

We now show that f is *not* Riemann–Stieltjes integrable with respect to g over $[0, 2]$. For each positive integer n, if

$$\mathfrak{n}_n = \{0, 1/n, 2/n, \ldots, 1 - 1/n, 1 - 1/2n, 1 + 1/2n,$$
$$1 + 1/n, \ldots, 2 - 1/n, 2\},$$

where the subintervals of $\mathfrak{n}_n$ are

$$I_{n,k} = [(k - 1)/n, k/n], \ 1 \leqslant k \leqslant n - 1,$$
$$I_{n,n} = [1 - 1/n, \ 1 - 1/2n],$$
$$I_{n,n+1} = [1 - 1/2n, \ 1 + 1/2n],$$
$$I_{n,n+2} = [1 + 1/2n, \ 1 + 1/n], \text{ and}$$
$$I_{n,k} = [(k - 2)/n, \ (k - 1)/n], \ n + 3 \leqslant k \leqslant 2n + 1,$$

then $\|\mathfrak{n}_n\| = 1/n$. Let

$$\xi_n = \{\xi_{n,k} : 1 \leqslant k \leqslant 2n + 1\}$$

and

$$\zeta_n = \{\zeta_{n,k} : 1 \leqslant k \leqslant 2n + 1\},$$

where $\xi_{n,k}$ and $\zeta_{n,k}$ are respectively the left and right endpoints of the subinterval $I_{n,k}$, where $1 \leqslant k \leqslant 2n + 1$. Then

$$R(f, g, \mathfrak{n}_n, \xi_n) = 0$$

and

$$R(f, g, \mathfrak{n}_n, \zeta_n) = 1.$$

Therefore

$$\lim_{||\mathfrak{n}|| \to 0} R(f, g, \mathfrak{n}, \xi)$$

does not exist, and so f is not Riemann–Stieltjes integrable with respect to g over $[0, 2]$.

The above example shows that the Darboux–Stieltjes and the Riemann–Stieltjes integrals are not equivalent. The following theorem indicates that any function f which is Riemann–Stieltjes integrable with respect to the increasing function g over $[a, b]$ is also Darboux–Stieltjes integrable with respect to g over $[a, b]$.

Theorem 11.3.2. *Let f be bounded and g be increasing on the compact interval $I = [a, b]$. If f is Riemann–Stieltjes integrable with respect to g over I, then f is Darboux–Stieltjes integrable with respect to g over I and*

$$\text{DS} \int_a^b f(x)dg(x) = \text{RS} \int_a^b f(x)dg(x).$$

Proof. We assume that f is Riemann–Stieltjes integrable with respect to g over $[a, b]$ and show that f is Darboux–Stieltjes integrable with respect to g over $[a, b]$. Assume that $\varepsilon > 0$ is given. Definitions 11.3.2 and 11.3.3 imply that there is $\delta > 0$ such that

$$S - \varepsilon/2 < R(f, g, \mathfrak{n}, \xi) < S + \varepsilon/2$$

whenever $||\mathfrak{n}|| < \delta$ and for any choice of ξ, where

$$S = \text{RS} \int_a^b f(x)dg(x).$$

If $g(a) = g(b)$, then g is constant and the result follows. So assume that $g(a) < g(b)$. Let

$$\mathfrak{n} = \{a = x_0 < x_1 < ... < x_n = b\}$$

be any net of I with $||\mathfrak{n}|| < \delta$, and let

$$I_k = [x_{k-1}, x_k], k = 1, 2, ..., n,$$

be the subintervals of $\mathfrak{n}$. Since $M(f, I_k)$ denotes the least upper bound of the set $\{f(x): x \in I_k\}$, there is a $\xi_k \in I_k$ such that

$$f(\xi_k) > M(f, I_k) - \varepsilon/2[g(b) - g(a)],$$

where $k = 1, 2, ..., n$. Then, for $\xi = \{\xi_k : 1 \leqslant k \leqslant n\}$,

$$R(f, g, \mathfrak{n}, \xi) = \sum_{k=1}^{n} f(\xi_k)[g(x_k) - g(x_{k-1})]$$

$$> \sum_{k=1}^{n} [M(f, I_k) - \varepsilon/2[g(b) - g(a)]][g(x_k) - g(x_{k-1})]$$

$$= \sum_{k=1}^{n} M(f, I_k)[g(x_k) - g(x_{k-1})]$$

$$- \left[\varepsilon/2[g(b) - g(a)] \sum_{k=1}^{n} [g(x_k) - g(x_{k-1})]\right]$$

$$= u(f, g, \mathfrak{n}) - \varepsilon/2.$$

Using the inequality
$$R(f, g, \mathfrak{n}, \xi) < S + \varepsilon/2,$$
we obtain
$$u(f, g, \mathfrak{n}) - \varepsilon/2 < R(f, g, \mathfrak{n}, \xi) < S + \varepsilon/2$$
or
$$u(f, g, \mathfrak{n}) < S + \varepsilon.$$

Since the upper integral is less than or equal to any upper sums, we have
$$\overline{\int} f \, dg \leqslant u(f, g, \mathfrak{n}) < S + \varepsilon.$$

The positive number ε was arbitrary, and so
$$\overline{\int} f \, dg \leqslant S.$$

By a similar argument, which is left as an exercise, we have
$$S \leqslant \underline{\int} f \, dg.$$
Hence
$$\overline{\int} f \, dg \leqslant S \leqslant \underline{\int} f \, dg \leqslant \overline{\int} f \, dg$$
and
$$\mathrm{DS} \int_a^b f(x) dg(x) = \overline{\int} f \, dg = \underline{\int} f \, dg = \mathrm{S} = \mathrm{RS} \int_a^b f(x) dg(x).$$

Therefore, f is Darboux–Stieltjes integrable with respect to g over $[a, b]$, and the two integrals are equal.

In Example 11.3.2, it was pointed out that not all functions f which are Darboux–Stieltjes integrable with respect to an increasing function g are

Riemann–Stieltjes integrable with respect to g. We now present a condition on the function g which guarantees that f is Riemann–Stieltjes integrable with respect to g whenever f is Darboux–Stieltjes integrable with respect to g.

Theorem 11.3.3. *Let g be an increasing continuous function on the compact interval $[a, b]$. If f is Darboux–Stieltjes integrable with respect to g over $[a, b]$, then f is Riemann–Stieltjes integrable with respect to g over $[a, b]$, and*

$$\text{DS} \int_a^b f(x)dg(x) = \text{RS} \int_a^b f(x)dg(x).$$

Proof. Denote the Darboux–Stieltjes integral by S, that is,

$$S = \text{DS} \int_a^b f(x)dg(x),$$

and assume that $\varepsilon > 0$ is given. Theorem 11.2.4 implies that there is a net $\mathfrak{n}^* = \{a = y_0 < y_1 < \ldots < y_k = b\}$ of $[a, b]$ such that

$$S - \varepsilon/2 < l(f, g, \mathfrak{n}^*) \leqslant u(f, g, \mathfrak{n}^*) < S + \varepsilon/2.$$

For the remainder of the proof we assume the net $\mathfrak{n}^*$ is fixed. Let

$$M = M(f, [a, b]),$$
$$m = m(f, [a, b]).$$

If

$$M = m,$$

then f is constant on $[a, b]$ and

$$\text{DS} \int_a^b f(x)dg(x) = \text{RS} \int_a^b f(x)dg(x) = M[g(b) - g(a)].$$

Thus we consider the case where $m < M$. Since g is continuous on $[a, b]$ it follows that g is uniformly continuous on $[a, b]$, and so there is a $\delta > 0$ such that $a \leqslant x \leqslant y \leqslant b$ and $y - x < \delta$ imply that

$$0 \leqslant g(y) - g(x) < \varepsilon/2k(M - m).$$

We shall show that if $\mathfrak{n}$ is any net of $[a, b]$ with $\|\mathfrak{n}\| < \delta$, then

$$S - \varepsilon < R(f, g, \mathfrak{n}, \xi) < S + \varepsilon,$$

and so f is Riemann–Stieltjes integrable with respect to g over $[a, b]$ and

$$\text{RS} \int_a^b f(x)dg(x) = S.$$

If $\mathfrak{n} = \{a = x_0 < x_1 < \ldots < x_n = b\}$ is any net of $[a, b]$ with $\|\mathfrak{n}\| < \delta$ and

$$\Delta = \max\{g(x_i) - g(x_{i-1}): 1 \leqslant i \leqslant n\},$$

then

$$0 \leqslant g(x_i) - g(x_{i-1}) \leqslant \Delta < \varepsilon/2k(M - m),$$

for $i = 1, 2, \ldots, n$.

Set $\mathfrak{n}_1 = \mathfrak{n} \cup \mathfrak{n}^*$. Then the refinement $\mathfrak{n}_1$ of $\mathfrak{n}$ has at most $k - 1$ points more than the net $\mathfrak{n}$, since $\mathfrak{n}^*$ has $k + 1$ points. Hence

$$l(f, g, \mathfrak{n}_1) - l(f, g, \mathfrak{n}) \geqslant 0$$

and

$$l(f, g, \mathfrak{n}_1) - l(f, g, \mathfrak{n}) \leqslant (k - 1)(M - m)\Delta < \varepsilon/2.$$

Also, $\mathfrak{n}_1$ is a refinement of $\mathfrak{n}^*$, and so

$$l(f, g, \mathfrak{n}^*) \leqslant l(f, g, \mathfrak{n}_1), \quad u(f, g, \mathfrak{n}_1) \leqslant u(f, g, \mathfrak{n}^*).$$

Combining these inequalities, we have

$$S - \varepsilon/2 < l(f, g, \mathfrak{n}^*) \leqslant l(f, g, \mathfrak{n}_1) < l(f, g, \mathfrak{n}) + \varepsilon/2$$

and

$$u(f, g, \mathfrak{n}) < S + \varepsilon.$$

Hence

$$S - \varepsilon < l(f, g, \mathfrak{n}) \leqslant R(f, g, \mathfrak{n}, \xi) \leqslant u(f, g, \mathfrak{n}) < S + \varepsilon,$$

which completes the proof.

Corollary 11.3.3. *If f is a monotone function on the compact interval $[a, b]$ and g is a continuous increasing function on $[a, b]$, then f is Riemann–Stieltjes integrable with respect to g over $[a, b]$.*

This corollary follows directly from Theorems 11.2.3 and 11.3.3. We now present the analogue of Theorem 11.2.2 for Riemann–Stieltjes integrals.

Theorem 11.3.4. *Let g be an increasing function on the compact interval $[a, b]$. If f is continuous on $[a, b]$, then f is Riemann–Stieltjes integrable with respect to g over $[a, b]$.*

Proof. Let $\varepsilon > 0$ be given. Since f is continuous on $[a, b]$, it follows from Theorem 11.2.2 that f is Darboux–Stieltjes integrable with respect to g over $[a, b]$. Denote by S the value of this integral, that is,

$$S = \mathrm{DS}\int_a^b f(x)dg(x).$$

Since g is increasing on $[a, b]$, we can choose $\lambda > 0$ such that

$$[g(b) - g(a)]\lambda < \varepsilon.$$

Since f is continuous on $[a, b]$, f is uniformly continuous on $[a, b]$, and so there is $\delta > 0$ such that $x, y \in [a, b]$ and $|x - y| < \delta$ imply that

$$|f(x) - f(y)| < \lambda.$$

Let $\mathfrak{n} = \{a = x_0 < x_1 < ... < x_n = b\}$ be any net of $[a, b]$ with $\|\mathfrak{n}\| < \delta$ and with subintervals I_k, where $k = 1, 2, ..., n$. Then

$$M(f, I_k) - m(f, I_k) < \lambda,$$

$k = 1, 2, ..., n$, and

$$u(f, g, \mathfrak{n}) - l(f, g, \mathfrak{n}) = \sum_{k=1}^{n} [M(f, I_k) - m(f, I_k)][g(x_k) - g(x_{k-1})]$$

$$< \lambda \sum_{k=1}^{n} [g(x_k) - g(x_{k-1})]$$

$$= \lambda[g(b) - g(a)]$$

$$< \varepsilon.$$

Since any Riemann–Stieltjes sum $R(f, g, \mathfrak{n}, \xi)$ satisfies the inequality

$$u(f, g, \mathfrak{n}) \geqslant R(f, g, \mathfrak{n}, \xi) \geqslant l(f, g, \mathfrak{n}),$$

and

$$u(f, g, \mathfrak{n}) \geqslant S \geqslant l(f, g, \mathfrak{n}),$$

it follows that

$$|R(f, g, \mathfrak{n}, \xi) - S| < \varepsilon.$$

Hence, f is Riemann–Stieltjes integrable with respect to g over $[a, b]$, and

$$RS \int_a^b f(x) dg(x) = S.$$

We now give an example to show that not all elementary properties of the Riemann integral are extendable to the Riemann–Stieltjes integral. In particular, if a bounded function f is Riemann integrable on the compact intervals $[a, c]$ and $[c, b]$, then f is Riemann integrable on the interval $[a, b]$ and

$$\int_a^b f(x) dx = \int_a^c f(x) dx + \int_c^b f(x) dx;$$

however, this is not true for Riemann–Stieltjes integrability.

Example 11.3.3. Consider the functions f and g and the interval $[0, 2]$ defined in Example 11.3.2. Since g is constant on the interval $[0, 1]$, f is Riemann–Stieltjes integrable with respect to g over $[0, 1]$ and

$$\text{RS} \int_0^1 f(x)dg(x) = 0.$$

The function f is constant on the interval $[1, 2]$, and so f is Riemann–Stieltjes integrable with respect to g over $[1, 2]$ and

$$\text{RS} \int_1^2 f(x)dg(x) = 1.$$

But f is *not* Riemann–Stieltjes integrable with respect to g over the interval $[0, 2]$.

The above example is a special case of the following theorem, the proof of which is left as Exercise 14 of Problem Set 11.3.

Theorem 11.3.5. *Let f be bounded and g be increasing on the compact interval $[a, b]$. If f and g have a common point of discontinuity $c \in [a, b]$, then f is not Riemann–Stieltjes integrable with respect to g over $[a, b]$.*

The following two results are extendable to Riemann–Stieltjes integration; the proofs of these are left as Exercises 16 and 17 of Problem Set 11.3.

Theorem 11.3.6. *If f is Riemann–Stieltjes integrable with respect to g over $[a, c]$, $[c, b]$, and $[a, b]$, then*

$$\text{RS} \int_a^b f(x)dg(x) = \text{RS} \int_a^c f(x)dg(x) + \text{RS} \int_c^b f(x)dg(x).$$

Theorem 11.3.7. *If f is Riemann–Stieltjes integrable with respect to g over $[a, b]$ and $c \in [a, b]$, then f is Riemann–Stieltjes integrable with respect to g over $[a, c]$ and $[c, b]$ and*

$$\text{RS} \int_a^b f(x)dg(x) = \text{RS} \int_a^c f(x)dg(x) + \text{RS} \int_c^b f(x)dg(x).$$

We now give a result relating differentiation and integration. It is useful in reducing Riemann–Stieltjes integrals to Riemann integrals.

Theorem 11.3.8. *Let f be bounded, and let g be increasing and differentiable on $[a, b]$. If f and g' are Riemann integrable on $[a, b]$,* *then f is Riemann–Stieltjes integrable with respect to g over [a, b] and*

$$\text{RS} \int_a^b f(x)dg(x) = \int_a^b f(x)g'(x)dx.$$

Proof. Since f is bounded, there is an $M > 0$ such that $|f(x)| \leqslant M$ for each $x \in [a, b]$. Let $\varepsilon > 0$ be given, and let $\lambda = \varepsilon/(2M + 1)$. The functions f and g' are each Riemann integrable on $[a, b]$; thus Theorem 6.3.5 implies that the function fg' is Riemann integrable on $[a, b]$. Since g' is Riemann integrable on $[a, b]$, it follows from Theorem 6.2.3 that there is a $\delta' > 0$ such that

$$|R(g', \mathfrak{n}, \xi) - \int_a^b g'(x)dx| < \lambda$$

whenever $\|\mathfrak{n}\| < \delta'$. Likewise, Theorem 6.2.3 implies that there is a $\delta'' > 0$ such that

$$|R(fg', \mathfrak{n}, \xi) - \int_a^b f(x)g'(x)dx| < \lambda$$

whenever $\|\mathfrak{n}\| < \delta''$. Let $\delta = \min(\delta', \delta'')$. Then the two inequalities above both hold whenever $\|\mathfrak{n}\| < \delta$.

Let $\mathfrak{n} = \{a = x_0 < x_1 < ... \leqslant x_n = b\}$ be any net of $[a, b]$ with $\|\mathfrak{n}\| < \delta$, and let ξ_k and ζ_k be any two points of the subinterval $I_k = [x_{k-1}, x_k]$ of the net $\mathfrak{n}$, where $k = 1, 2, ..., n$. Then we have

$$\left| \sum_{k=1}^n g'(\xi_k)(x_k - x_{k-1}) - \int_a^b g'(x)dx \right| < \lambda$$

and

$$\left| \sum_{k=1}^n g'(\zeta_k)(x_k - x_{k-1}) - \int_a^b g'(x)dx \right| < \lambda.$$

Combining these inequalities, we obtain

$$\left| \sum_{k=1}^n g'(\xi_k)(x_k - x_{k-1}) - \sum_{k=1}^n g'(\zeta_k)(x_k - x_{k-1}) \right|$$

$$\leqslant \left| \sum_{k=1}^n g'(\xi_k)(x_k - x_{k-1}) - \int_a^b g'(x)dx \right|$$

$$+ \left| \sum_{k=1}^n g'(\zeta_k)(x_k - x_{k-1}) - \int_a^b g'(x)dx \right|$$

$$< \lambda + \lambda = 2\lambda.$$

These inequalities are independent of how the points ξ_k and ζ_k are chosen in $[x_{k-1}, x_k]$, where $k = 1, 2, ..., n$.

Let $\xi_k \in [x_{k-1}, x_k]$, $k = 1, 2, ..., n$. By the Mean Value Theorem (Theorem 5.3.4), there is a $\zeta_k \in [x_{k-1}, x_k]$ such that

$$g(x_k) - g(x_{k-1}) = g'(\zeta_k)(x_k - x_{k-1}),$$

$k = 1, 2, ..., n$. Thus

$$\left| \sum_{k=1}^{n} f(\xi_k)[g(x_k) - g(x_{k-1})] - \int_a^b f(x)g'(x)dx \right|$$

$$= \left| \sum_{k=1}^{n} f(\xi_k)g'(\zeta_k)(x_k - x_{k-1}) - \int_a^b f(x)g'(x)dx \right|$$

$$= \left| \sum_{k=1}^{n} f(\xi_k)g'(\xi_k)(x_k - x_{k-1}) - \int_a^b f(x)g'(x)dx \right.$$

$$\left. + \sum_{k=1}^{n} f(\xi_k)[g'(\zeta_k) - g'(\xi_k)](x_k - x_{k-1}) \right|$$

$$\leqslant \left| \sum_{k=1}^{n} f(\xi_k)g'(\xi_k)(x_k - x_{k-1}) - \int_a^b f(x)g'(x)dx \right|$$

$$+ \left| \sum_{k=1}^{n} f(\xi_k)[g'(\zeta_k) - g'(\xi_k)](x_k - x_{k-1}) \right|$$

$$< \lambda + \sum_{k=1}^{n} |f(\xi_k)||g'(\zeta) - g'(\xi_k)|(x_k - x_{k-1})$$

$$< \lambda + M(2\lambda) = \lambda(1 + 2M) = \varepsilon.$$

Therefore,

$$|R(f, g, \mathfrak{n}, \xi) - \int_a^b f(x)g'(x)dx| < \varepsilon$$

whenever $\|\mathfrak{n}\| < \delta$, the function f is Riemann–Stieltjes integrable with respect to g over $[a, b]$, and

$$RS \int_a^b f(x)dg(x) = \int_a^b f(x)g'(x)dx.$$

We now give an example to demonstrate the use of Theorem 11.3.7.

Example 11.3.4. Evaluate the Riemann–Stieltjes integral

$$RS \int_0^{\pi/2} \cos x \, d(\sin x).$$

Since the cosine function is integrable on $[0, \pi/2]$ and the sine function is differentiable on $[0, \pi/2]$, we may apply Theorem 11.3.7 and obtain

$$RS \int_0^{\pi/2} \cos x \, d(\sin x) = \int_0^{\pi/2} \cos^2 x \, dx$$

$$= \int_0^{\pi/2} \frac{(\cos 2x) + 1}{2} \, dx$$

$$= \frac{\sin 2x}{4} + \frac{x}{2} \Big]_0^{\pi/2}$$

$$= \pi/4.$$

Problem Set 11.3

In Exercises 1–10, evaluate the given Riemann–Stieltjes integral.

1. $RS \displaystyle\int_0^{\pi/2} x \, d(\sin x).$

2. $RS \displaystyle\int_\pi^{2\pi} \sin x \, d(\cos x).$

3. $RS \displaystyle\int_{-\pi/2}^{0} e^{|x|} d(\cos x).$

4. $RS \displaystyle\int_0^1 x \, d(e^x).$

5. $RS \displaystyle\int_0^4 (x^2 + 1) d([x]),$ where $[x]$ denotes the bracket function.

6. $RS \displaystyle\int_0^4 e^x d(x + [x]).$

7. $RS \displaystyle\int_0^3 [x] \, d(e^x).$

8. $RS \displaystyle\int_{-10}^{10} (x^2 + e^x) d(\text{sgn } x)$.

9. $RS \displaystyle\int_{0}^{10} (1/\sqrt{x + x^2}) d(\sqrt{1 + x^2})$.

10. $RS \displaystyle\int_{0}^{1} x \, dg(x)$, where $g(x) = \begin{cases} -1 & \text{if } x = 0, \\ x & \text{if } 0 < x < 1, \\ 2 & \text{if } x = 1. \end{cases}$

11. Let f be continuous on the closed interval $[0, n]$. Prove that

$$RS \int_0^n f(x) d([x]) = \sum_{k=1}^n f(n).$$

12. Prove Theorem 11.3.1.

13. Show that if the limit in Definition 11.3.1 exists, then it is unique.

14. Prove Theorem 11.3.5.

15. Prove the following **Cauchy Criterion for Riemann–Stieltjes Integrability**: The bounded function f is Riemann–Stieltjes integrable with respect to the increasing function g over $[a, b]$ if and only if to each $\varepsilon > 0$ there is $\delta = \delta(\varepsilon) > 0$ such that

$$|R(f, g, \mathfrak{n}, \xi) - R(f, g, \mathfrak{n}', \xi')| < \varepsilon$$

for any nets $\mathfrak{n}$ and $\mathfrak{n}'$ with $||\mathfrak{n}|| < \delta$ and $||\mathfrak{n}'|| < \delta$.

16. Prove Theorem 11.3.6.

17. Prove Theorem 11.3.7.

For Exercises 18–26, assume that f_1 and f_2 are Riemann–Stieltjes integrable with respect to the increasing functions g_1 and g_2 over $[a, b]$.

18. If c is a constant, then prove that the function cf_1 is Riemann–Stieltjes integrable with respect to the g_1 over $[a, b]$ and that

$$RS \int_a^b cf_1(x) dg_1(x) = c\left(RS \int_a^b f_1(x) dg_1(x)\right).$$

19. Prove that the function $f_1 + f_2$ is Riemann–Stieltjes integrable with respect to g_1 over $[a, b]$ and that

$$RS \int_a^b [f_1(x) + f_2(x)] dg_1(x) = RS \int_a^b f_1(x) dg_1(x) + RS \int_a^b f_2(x) dg_1(x).$$

20. If c_1 and c_2 are constants, then prove that the function $c_1 f_1 + c_2 f_2$ is Riemann–Stieltjes integrable with respect to g_1 over $[a, b]$, and that

$$RS \int_a^b [c_1 f_1(x) + c_2 f_2(x)] dg_1(x)$$
$$= c_1\left(RS \int_a^b f_1(x) dg_1(x)\right) + c_2\left(RS \int_a^b f_2(x) dg_1(x)\right).$$

21. If $m \leqslant f_1(x) \leqslant M$ for $x \in [a, b]$, then prove that

$$m[g_1(b) - g_1(a)] \leqslant RS \int_a^b f_1(x) dg_1(x) \leqslant M[g_1(b) - g_1(a)].$$

22. If $c \geqslant 0$ is constant, then prove that the function f_1 is Riemann–Stieltjes integrable with respect to cg_1 over $[a, b]$ and that

$$RS \int_a^b f_1(x) d[cg_1(x)] = c\left(RS \int_a^b f_1(x) dg_1(x)\right).$$

23. Prove that the function f_1 is Riemann–Stieltjes integrable with respect to $g_1 + g_2$ over $[a, b]$ and that

$$RS \int_a^b f_1(x) d[g_1(x) + g_2(x)] = RS \int_a^b f_1(x) dg_1(x) + RS \int_a^b f_2(x) dg_2(x).$$

24. If $c_1 \geqslant 0$ and $c_2 \geqslant 0$ are constants, then prove that the function f_1 is Riemann–Stieltjes integrable with respect to $c_1 g_1 + c_2 g_2$ over $[a, b]$ and that

$$RS \int_a^b f_1(x) d[c_1 g_1(x) + c_2 g_2(x)] = c_1\left(RS \int_a^b f_1(x) dg_1(x)\right)$$

$$+ c_2\left(RS \int_a^b f_1(x) dg_2(x)\right).$$

25. Prove that the function $f_1 f_2$ is Riemann–Stieltjes integrable with respect to g_1 over $[a, b]$.

26. Prove that the function $|f_1|$ is Riemann–Stieltjes integrable with respect to g_1 over $[a, b]$.

27. Let f be Riemann–Stieltjes integrable with respect to the increasing function g over $[a, b]$, and let

$$F(x) = RS \int_a^x f(\zeta) dg(\zeta)$$

for each $x \in [a, b]$. If g is continuous at $x_0 \in [a, b]$, then prove that F is continuous at x_0.

28. Let f be Riemann–Stieltjes integrable with respect to the increasing function g over $[a, b]$. If $f(x) \geqslant 0$ for each $x \in [a, b]$, then prove that

$$RS \int_a^b f(x) dg(x) \geqslant 0.$$

29. Let f_1 and f_2 be Riemann–Stieltjes integrable with respect to the increasing function g over $[a, b]$. If $f_1(x) \leqslant f_2(x)$ for each $x \in [a, b]$, then prove that

$$RS \int_a^b f_1(x) dg(x) \leqslant RS \int_a^b f_2(x) dg(x).$$

30. Let f be continuous and nonnegative on $[a, b]$, and assume that there is an

$x_0 \in [a, b]$ such that $f(x_0) \neq 0$. If $g(x)$ is strictly increasing on $[a, b]$, then prove that

$$\text{RS} \int_a^b f(x)dg(x) > 0.$$

31. Let f_1 and f_2 be continuous on $[a, b]$, let $f_1(x) \leq f_2(x)$ for each $x \in [a, b]$, and assume that there is an $x_0 \in [a, b]$ such that $f_1(x_0) \neq f_2(x_0)$. If g is strictly increasing on $[a, b]$, then prove that

$$\text{RS} \int_a^b f_1(x)dg(x) < \text{RS} \int_a^b f_2(x)dg(x).$$

32. Let f be continuous and positive on $[a, b]$. If g is increasing and nonconstant on $[a, b]$, then prove that

$$\text{RS} \int_a^b f(x)dg(x) > 0.$$

33. Let f_1 and f_2 be continuous on $[a, b]$ and $f_1(x) < f_2(x)$ for each $x \in [a, b]$. If g is increasing and nonconstant on $[a, b]$, then prove that

$$\text{RS} \int_a^b f_1(x)dg(x) < \text{RS} \int_a^b f_2(x)dg(x).$$

34. Prove the following form of **First Mean Value Theorem for Riemann–Stieltjes Integrals**: If f is continuous and g is increasing on $[a, b]$, then there is a $c \in [a, b]$ such that

$$\text{RS} \int_a^b f(x)dg(x) = f(c)[g(b) - g(a)].$$

[*Hint*: Use the fact that
$$m(f, [a, b])[g(b) - g(a)] \leq \text{RS} \int_a^b f(x)dg(x) \leq M(f, [a, b])[g(b) - g(a)].]$$

35. Prove the following form of the **First Mean Value Theorem for Riemann–Stieltjes Integrals**: If f is continuous and g is strictly increasing on $[a, b]$, then there is $c \in (a, b)$ such that

$$\text{RS} \int_a^b f(x)dg(x) = f(c)[g(b) - g(a)].$$

36. Give an example to show that the equation in Exercise 35 need not hold for $c \in (a, b)$ if g is not strictly increasing. [*Hint*: Let $g(x) = [x]$ for each $x \in [0, 1]$.]

37. Prove the following form of the **Second Mean Value Theorem for Riemann–Stieltjes Integrals**. If f is increasing and g is continuous on $[a, b]$, then there exists a point $c \in [a, b]$ such that

$$\text{RS} \int_a^b f(x)dg(x) = f(a)[g(c) - g(a)] + f(b)[g(b) - g(c)].$$

[*Hint*: Use Exercise 34 and integration by parts.]

38. Prove the following form of the **Second Mean Value Theorem for Riemann–Stieltjes Integrals**: If f is strictly increasing and g is continuous on $[a, b]$, then there exists a point $c \in (a, b)$ such that

$$\text{RS} \int_a^b f(x)dg(x) = f(a)[g(c) - g(a)] + f(b)[g(b) - g(c)].$$

39. Give an example to show that the equation in Exercise 38 need not hold for $c \in (a, b)$ if f is not strictly increasing.

40. Prove the following form of the **Second Mean Value Theorem for Riemann Integrals**: If f is increasing and h is continuous on $[a, b]$, then there exists $c \in [a, b]$ such that

$$\int_a^b f(x)h(x)dx = f(a)\int_a^c h(x)dx + f(b)\int_c^b h(x)dx.$$

[*Hint*: Let
$$g(x) = \int_a^x h(\zeta)d\zeta$$

for each $x \in [a, b]$ and use Exercise 37.]

41. Prove the following form of the **Second Mean Value Theorem of Riemann Integrals**: If f is strictly increasing and h is continuous on $[a, b]$, then there exists $c \in (a, b)$ such that

$$\int_a^b f(x)h(x)dx = f(a)\int_a^c h(x)dx + f(b)\int_c^b h(x)dx.$$

42. Give an example to show that the equation in Exercise 41 need not hold for $c \in (a, b)$ if f is not strictly increasing.

43. Prove the following form of the **Second Mean Value Theorem for Riemann Integrals** due to Bonnet: If f is nonnegative and decreasing (or f is nonpositive and increasing) and h is continuous on $[a, b]$, then there exists $c \in [a, b]$ such that

$$\int_a^b f(x)h(x)dx = f(a)\int_a^c h(x)dx.$$

[*Hint*: Use Exercise 41 and redefine f to be 0 at either a or b.]

44. Prove the following form of the **Second Mean Value Theorem for Riemann Integrals** due to Bonnet: If f is nonnegative and increasing (or f is nonpositive and decreasing) and h is continuous on $[a, b]$, then there is a $c \in [a, b]$ such that

$$\int_a^b f(x)h(x)dx = f(b)\int_c^b h(x)dx.$$

45. Let ϕ be continuous and strictly increasing on $[a, b]$, let $\phi(a) = c$, and let $\phi(b) = d$. If either of the integrals

$$\int_c^d f(x)dg(x) \quad \text{or} \quad \int_a^b f(\phi(t))dg(\phi(t))$$

exists, prove that the other integral exists and that the two integrals are equal.

46. Let ϕ be continuous and increasing on $[a, b]$, let $\phi(a) = c$, and let $\phi(b) = d$. If the integral

$$\int_c^d f(x)dg(x)$$

exists, prove that the integral

$$\int_a^b f(\phi(t))dg(\phi(t))$$

exists and that the two integrals are equal.

47. Let

$$f(x) = \begin{cases} 0 & \text{if} \quad 0 \leqslant x < 1/2, \\ 1 & \text{if} \quad 1/2 \leqslant x \leqslant 1, \end{cases}$$

$$g(x) = \begin{cases} 0 & \text{if} \quad 0 \leqslant x \leqslant 1/2, \\ 1 & \text{if} \quad 1/2 < x \leqslant 1, \end{cases}$$

$$\phi(t) = \begin{cases} 3/2t & \text{if} \quad 0 \leqslant t \leqslant 1/3, \\ 1/2 & \text{if} \quad 1/3 < t < 2/3, \\ (3t - 1)/2 & \text{if} \quad 2/3 \leqslant t \leqslant 1. \end{cases}$$

Using the functions f, g, and ϕ, show that, under the hypotheses of Exercise 46, the existence of the second integral does not imply the existence of the first integral.

11.4. Riemann-Stieltjes Integrals of Functions of Bounded Variation

In this section, we extend the definition of Riemann–Stieltjes integration to include functions of bounded variation. Let g be a function of bounded variation on the compact interval $[a, b]$, and let f be bounded on $[a, b]$. Theorem 11.1.9 implies that g may be written as the difference of two increasing functions, say,

$$g = h - k.$$

Then it seems reasonable to say that the bounded function f is Riemann–Stieltjes integrable with respect to the function of bounded variation g over $[a, b]$ if and only if f is Riemann–Stieltjes integrable with respect to each of the increasing functions h and k over $[a, b]$ and to define

$$\int_a^b f(x)dg(x) = \int_a^b f(x)dh(x) - \int_a^b f(x)dk(x).$$

The problem that arises with this definition is that there are many ways of writing the function g as the difference of two increasing functions (cf. the discussion following Theorem 11.1.9). Therefore, to use this definition we must show that it is independent of the decomposition of the function g into the difference of two increasing functions.

Theorem 11.4.1. *Let* $g_1, g_2, h_1,$ *and* h_2 *be increasing functions on* $[a, b]$ *such that* $g_1 - h_1 = g_2 - h_2$. *If the function* f *is Riemann–Stieltjes integrable with respect to each of the functions* $g_1, g_2, h_1,$ *and* h_2 *over* $[a, b]$, *then*

$$\text{RS} \int_a^b f(x)dg_1(x) - \text{RS} \int_a^b f(x)dh_1(x)$$

$$= \text{RS} \int_a^b f(x)dg_2(x) - \text{RS} \int_a^b f(x)dh_2(x).$$

The proof of this theorem depends on the results of Exercise 23 of Problem Set 11.3 and is left as Exercise 11 of Problem Set 11.4.

Theorem 11.4 1 indicates that the value of the integral is independent of the representation of function of bounded variation, assuming that the integral exists. The following example shows that the existence of the integral is not independent of the representation of the function of bounded variation.

Example 11.4.1. Let $f = g_1 = h_1$, where f is defined as in Example 11.2.4. Let $g_2 = h_2 = 0$. Then the integrals

$$\text{RS} \int_0^2 f(x)dg_1(x) \quad \text{and} \quad \text{RS} \int_0^2 f(x)dh_1(x)$$

do not exist, while the integrals

$$\text{RS} \int_0^2 f(x)dg_2(x) = \text{RS} \int_0^2 f(x)dh_2(x) = 0.$$

We now define Riemann–Stieltjes integrability with respect to a function of bounded variation.

Definition 11.4.1. *Let* f *be bounded and* g *be of bounded variation on* $[a, b]$. *If there are increasing functions* h *and* k *on* $[a, b]$ *such that* $g = h - k$, *and if* f *is Riemann–Stieltjes integrable with respect to* h *and* k *over* $[a, b]$, *then* f *is said to be **Riemann–Stieltjes integrable** with respect to* g *over* $[a, b]$, *and we define*

$$\text{RS} \int_a^b f(x)dg(x) = \text{RS} \int_a^b f(x)dh(x) - \text{RS} \int_a^b f(x)dk(x).$$

If g is an increasing function on $[a, b]$, then we write $g = g - 0$, where 0 denotes the zero function.

Theorem 11.4.2. *Let f be bounded and g be a function of bounded variation on $[a, b]$. Then f is Riemann–Stieltjes integrable with respect to g over $[a, b]$ and has Riemann–Stieltjes integral $\mathfrak{s}$ if and only if*

$$\lim_{||\mathfrak{n}|| \to 0} R(f, g, \mathfrak{n}, \xi) = \mathfrak{s}.$$

The proof is left as Exercise 12 of Problem Set 11.4.

Here the definition of a Riemann–Stieltjes sum has been enlarged to include functions g of bounded variation. Essentially all the results that apply to Riemann–Stieltjes integrable functions with respect to increasing functions can be extended to Riemann–Stieltjes integrable functions with respect to functions of bounded variation. The reader should try making some of these extensions.

Thus far, no mention has been made of interchanging the roles of the functions f and g; this is discussed in the following result and is called *integration by parts*.

Theorem 11.4.3. *(Integration by Parts) Let f and g be of bounded variation over $[a, b]$. The function f is Riemann–Stieltjes integrable with respect to the function g over $[a, b]$ if and only if g is Riemann–Stieltjes integrable with respect to f over $[a, b]$. In either case*

$$\mathrm{RS} \int_a^b f(x) dg(x) + \mathrm{RS} \int_a^b g(x) dg(x) = f(b)g(b) - f(a)g(a).$$

Proof. Let f be Riemann–Stieltjes integrable with respect to g over $[a, b]$, and let $\varepsilon > 0$ be given. Then Theorem 11.4.2 implies that there exists a $\delta > 0$ such that

$$|R(f, g, \mathfrak{n}, \xi) - \int_a^b f(x) dg(x)| < \varepsilon$$

whenever $||\mathfrak{n}|| < \delta$.

Let $\mathfrak{n}' = \{a = x_0 < x_1 < \ldots < x_n = b\}$ be any net of $[a, b]$ with $||\mathfrak{n}|| < \delta/2$, and let $\xi' = \{\xi_1, \xi_2, \ldots, \xi_n\}$ be any set of points such that $\xi_k \in [x_{k-1}, x_k]$, $k = 1, 2, \ldots, n$. We consider the Riemann–Stieltjes sum

$$R(g, f, \mathfrak{n}', \xi') = \sum_{k=1}^n g(\xi_k)[f(x_k) - f(x_{k-1})].$$

$$= -g(\xi_1)f(x_0)$$

$$- \sum_{k=2}^n f(x_{k-1})[g(\xi_k) - g(\xi_{k-1})]$$

$$+ f(x_n)g(\xi_n).$$

Let $\xi_0 = x_0$ and $\xi_{n+1} = x_n$. Then adding and subtracting $g(x_0)f(x_0)$ and $g(x_n)f(x_n)$ gives

$$R(g, f, \mathfrak{n}, ' \xi') = f(x_n)\,g(x_n) - f(x_0)\,g(x_0)$$

$$- \sum_{k=1}^{n+1} f(x_{k-1})[g(\xi_k) - g(\xi_k)].$$

Now,

$$\sum_{k=1}^{n+1} f(x_{k-1})[g(\xi_k) - g(\xi_{k-1})]$$

is a Riemann–Stieltjes sum of the form $R(f, g, \mathfrak{n}, \xi)$, where $\mathfrak{n} = \{a = \xi_0 < \xi_1 < \ldots < \xi_{n+1} = b\}$ is a net of $[a, b]$ and

$$\xi = \{x_0, x_1, \ldots, x_n\}.$$

Since $\|\mathfrak{n}'\| < \delta/2$, we have $\|\mathfrak{n}\| < \delta$ and

$$|R(f, g, \mathfrak{n}, \xi) - \int_a^b f(x)dg(x)| < \varepsilon.$$

Thus

$$|R(g, f, \mathfrak{n}', \xi') - \{f(b)g(b) - f(a)g(a) - \int_a^b f(x)dg(x)\}|$$

$$= \left| \int_a^b f(x)dg(x) - R(f, g, \mathfrak{n}, \xi) \right|$$

$$< \varepsilon$$

whenever $\|\mathfrak{n}'\| < \delta/2$. Therefore, Theorem 11.4.2 implies that g is Riemann–Stieltjes integrable with respect to f and

$$\text{RS} \int_a^b g(x)df(x) = f(b)g(b) - f(a)g(a) - \text{RS} \int_a^b f(x)dg(x),$$

as desired.

We demonstrate this theorem with the following example.

Example 11.4.2. Let $f(x) = x^2$ for $0 \leqslant x \leqslant 2$, and let

$$g(x) = \begin{cases} x & \text{if } 0 \leqslant x \leqslant 1, \\ x + 2 & \text{if } 1 < x \leqslant 2. \end{cases}$$

Evaluate the Riemann–Stieltjes integral

$$\int_0^2 f(x)dg(x).$$

Since f is continuous on $[0, 2]$ and g is increasing on $[0, 2]$, Theorem 11.3.4 implies that f is Riemann–Stieltjes integrable with respect to g over $[0, 2]$. Since f is of bounded variation, it follows from Theorem 11.4.3 that g is Riemann–Stieltjes integrable with respect to f over $[0, 2]$ and that

$$\int_0^2 f(x)dg(x) = f(2)g(2) - f(0)g(0) - \int_0^2 g(x)df(x)$$

$$= (4)(4) - (0)(0) - \int_0^2 g(x)df(x).$$

Since f is differentiable on $[0, 2]$, we may write $df(x) = f'(x)dx$, and so

$$\int_0^2 g(x)df(x) = \int_0^2 g(x)(2x)dx$$

$$= \int_0^1 x(2x)dx + \int_1^2 (x + 2)(2x)dx$$

$$= \frac{2x^3}{x}\bigg|_0^1 + \frac{2x^3}{3} + 2x^2\bigg|_1^2$$

$$= \frac{2}{3} + \frac{16}{3} + 8 - \frac{2}{3} - 2$$

$$= 34/3.$$

Therefore,

$$\int_0^2 f(x)dg(x) = 16 - 34/3 = 14/3.$$

Problem Set 11.4

In Exercises 1–10, evaluate the given Riemann–Stieltjes integrals.

1. RS $\int_0^\pi x\, d(\sin x)$.

2. RS $\int_0^{2\pi} \sin x\, d(\cos x)$.

3. RS $\displaystyle\int_{-\pi}^{\pi} e^{|x|} d(\cos x)$.

4. RS $\displaystyle\int_{-1}^{1} x\, d(e^x - e^{-x})$.

5. RS $\displaystyle\int_{-4}^{0} (x^2 + 1)\, d([x])$,   where $[x]$ denotes the bracket function.

6. RS $\displaystyle\int_{0}^{4} e^x\, d([x] - x)$.

7. RS $\displaystyle\int_{-3}^{3} [x]\, d(|x|)$,   where $|x|$ denotes the absolute value function.

8. RS $\displaystyle\int_{0}^{5} (x^2 + e^x)\, d(|2 - x|)$.

9. RS $\displaystyle\int_{-1}^{1} x\, dg(x)$,   where $g(x) = \begin{cases} -1 & \text{if } x = 0, \\ |x| & \text{if } -1 < x < 0 \quad \text{or} \quad 0 < x < 1, \\ 2 & \text{if } x = -1 \quad \text{or} \quad 1. \end{cases}$

10. RS $\displaystyle\int_{-1}^{1} e^x\, d(|x|)$.

11. Prove Theorem 11.4.1.

12. Prove Theorem 11.4.2.

     For Exercises 13–21, assume that f_1 and f_2 are Riemann–Stieltjes integrable with respect to the function of bounded variation g_1 and g_2 over $[a, b]$.

13. If c is a constant, then prove that the function cf_1 is Riemann–Stieltjes integrable with respect to the g_1 over $[a, b]$ and that

$$\text{RS} \int_a^b cf_1(x)dg_1(x) = c\left(\text{RS} \int_a^b f_1(x)dg_1(x) \right).$$

14. Prove that the function $f_1 + f_2$ is Riemann–Stieltjes integrable with respect to g_1 over $[a, b]$ and that

$$\text{RS} \int_a^b [f_1(x) + f_2(x)]dg_1(x) = \text{RS} \int_a^b f_1(x)dg_1(x) + \text{RS} \int_a^b f_2(x)dg_1(x).$$

15. If c_1 and c_2 are constants, then prove that the function $c_1 f_1 + c_2 f_2$ is Riemann–Stieltjes integrable with respect to g_1 over $[a, b]$ and that

$$\text{RS} \int_a^b [c_1 f_1(x) + c_2 f_2(x)]dg_1(x) = c_1\left(\text{RS} \int_a^b f_1(x)dg_1(x) \right) + c_2\left(\text{RS} \int_a^b f_2(x)dg_1(x) \right).$$

16. If $m \leqslant f_1(x) \leqslant M$ for $x \in [a, b]$, then prove that

$$m[V(g_1, [a, b])] \leqslant \left| \text{RS} \int_a^b f_1(x) dg_1(x) \right| \leqslant M[V(g_1, [a, b])].$$

17. If c is a constant, then prove that the function f_1 is Riemann–Stieltjes integrable with respect to cg_1 over $[a, b]$ and that

$$\text{RS} \int_a^b f_1(x) d[cg_1(x)] = c \left(\text{RS} \int_a^b f_1(x) dg_1(x) \right).$$

18. Prove that the function f_1 is Riemann–Stieltjes integrable with respect to $g_1 + g_2$ over $[a, b]$ and that

$$\text{RS} \int_a^b f_1(x) d[g_1(x) + g_2(x)] = \text{RS} \int_a^b f_1(x) dg_1(x) + \text{RS} \int_a^b f_2(x) dg_2(x).$$

19. If c_1 and c_2 are constants, then prove that the function f_1 is Riemann–Stieltjes integrable with respect to $c_1 g_1 + c_2 g_2$ over $[a, b]$ and that

$$\text{RS} \int_a^b f_1(x) d[c_1 g_1(x) + c_2 g_2(x)]$$

$$= c_1 \left(\text{RS} \int_a^b f_1(x) dg_1(x) \right) + c_2 \left(\text{RS} \int_a^b f_1(x) dg_2(x) \right).$$

20. Prove that the function $f_1 f_2$ is Riemann–Stieltjes integrable with respect to g_1 over $[a, b]$.

21. Prove that the function $|f_1|$ is Riemann–Stieltjes integrable with respect to g_1 over $[a, b]$ and that

$$\left| \text{RS} \int_a^b f_1(x) dg_1(x) \right| \leqslant \text{RS} \int_a^b |f_1(x)| dv_{g_1}(x).$$

(Cf. Definition 11.1.2.)

22. Let g be of bounded variation on $[a, b]$ and f be Riemann–Stieltjes integrable with respect to g over $[a, b]$. Prove that

$$\left| \text{RS} \int_a^b f(x) dg(x) \right| \leqslant M(f, [a, b]) \cdot V(g, [a, b]).$$

23. Prove a **Cauchy Criterion for Riemann–Stieltjes integrability** for functions of bounded variation. (Cf. Exercise 15 of Problem set 11.3.)

24. Let f be continuous and g be of bounded variation on $[a, b]$, and let

$$F(x) = \int_a^x f(\zeta) dg(\zeta)$$

for each $x \in [a, b]$. Prove that F is of bounded variation on $[a, b]$.

Selected Answers

1. $\varnothing$, $\{a\}$, $\{b\}$, $\{c\}$, $\{d\}$, $\{a, b\}$, $\{a, c\}$, $\{a, d\}$, $\{b, c\}$, $\{b, d\}$, $\{c, d\}$, $\{a, b, d\}$, $\{a, b, c\}$, $\{a, c, d\}$, $\{b, c, d\}$, $A = \{a, b, c, d\}$.

2. $A = \{a, b\}$, $B = \{b, c\}$; $A \nsubseteq B$ and $B \nsubseteq A$.

3. No. Since $\varnothing$ contains no elements and $\{\varnothing\}$ contains exactly one element, namely $\varnothing$.

5. (a) $A \cup B = B \cup A = \{a, b, d, 1, 2, 3, 5\}$.
   (b) $A \cap B = B \cap A = \{b, 1\} = C$.
   (c) $A \cup (B \cap C) = (A \cup B) \cap (A \cup C) = A$.
   (d) $A \cap (B \cup C) = (A \cap B) \cup (A \cap C) = C$.

14. No. Let A be a proper subset of B, and let $B = C$. Then $C \cup A = B \cup A = C \cup B = B = C$ and $C \cup A$ is *not* a proper subset of $C \cup B$.

18. $A \setminus B = A \setminus C = \{a, 2\}$, $B \setminus A = B \setminus C = \{d, 3, 5\}$, and $C \setminus A = C \setminus B = \varnothing$.

Problem Set 1.2.

1. (a) $A \times B = \{(1, a),\ (1, b),\ (1, c),\ (1, d),\ (2, a),\ (2, b),\ (2, c),\ (2, d),\ (3, a),$ $(3, b),\ (3, c),\ (3, d),\ (4, a),\ (4, b),\ (4, c),\ (4, d)\}$.
   (b) $B \times A = \{(a, 1),\ (a, 2),\ (a, 3),\ (a, 4),\ (b, 1),\ (b, 2),\ (b, 3),\ (b, 4),\ (c, 1),\ (c, 2),$ $(c, 3),\ (c, 4),\ (d, 1),\ (d, 2),\ (d, 3),\ (d, 4)\}$.
   (c) No. Since $A \times B \neq B \times A$, unless $A = B$.
   (d) $(A \times B) \cup C = \{(1, a), (1, b), (1, c), (1, d), (2, a), (2, b), (2, c), (2, d), (3, a),$ $(3, b), (3, c), (3, d), (4, a), (4, b), (4, c), (4, d), -3, -2, -1\}$.

$(A \cup C) \times (B \cup C) = \{1, 2, 3, 4, -3, -2, -1\} \times \{a, b, c, d, -3, -2, -1\}$
$= \{(1, a), (1, b), (1, c), (1, d), (1, -3), (1, -2), (1, -1), (2, a), (2, b), (2, c),$
$(2, d), (2, -3), (2, -2), (2, -1), (3, a), (3, b), (3, c), (3, d), (3, -3), (3, -2),$
$(3, -1), (4, a), (4, b), (4, c), (4, d), (4, -3), (4, -2), (4, -1), (-3, a),$
$(-3, b), (-3, c), (-3, d), (-3, -3), (-3, -2), (-3, -1), (-2, a),$
$(-2, b), (-2, c), (-2, d), (-2, -3), (-2, -2), (-2, -1), (-1, a),$
$(-1, b), (-1, c), (-1, d), (-1, -3), (-1, -2), (-1, -1)\}.$

(e) $(A \times B) \cap C = \emptyset$, $(A \cap C) \times (B \cap C)$ is not defined, since $A \cap C = B \cap C = \emptyset$.

2. No. See Exercise 1(e) for a counterexample.

5. $A \times A = \{(1, 1), (1, 2), (1, 3), (1, 4), (2, 1), (2, 2), (2, 3), (2, 4), (3, 1), (3, 2), (3, 3), (3, 4), (4, 1), (4, 2), (4, 3), (4, 4)\}$.
   (a) $R = \{(2, 1), (3, 1), (4, 1), (3, 2), (4, 2), (4, 3)\}$.
   (b) $R = \{(1, 1), (2, 4)\}$.

6. $A \times B = \{(a, 1), (a, 2), (a, 3), (b, 1), (b, 2), (b, 3)\}$.
   $F_1 = \{(a, 1), (b, 1)\}, F_2 = \{(a, 2), (b, 2)\}, F_3 = \{(a, 3), (b, 3)\},$
   $F_4 = \{(a, 1), (b, 2)\}, F_5 = \{(a, 1), (b, 3)\}, F_6 = \{(a, 2), (b, 1)\},$
   $F_7 = \{(a, 2), (b, 3)\}, F_8 = \{(a, 3), (b, 1)\}, F_9 = \{(a, 3), (b, 2)\}.$
   The functions $F_4, F_5, F_6, F_7, F_8,$ and F_9 are 1-1. None of these functions are onto.

7. $A \times B = \{(a, 1), (a, 2), (b, 1), (b, 2), (c, 1), (c, 2)\}$.
   $F_1 = \{(a, 1), (b, 1), (c, 1)\}, F_2 = \{(a, 1), (b, 1), (c, 2)\}, F_3 = \{(a, 1), (b, 2), (c, 1)\},$
   $F_4 = \{(a, 1), (b, 2), (c, 2)\}, F_5 = \{(a, 2), (b, 1), (c, 1)\}, F_6 = \{(a, 2), (b, 1), (c, 2)\},$
   $F_7 = \{(a, 2), (b, 2), (c, 1)\}, F_8 = \{(a, 2), (b, 2), (c, 2)\}.$ None of the functions are 1-1. The functions $F_2, F_3, F_4, F_5, F_6,$ and F_7 are onto.

8. $A \times B = \{(a, 1), (a, 2), (a, 3), (b, 1), (b, 2), (b, 3), (c, 1), (c, 2), (c, 3)\}$.
   $F_1 = \{(a, 1), (b, 1), (c, 1)\}, F_2 = \{(a, 1), (b, 1), (c, 2)\}, F_3 = \{(a, 1), (b, 1), (c, 3)\},$
   $F_4 = \{(a, 1), (b, 2), (c, 1)\}, F_5 = \{(a, 1), (b, 2), (c, 2)\}, F_6 = \{(a, 1), (b, 2), (c, 3)\},$
   $F_7 = \{(a, 1), (b, 3), (c, 1)\}, F_8 = \{(a, 1), (b, 3), (c, 2)\}, F_9 = \{(a, 1), (b, 3), (c, 3)\},$
   $F_{10} = \{(a, 2), (b, 1), (c, 1)\}, F_{11} = \{(a, 2), (b, 1), (c, 2)\}, F_{12} = \{(a, 2), (b, 1), (c, 3)\},$
   $F_{13} = \{(a, 2), (b, 2), (c, 1)\}, F_{14} = \{(a, 2), (b, 2), (c, 2)\}, F_{15} = \{(a, 2), (b, 2), (c, 3)\},$
   $F_{16} = \{(a, 2), (b, 3), (c, 1)\}, F_{17} = \{(a, 2), (b, 3), (c, 2)\}, F_{18} = \{(a, 2), (b, 3), (c, 3)\},$
   $F_{19} = \{(a, 3), (b, 1), (c, 1)\}, F_{20} = \{(a, 3), (b, 1), (c, 2)\}, F_{21} = \{(a, 3), (b, 1), (c, 3)\},$
   $F_{22} = \{(a, 3), (b, 2), (c, 1)\}, F_{23} = \{(a, 3), (b, 2), (c, 2)\}, F_{24} = \{(a, 3), (b, 2), (c, 3)\},$
   $F_{25} = \{(a, 3), (b, 3), (c\ 1)\}, F_{26} = \{(a, 3), (b, 3), (c, 2)\}, F_{27} = \{(a, 3), (b, 3), (c, 3)\}.$
   The functions $F_6, F_8, F_{12}, F_{16}, F_{20}, F_{22}$ are both 1-1 and onto. These are the same since A and B have the same number of elements.

9. (a), (b), (c) are functions; (a) is 1-1; and (a) is onto.

10. (a) $F_1 = \{(x, x): x \in Z\}$.
    (b) $F_2 = \{(x, x^2): x \in Z\}$.
    (c) $F_3 = \{(x, 2x): x \in Z\}$.
    (d) $F_4 = \{(x, x): x \in Z^+\} \cup \{(x, x + 1): x \in \mathbf{C}(Z^+)\}$, where Z^+ denotes the set of positive integers and $\mathbf{C}(Z^+)$ denotes the set of nonpositive integers.

11. (b) Let $A = \{a, b\}$, $C = \{a, b, c\}$, $D = \{1, 2, 3\}$, $g = \{(a, 1), (b, 2), (c, 3)\}$, and $f = g/A = \{(a, 1), (b, 2)\}$.

12. (b) Let $A = \{a, b\}$, $B = \{1, 2\}$, $C = \{a, b, c\}$, $f = \{(a, 1), (b, 2)\}$, and $g = \{(a, 1), (b, 2), (c, 2)\}$.

13. (a) Let $A = \{-1, 1\}$ and $F: A \to A$ such that $F = \{(-1, 1), (1, 1)\}$. Let $D = \{1\}$ and $E = \{-1\}$. Then $D \cap E = \emptyset$, $F[D \cap E] = \emptyset$, $F[D] = F[E] = F[D] \cap F[E] = \{1\}$.

Problem Set 1.3.

1. $f \circ g: R \to R$ such that $(f \circ g)(x) = \sin(2x + \pi)$. $\operatorname{Dom} f \circ g = R$ and $\operatorname{Ran} f \circ g = \{y : y \in R \text{ and } -1 \leqslant y \leqslant 1\}$.
$g \circ f: R \to R$ such that $(g \circ f)(x) = 2(\sin x) + \pi$. $\operatorname{Dom} g \circ f = R$ and $\operatorname{Ran} g \circ f = \{y : y \in R \text{ and } -2 + \pi \leqslant y \leqslant 2 + \pi\}$.

2. $f \circ g: R \to R$ such that $(f \circ g)(x) = 3x^2 - 4x + 1$. $\operatorname{Dom} f \circ g = R$ and $\operatorname{Ran} f \circ g = \{y : y \in R \text{ and } y \geqslant -1/3\}$.
$g \circ f: R \to R$ such that $(g \circ f)(x) = 3x^2 + 2x - 1$. $\operatorname{Dom} g \circ f = R$ and $\operatorname{Ran}(g \circ f) = \{y : y \in R \text{ and } y \geqslant -4/3\}$.

3. $g \circ f: A \to C$ such that $g \circ f = \{(a, w), (b, w), (c, w), (d, u)\}$. $\operatorname{Ran} g \circ f = \{w, u\}$.

4. $g \circ f: A \to B$ such that $(g \circ f)(x) = (2 - x)/(3 - x)$. $\operatorname{Ran} g \circ f = \{y : y \in R \text{ and } 1 > y > 2/3\}$.

5. $f \circ g: R \to R$ such that $(f \circ g)(x) = x^{46} + 1$.
$\operatorname{Ran} f \circ g = \{y : y \in R \text{ and } y \geqslant 1\}$.
$g \circ f: R \to R$ such that $(g \circ f)(x) = (x^2 + 1)^{23}$.
$\operatorname{Ran} g \circ f = \{y : y \in R \text{ and } y \geqslant 1\}$.

6. $f^{-1}: R \to R$ such that $f^{-1}(x) = (x + 5)/3$.

7. Assume that $f(x) = f(y)$. Then $x^2 - 4x = y^2 - 4y$ or $(x^2 - y^2) - 4(x - y) = (x - y)(x + y - 4) = 0$. Thus $x = y$ or $x + y = 4$. Since $x, y \in A$, this implies that $x = y$.
Let $y \in B$. Then $y \geqslant -4$. From the equation $y = f(x) = x^2 - 4x$, we can solve for x to get $x = 2 + \sqrt{4 + y} \in A$. Thus $f(2 + \sqrt{4 + y}) = y$ and f is onto B.
$f^{-1}: B \to A$ such that $f^{-1}(x) = 2 + \sqrt{4 + x}$.

8. $f^{-1}: B \to A$ such that $f^{-1}(x) = x/(1 - x)$.

13. Let $A = B = C = Z$, the set of integers.
   (a) Let $f: Z \to Z$ and $g: Z \to Z$ such that $f(x) = x^2$ and $g(x) = x$. Then $(g \circ f): Z \to Z$ such that $(g \circ f)(x) = x^2$. The function g is onto Z, but $g \circ f$ is not onto Z.
   (b) Let $f: Z \to Z$ and $g: Z \to Z$ such that $f(x) = x$ and $g(x) = x^2$. Then $(g \circ f): Z \to Z$ such that $(g \circ f)(x) = x^2$. The function f is 1-1, but $g \circ f$ is not 1-1.

(c) Let $f: Z \to Z$ and $g: Z \to Z$ such that $f(x) = 2x$ and
$$g(x) = \begin{cases} x/2 & \text{if } x \text{ is even,} \\ (x-1)/2 & \text{if } x \text{ is odd.} \end{cases}$$
Then $(g \circ f): Z \to Z$ such that $(g \circ f)(x) = x$. The function f is not onto Z, but $g \circ f$ is onto Z.

(d) See part (c).

Problem Set 1.4.

1. Let $f: \{\ldots, -3, -2, -1\} \to N$ such that $f(x) = -x$.

2. Let $f: Z \to N$ such that $f(x) = 2x$ if $x \geqslant 1$ and $f(x) = -2x + 1$ if $x \leqslant 0$.

3. Let $f: \{3n: n \text{ is a natural number}\} \to N$ such that $f(x) = x/3$.

6. Let $A_n = \{1\}$ and $B_n = \{n\}$ for each natural number n. Then
$$\bigcup_{k=1}^{\infty} A_k = \{1\} \quad \text{and} \quad \bigcup_{k=1}^{\infty} B_k = N.$$

Problem Set 2.2.

1. Let Z denote the set of integers and define the binary operation $\circ: Z \times Z \to Z$ by the equation $x \circ y = x - x^2 y^2 + y$. Then $\circ$ is commutative but *not* associative.

2. Let Z denote the set of integers and define the binary operation $\circ: Z \times Z \to Z$ by the equation $x \circ y = x$. Then $\circ$ is associative but *not* commutative.

3. (a) Yes
(b) No. $(2^2)^3 = 64 \neq 256 = 2^{(2^3)}$.
(c) No. $2^3 = 8 \neq 9 = 3^2$.

8. (a) The set Z^+ is *not* an additive group, since it has no identity.
(b) The set E is an additive group.
(c) The set of multiples of 5 is an additive group.
(d) The set $\{0\}$ is an additive group.
(e) The set $\{1\}$ is *not* an additive group, since it has no identity.
(f) The set R is an additive group.

9. (e) The set $\{1\}$ is a multiplicative group.

17.

$\circ$	a	b	c
a	a	b	c
b	b	c	a
c	c	a	b

18. The group given in Example 2.2.6(c).

Problem Set 2.3.

9.

+	a	b	c	d	e
a	a	b	c	d	e
b	b	c	d	e	a
c	c	d	e	a	b
d	d	e	a	b	c
e	e	a	b	c	d

·	a	b	c	d	e
a	a	a	a	a	a
b	a	b	c	d	e
c	a	c	e	b	d
d	a	d	b	e	c
e	a	e	d	c	b

Problem Set 2.4.

15.
$$x < y$$
$$x + x < x + y$$
$$2x < x + y$$
$$x < (x + y)2^{-1} = (x + y)/2$$

$$x < y$$
$$x + y < y + y$$
$$x + y < 2y$$
$$(x + y)/2 = (x + y)2^{-1} < y$$

17. No. See Exercise 16. **18.** No. See Exercise 16.

33. Assume $x \neq 0$. If $x > 0$, then $x + x > 0$ and if $x < 0$, then $x + x < 0$. In either case, we have a contradiction. Thus $x = 0$.

38. (a), (f), (g), (h), (i), (j), (k).

Problem Set 2.5.

4. Let $S = \{n: n \in N$ and $s_n = n(n + 1)(2n + 1)/6\}$. We need to show that $S = N$.
(I). Since $s_1 = 1 = 1(1 + 1)(2 \cdot 1 + 1)/6$, we have that $1 \in S$.
(II). Assume that $k \in S$, that is, $s_k = k(k + 1)(2k + 1)/6$. Then
$$s_{k+1} = s_k + (k + 1)^2 = k(k + 1)(2k + 1)/6 + (k + 1)^2$$
$$= (k + 1)[(k + 1) + 1][2(k + 1) + 1]/6,$$
and so $(k + 1) \in S$. The Principle of Mathematical Induction implies that $S = N$.

22. Let $S = \{n: n \in N$ and $s_n = 2^n - 1\}$. We need to show that $S = N$. Assume that, for each $m < k$, $m \in S$; that is, $s_m = 2^m - 1$ for each $m < k$. Then
$$s_k = 3s_{k-2} - 2s_{k-1} = 3(2^{k-1} - 1) - 2(2^{k-2} - 1) = 2^k - 1.$$
Thus $k \in S$ and the Second Principle of Mathematical Induction implies that $S = N$.

30. Let x and y be natural numbers and let $S = \{nx: n \in N\} \subset N$. Assume that $nx \leqslant y$ for each $n \in N$. Then y is an upper bound for S (cf. Exercise 23) and so S has a greatest element, say y_0. Then $y_0 \in S$ and $nx \leqslant y_0$ for every $n \in N$. Since $y_0 \in S$, there is $n_0 \in S$ such that $y_0 = n_0 x$. But then $(n_0 + 1)x > y_0$ and y_0 is not an upper bound of S. This is a contradiction, and so there is a natural number n_1 such that $n_1 x > y$.

35. Let $f(n) = n^3$. Using Exercise 34, we have

$$\sum_{k=1}^{n} [(k+1)^2 - k^2] = \sum_{k=1}^{n} (2k+1) = (n+1)^2 - 1^2 = n^2 + 2n.$$

Problem Set 2.6.

25. Let p be a prime number and assume that $\sqrt{p}$ is a rational number. Then we can write $\sqrt{p} = m/n$, where m and n are integers whose only common factors are $+1$ and -1. Then $p = m^2/n^2$, and so $pn^2 = m^2$. Since the only common factors of m and n are $+1$ and -1, p must divide m, that is, $m = pk$, for some integer k. Then $pn^2 = m^2 = p^2 k^2$, and so $n^2 = pk^2$. Now p must divide n, that is, $n = pg$, where g is an integer. Hence $\sqrt{p} = m/n = pk/pg = k/g$. But this is a contradiction since the only common factors of m and n were assumed to be $+1$ and -1. Therefore $\sqrt{p}$ is *not* a rational number.

30. Let η be an irrational number, and let $r = p/g \neq 0$ be a rational number. If $r\eta = (p/g)\eta$ is a rational number then there are integers m and n such that $r\eta = (p/g)\eta = m/n$. Then $\eta = mq/pn$ and η is a rational number. This is a contradiction, and so $r\eta$ is an irrational number.

Problem Set 2.7.

1. (a) $\sup S = 13$, $\inf S = 3$.
   (b) $\sup S = 0$, $\inf = -1$.
   (c) $\sup S = 2$, $\inf S = -2$.
   (d) $\sup S = +\infty$ (does not exist), $\inf S = 0$.
   (e) $\sup S = 4$, $\inf S = -2$.
4. $S = \{x: |x| < \sqrt{2}\}$.
7. Consider the set $\{x - 1/n: n \in N\}$. This set has no least upper bound in $S = R \setminus \{x\}$.

Problem Set 3.1.

1. Converges to 2. **2.** Converges to 1. **3.** Converges to 0. **4.** Diverges.
5. Converges to 0. **6.** Converges to 1 if $0 < h \leqslant 1$ and diverges if $h > 1$. **8.** 0.
  9. 0. **10.** 1. **11.** 0. **12.** 0. **13.** 0.

Problem Set 3.2.

1. $\{1 + 1/n\}$
2. Let $a_n = 1 + 1/n$ and $b_n = 1$ for each natural number n.
7. Let $a_n = n$ and $b_n = -n$ for each positive integer n. Then $a_n + b_n = 0$ for each positive integer n.
8. Let $a_n = (-1)^n$ and $b_n = (-1)^{n+1}$ for each positive integer n. Then $a_n b_n = -1$ for each positive integer n.
14. Let $a_n = (-1)^{n+1}$ for each positive integer n.

Problem Set 3.3.

3. Let $d = |x - y|$, $T_x = \{z: z \in R \text{ and } |x - z| < d/2\}$, and $T_y = \{z: z \in R \text{ and } |y - z| < d/2\}$. Then $T_x \cap T_y = \emptyset$.

4. If $x \neq z$ then let $d = |x - z|$ and let $T_x = \{u: u \in R \text{ and } |x - u| < d/2\}$. If $x = z$ then use the neighborhood T_x defined in Exercise 3.

Problem Set 3.4.

8. Let $a_n = 2n$ and $b_n = -n$ for each positive integer n. Then $\{a_n + b_n\} = \{n\} \to +\infty$.

9. Let $a_n = n + A$ and $b_n = -n$ for each positive integer n. Then $a_n + b_n = A$ for each positive integer n.

10. Let $a_n = n$ and $b_n = -2n$ for each positive integer n. Then $\{a_n + b_n\} \to -\infty$.

11. Let $a_n = \left(\dfrac{-1 + (-1)^n}{2}\right)^{(n-1)/2}$ for each positive integer n. Then $\{a_n\}$ has $-1, 0, 1$ at its only limit points.

13. Let $a_n = n$ for each positive integer n.

14. Let $a_n = -n$ for each positive integer n.

15. Let $a_n = (-n)^n$ for each positive integer n.

16. Let $a_n = (-1 + (-1)^n)^{(n-1)/2}$ for each positive integer n.

18. $\lim_n \sup a_n = 3$, $\lim_n \inf a_n = -1$.

19. $\lim_n \sup a_n = 1$, $\lim_n \inf a_n = -1$.

20. $\lim_n \sup a_n = \lim_n \inf a_n = 1$.

21. $\lim_n \sup a_n = 1$, $\lim_n \inf a_n = -1$.

31. Let $a_n = (-1)^n$ and $b_n = (-1)^{n+1}$ for each positive integer n. Then $a_n + b_n = 0$ for each positive integer n and $\lim_n \sup a_n = 1$, $\lim_n \sup b_n = 1$, but $\lim_n \sup (a_n + b_n) = 0$.

32. Same as Exercise 31. **33.** Same as Exercise 31.

34. Let $a_n = (-1)^n$ and $b_n = (-1)^{n+1}$ for each positive integer n. Then $(\lim_n \inf a_n)(\lim_n \inf b_n) = 1$ and $\lim_n \inf (a_n b_n) = -1$.

Problem Set 3.5.

6. Let $I_n = (1 - 1/n, 1 + n)$ for each positive integer n. Then $\bigcap_{n=1}^{\infty} I_n = \{1\}$.

12. Let $I_n = [1 + 1/n, 1 + n]$ for each positive integer n. Then $\bigcup_{n=1}^{\infty} I_n = (1, +\infty)$.

16. Let $S = [-1, 1]$.

Problem Set 4.1.

1. 9. **2.** a^2. **3.** 12. **4.** $a^2 + a$. **5.** 15/2. **6.** 0. **7.** 243. **8.** 0. **9.** 0. **10.** 1. **11.** 3. **12.** 1/2. **13.** 20. **14.** $-1/2$. **15.** $a/2$.

23. If $n \leqslant a < n+1$, where n is an integer, then $\lim\limits_{x \to a^+} f(x) = a - n$.

If $n < a \leqslant n+1$, where n is an integer, then $\lim\limits_{x \to a^-} f(x) = a - n$.

The limit, $\lim\limits_{x \to a} f(x)$, exists for all noninteger values of a.

25. If $a > 0$, then $\lim\limits_{x \to a^+} \{\mathrm{sgn}\, x\} = \lim\limits_{x \to a^-} \{\mathrm{sgn}\, x\} = \lim\limits_{x \to a} \{\mathrm{sgn}\, x\} = 1$. If $a < 0$, then $\lim\limits_{x \to a^+} (\mathrm{sgn}\, x\} = \lim\limits_{x \to a^-} \{\mathrm{sgn}\, x\} = \lim\limits_{x \to a} \{\mathrm{sgn}\, x\} - 1$. If $a = 0$, then $\lim\limits_{x \to 0^+} \{\mathrm{sgn}\, x\} = 1$, $\lim\limits_{x \to 0^-} \{\mathrm{sgn}\, x\} = -1$, and $\lim\limits_{x \to 0} \{\mathrm{sgn}\, x\}$ does not exist.

27. If $a < 0$ or $a > 1$, then $\lim\limits_{x \to a^+} \chi_S(x) = \lim\limits_{x \to a^-} \chi_S(x) = \lim\limits_{x \to a} \chi_S(x) = 0$. If $0 < a < 1$, then $\lim\limits_{x \to a^+} \chi_S(x) = \lim\limits_{x \to a^-} \chi_S(x) = \lim\limits_{x \to a} \chi_S(x) = 1$. Also $\lim\limits_{x \to 0^-} \chi_S(x) = 0$, $\lim\limits_{x \to 0^+} \chi_S(x) = 1$, $\lim\limits_{x \to 1^-} \chi_S(x) = 1$, $\lim\limits_{x \to 1^+} \chi_S(x) = 0$, and $\lim\limits_{x \to 0} \chi_S(x)$ and $\lim\limits_{x \to 1} \chi_S(x)$ do not exists.

28. For any real number a, $\lim\limits_{x \to a^+} \chi_Z(x) = \lim\limits_{x \to a^-} \chi_Z(x) = \lim\limits_{x \to a} \chi_Z(x) = 0$.

29. The limits, $\lim\limits_{x \to a^+} \chi_Q(x)$, $\lim\limits_{x \to a^-} \chi_Q(x)$, and $\lim\limits_{x \to a} \chi_Q(x)$, do not exist for any real number a.

30. $\lim\limits_{x \to 2} (1/(x-2)^2) = +\infty$, $\lim\limits_{x \to -2} (-1/(x+2)^2) = -\infty$.

31. If $a \neq 1$, then $\lim\limits_{x \to a} \left| \dfrac{x+1}{x-1} \right| = \left| \dfrac{a+1}{a-1} \right|$. For $a = 1$, we have $\lim\limits_{x \to 1} \left| \dfrac{x+1}{x-1} \right| = +\infty$.

34. $\lim\limits_{x \to +\infty} \left(\dfrac{2x^2 + x + 1}{x^2 + 1} \right) = 2$, $\lim\limits_{x \to -\infty} \left(\dfrac{-2x+1}{x} \right) = -2$.

36. $\lim\limits_{x \to +\infty} x = +\infty$, $\lim\limits_{x \to +\infty} (-x) = -\infty$, $\lim\limits_{x \to -\infty} \left(\dfrac{-x^2+1}{x} \right) = +\infty$, and

$\lim\limits_{x \to -\infty} \left(\dfrac{x^2+1}{x} \right) = -\infty$.

37. $\lim\limits_{x \to 2+} \left(\dfrac{1}{x-2} \right) = +\infty$, $\lim\limits_{x \to 2-} \left(\dfrac{1}{2-x} \right) = +\infty$, $\lim\limits_{x \to 2+} \left(\dfrac{1}{2-x} \right) = -\infty$, and

$\lim\limits_{x \to 2-} \left(\dfrac{1}{x-2} \right) = -\infty$.

38. $\sin 0 = 0$, $\cos 0 = 1$, $\sin(\pi/6) = 1/2$, $\cos(\pi/6) = \sqrt{3}/2$, $\sin(\pi/4) = \sqrt{2}/2$, $\cos(\pi/4) = \sqrt{2}/2$, $\sin(\pi/3) = \sqrt{3}/2$, $\cos(\pi/3) = 1/2$, $\sin(\pi/2) = 1$, $\cos(\pi/2) = 0$, $\sin(\pi) = 0$, and $\cos(\pi) = -1$.

Problem Set 4.2.

5. No. Let $f_1: R \to R$ such that $f_1(x) = -x$, and let $L = 1$. Then $\lim_{x \to 1} |f_1(x)| = 1 = |L|$, but $\lim_{x \to 1} f_1(x) = -1 \neq L$.

Yes. Let $f_2: R \to R$ such that

$$f_2(x) = \begin{cases} -1 & \text{if } x \text{ is an irrational number,} \\ 1 & \text{if } x \text{ is a rational number.} \end{cases}$$

Then $\lim_{x \to a} |f_2(x)| = 1$ for every real number a, but $\lim_{x \to a} f_2(x)$ does not exist.

10. Let $f: R \to R$ such that $f(x) = \chi_Q(x)$, where Q denotes the set of rational numbers. Let $g: R \to R$ such that $g(x) = -f(x)$. Then $\lim_{x \to a} f(x)$, and $\lim_{x \to a} g(x)$ do not exist for any real number a, but $\lim_{x \to a} [f(x) + g(x)] = 0$.

11. Let $f: R \to R$ such that $f(x) = \chi_Q(x)$, and let $g: R \to R$ such that $g(x) = 1 - f(x)$. Then $\lim_{x \to a} f(x)$ and $\lim_{x \to a} g(x)$ do not exist for any real number a, but $\lim_{x \to a} f(x)g(x) = 0$.

27. Let $f: R \setminus \{a\} \to R$ such that $f(x) = 1/|x - a|$ and let $g: R \setminus \{a\} \to R$ such that $g(x) = -1/|x - a|^2$. Then $\lim_{x \to a} f(x) = +\infty$, $\lim_{x \to a} g(x) = -\infty$, and $\lim_{x \to a} [f(x) + g(x)] = -\infty$

28. Let $f: R \setminus \{a\} \to R$ such that $f(x) = 1/|x - a| + L$ and let $g: R \setminus \{a\} \to R$ such that $g(x) = -1/|x - a|$. Then $\lim_{x \to a} f(x) = +\infty$, $\lim_{x \to a} g(x) = -\infty$, and $\lim_{x \to a} [f(x) + g(x)] = L$.

29. Let $f: R \setminus \{a\} \to R$ such that $f(x) = 1/|x - a|^2$ and $g: R \setminus \{a\} \to R$ such that $g(x) = -1/|x - a|$. Then $\lim_{x \to a} f(x) = +\infty$, $\lim_{x \to a} g(x) = -\infty$, and $\lim_{x \to a} [f(x) + g(x)] = +\infty$.

30. Let $f: R \setminus \{a\} \to R$ such that $f(x) = 1/|x - a| + \chi_Q(x)$, and let $g: R \setminus \{a\} \to R$ such that $g(x) = -1/|x - a|$. Then $\lim_{x \to a} f(x) = +\infty$, $\lim_{x \to a} g(x) = -\infty$, and $\lim_{x \to a} [f(x) + g(x)] = \lim_{x \to a} \chi_Q(x)$ does not exist. (Cf. Exercise 29 of Problem Set 4.1.)

32. Assume that a is a limit point of D and $b \neq 0$ is a limit point of E. Define $f: D \to R$ such that $f(x) = b$ and define $g: E \to R$ such that

$$g(y) = \begin{cases} y & \text{if } y \neq b, \\ 2b & \text{if } y = b. \end{cases}$$

Then $\lim_{x \to a} f(x) = b$, $\lim_{y \to b} g(y) = b$, and $\lim_{x \to a} (g \circ f)(x) = \lim_{x \to a} g(f(x)) = \lim_{x \to a} g(b) = 2b$.

Problem Set 4.3.

1. The function f is continuous for all noninteger values of x.

12. Let $f_1: R \to R$ such that $f_1(x) = \chi_{\{2\}}(x)$. The function f_1 has a removable discontinuity at $x = 2$.

Let $f_2: R \to R$ such that $f_2(x) = \chi_Z(x)$. Then f_2 has a removable discontinuity at each integer.

Let $f_3: R \to R$ such that

$$f_3(x) = \begin{cases} x^2 & \text{if } x \neq 0, \\ 2 & \text{if } x = 0. \end{cases}$$

Then f_3 has a removable discontinuity at $x = 0$.

13. Let $f_1: R \to R$ such that $f_1(x) = [x]$, where $[x]$ denotes the bracket function. Then f_1 has a jump discontinuity at each integer.

Let $f_2: R \to R$ such that $f_2(x) = \operatorname{sgn} x$. (Cf. Exercise 25 of Problem Set 4.1.) Then f_2 has a jump discontinuity at $x = 0$.

Let $I = (0, 1)$ and $f_3: R \to R$ such that $f_3(x) = \chi_I(x)$. The function f_3 has jump discontinuities at $x = 0$ and $x = 1$.

Problem Set 4.4.

16. The bracket function is continuous from the left for all nonintegral real numbers.

27. The function χ_Q is upper-semicontinuous on the set Q.

28. The function χ_Q is lower-semicontinuous on the set of irrational numbers, $R \setminus Q$.

29. The function f is upper-semicontinuous on the set R.

30. The function f is lower-semicontinuous on the set of irrational numbers, $R \setminus Q$.

31. The function f is continuous on the set $R \setminus Q$.

33. Let $f: R \to R$ such that $f(x) = \chi_Q(x)$. Let $g: Q \to R$ such that $g(x) = 1$. Then $g(x) = f(x)$ for each $x \in Q$. Also, g is continuous for each rational number, but f is discontinuous for each rational number.

Problem 4.6.

13. Let $f: R \to R$, $g: R \to R$ such that $f(x) = g(x) = x$. Then f and g are uniformly continuous on R but fg is *not* uniformly continuous on R.

Problem Set 5.1.

5.
$$f'(x) = \begin{cases} 3x^2 \sin(1/x) - x \cos(1/x) & \text{if } x \neq 0, \\ 0 & \text{if } x = 0. \end{cases}$$

The function f' is continuous for all real numbers and is differentiable for all nonzero real numbers. The derivative of f' is given by $(f')'(x) = (6x - 1/x) \sin(1/x) - 4 \cos(1/x)$, if $x \neq 0$.

6. Let $f_1: R \to R$ such that

$$f_1(x) = \begin{cases} 0 & \text{if } x = r_1 \quad \text{or} \quad x = r_2, \\ \dfrac{1}{(x - r_1)(x - r_2)} & \text{if } x \neq r_1 \quad \text{and} \quad x \neq r_2. \end{cases}$$

Then f_1 is differentiable on $R \setminus \{r_1, r_2\}$

Let $d = |r_1 - r_2|$ and let $f_2: R \to R$ such that

$$f_2(x) = \begin{cases} 0 & \text{if } x \text{ is an irrational number,} \\ |x - r_1|^2 & \text{if } x \text{ is a rational number and } |x - r_1| < d/2, \\ |x - r_2|^2 & \text{if } x \text{ is a rational number and } |x - r_2| < d/2, \\ 1 & \text{otherwise.} \end{cases}$$

Then the function f_2 is differentiable only at r_1 and r_2.

7. Let $S = \{r_1, r_2, \ldots, r_n\}$. Define $f: R \to R$ by

$$f(x) = \begin{cases} 0 & \text{if } x \in S. \\ \displaystyle\sum_{k=1}^{n} \dfrac{1}{(x - r_k)} & \text{if } x \in R \setminus S. \end{cases}$$

Then the function f is differentiable on $R \setminus S$ and not differentiable on S.

8. $R \setminus \{0\}$. **9.** $R \setminus Z$.

12. Let $a = 0$ and $f: R \to R$ such that $f(x) = |x|$. Then

$$\lim_{h \to 0} \frac{f(0 + h) - f(0 - h)}{2h} = 0,$$

but f is *not* differentiable at 0.

15. Let $f: R \to R$ such that

$$f(x) = \begin{cases} x & \text{if } x < 0, \\ x^2 & \text{if } x \geq 0. \end{cases}$$

Then $f_-'(0) = 1$ and $f_+'(0) = 0$.

Problem Set 5.2.

6. Let $f: R \to R$ such that

$$f(x) = \begin{cases} x & \text{if } x \geq 0, \\ 0 & \text{if } x < 0. \end{cases}$$

Let $g: (-\infty, 0] \to R$ such that $g(x) = 0$. Then $f(x) = g(x)$ for each $x \in (-\infty, 0]$ and g is differentiable on $(-\infty, 0]$, but f is *not* differentiable at $x = 0$.

10. The function f is differentiable on R if $n > 1$. The function f' is continuous on R if $n > 1$. The function f' is differentiable on R if $n > 2$.

11.
$$f'(x) = \begin{cases} 2x \sin(1/x) - \cos(1/x) & \text{if } x \neq 0, \\ 0 & \text{if } x = 0. \end{cases}$$

The function f is differentiable on R and f' continuous on $R \setminus \{0\}$. The function f' is differentiable on $R \setminus \{0\}$.

Problem Set 5.3.

1.
$$f'(x) = \begin{cases} 2x \cos(1/x) + \sin(1/x) & \text{if } x \neq 0, \\ 0 & \text{if } x = 0. \end{cases}$$

The function f' is continuous and differentiable on $R \setminus \{0\}$.

23. Let $f: (0, 1) \to R$ such that $f(x) = x^{1/2}$. Then f is differentiable on $(0, 1)$ and uniformly continuous on $(0, 1)$, but $f'(x) = 1/2x^{1/2}$, which is unbounded on $(0, 1)$.

24. Let $f: [0, 1] \to R$ such that $f(x) = x^{1/2}$. Then f does not satisfy a Lipschitz condition at 0.

26. Cf. Exercise 24.

28. Let $f: [0, 1] \to R$ such that

$$f(x) = \begin{cases} x & \text{if } x \text{ is a rational number,} \\ 1 - x & \text{if } x \text{ is an irrational number.} \end{cases}$$

Then f is continuous only at $x = 1/2$ and f has the intermediate value property.

Problem Set 5.4.

4. Let $f: R \to R$ such that

$$f(x) = \begin{cases} x^2 \sin(1/x) & \text{if } x \neq 0, \\ 0 & \text{if } x = 0. \end{cases}$$

Then f has a generalized second derivative at 0 whose value is zero, but f does not have a second derivative at zero.

8. Let $f: R \to R$ such that

$$f(x) = \begin{cases} e^{-1/(x-a)^2(b-x)^2} & \text{if } a < x < b, \\ 0 & \text{otherwise.} \end{cases}$$

Problem Set 6.1.

7. $c(b - a)$.

8. (a) $u(f, \mathfrak{n}_4) = 5/8$, $l(f, \mathfrak{n}_4) = 3/8$.

    (b) $u(f, \mathfrak{n}_n) = (n + 1)/2n$, $l(f, \mathfrak{n}_n) = (n - 1)/2n$.

9. (a) $u(f, \mathfrak{n}_3) = (\sin(1/3) + \sin(2/3) + \sin(1))/3$.

    $l(f, \mathfrak{n}_3) = (\sin(1/3) + \sin(2/3))/3$.

    (b) $u(f, \mathfrak{n}_n) = \left(\sum_{k=1}^{n} \sin(k/n) \right)/3,$

    $l(f, \mathfrak{n}_n) = \left(\sum_{k=1}^{n-1} \sin(k/n) \right)/3.$

10. (a) $u(f, \mathfrak{n}_4) = (1 + e^{1/4} + e^{1/2} + e^{3/4})e^{1/4}/4,$

    $l(f, \mathfrak{n}_4) = (1 + e^{1/4} + e^{1/2} + e^{3/4})/4.$

(b) $u(f, \mathfrak{n}_n) = e^{1/n}(1 - e)/n(1 - e^{1/n})$,
$l(f, \mathfrak{n}_n) = (1 - e)/n(1 - e^{1/n})$.

13. (a) $1/2$

(b) $1 - \cos$ (1).

(c) $e - 1$.

15. $n(n - 1)/2$.   **18.** $\bar{\int} f = 3/4, \underline{\int} f = 1/4$.

Problem Set 6.2.

1. $\displaystyle\int_0^1 \frac{dx}{1 + x}$.   **2.** $\displaystyle\int_0^1 \frac{x\, dx}{1 + x^2}$.   **3.** $\displaystyle\int_0^1 \frac{dx}{1 + x^2}$.

9. $\displaystyle\int_0^1 \frac{dx}{1 + x}$.   **10.** $\displaystyle\int_0^1 \frac{x\, dx}{(1 + x)^2}$.   **11.** $\displaystyle\int_0^1 \frac{dx}{\sqrt{1 + x}}$.

Problem Set 6.3.

4. (a) $4 - 3\cos(1)$.

(b) $e + 1 - 2\cos(1)$.

(c) $a + b/2 - c/3$.

11. Let $f: [0, 1] \to R$ such that
$$f(x) = \begin{cases} -1 & \text{if } x \text{ is an irrational number,} \\ 1 & \text{if } x \text{ is a rational number.} \end{cases}$$
Then $[f(x)]^2 = 1$ and $\displaystyle\int_0^1 [f(x)]^2 dx = 1$, but f is not integrable on $[0, 1]$.

(Cf. Example 6.1.4.)

12. Let $f: [0, 1] \to R$ be defined as in Exercise 11, and let $g: [0, 1] \to R$ such that $g(x) = -f(x)$. Then $(fg)(x) = f(x)g(x) = -1$ for each $x \in [0, 1]$ and fg is integrable on $[0, 1]$. The functions f and g are not integrable on $[0, 1]$. (Cf. Example 6.1.4.)

14. Let $f: I \to R$ such that
$$f(x) = \begin{cases} 1/3 & \text{if } x \text{ is a rational number,} \\ -1/2 & \text{if } x \text{ is an irrational number.} \end{cases}$$
Then $[M(f, I)]^2 = 1/9$, $M(f^2, I) = 1/4$, $[m(f, I)]^2 = 1/4$, and $m(f^2, I) = 1/9$.

Problem Set 6.4.

1. $-2/3$.   **2.** $1/2$.   **3.** $52/9$.   **4.** 1.   **5.** 0.   **6.** 0.

7. 0.   **8.** $\cos(x^2)$.   **9.** $-\sin(x^2)$.

10. $2x\cos(x^4)$.   **11.** $-2x/(1 + x^8)$.   **12.** $5x^4(1 + x^{10})^{1/2} - 3x^2(1 + x^6)^{1/2}$.

16. $-f(x)$.   **19.** $f \equiv 0$.

Problem Set 6.5.

1. $(e^a - 1)/a.$ **2.** 0.   **3.** $(1 - \cos(a\pi))/a.$

4. $\pi/2.$ **5.** $-2.$ **6.** 1.

7. $(e^{\pi/2} + 1)/2.$ **25.** $2x/(x^2 + 1).$ **26.** $3[\cos(3x)/\sin(3x)] = 3\cot(3x).$

27. $2(1 - \ln x)/x^2.$ **28.** $-4x(\sin(x^2))/(\cos(x^2)) = -4x\tan(x^2).$

29. $(3x + 2)/(x^2 + x).$ **30.** $\ln x.$ **31.** $(\ln 2)/2.$

32. 1.   **33.** $2(2 - \sqrt{e}).$ **34.** $\sin(1).$

35. $(1 + 2e^2)/4.$ **36.** $1/3.$ **37.** $\ln 2.$

Problem Set 7.1.

4. (a) Converges to 1.

(b) Converges to 1/2.

(c) Converges to 0.

(d) Converges to 1.

(e) Diverges.

6. Let $a_n = (-1)^{n+1}/n$ and $b_n = (-1)^{n+1}$ for each positive integer n.

11. Let $a_n = (-1)^{n+1}$ for each positive integer n. Then $a_{2n-1} + a_{2n} = 0$ for each positive integer n, and $\sum_{n=1}^{\infty} (a_{2n-1} + a_{2n})$ converges to 0, while $\sum_{n=1}^{\infty} a_n$ diverges.

14. Let

$$a_n = \begin{cases} 1/n^2 & \text{if } n \neq k^2, \, k \text{ a positive integer}, \\ 1/k^2 & \text{if } n = k^2, \, k \text{ a positive integer}. \end{cases}$$

Then $\sum_{n=1}^{\infty} a_n$ converges, but $\lim_{k^2 \to +\infty} k^2 a_{k^2} = 1.$

15. (b) $1/p_1.$

Problem Set 7.2.

1. Diverges.   **2.** Converges.   **3.** Converges.

4. Converges.   **5.** Diverges.   **6.** Diverges.

7. Diverges.   **8.** Diverges.   **9.** Diverges.

10. Converges.   **11.** Diverges.   **12.** Converges.

13. Converges.   **19.** Converges for $p > 1$ and diverges for $p \leq 1.$

20. Converges for $p > 1$ and diverges for $p \leq 1.$

22. If $p > 1$, then the series converges for all values of q. If $p = 1$, then the series converges for all values of $q < -1$, and diverges for all values of $q \geq -1$. If $p < 1$, the series diverges for all values of q.

25. For $x \in [n, n + 1)$, where n is a positive integer, define f as follows:

$$f(x) = \begin{cases} a_n + (4 - 2a_n)(x - n) & \text{if } n \leq x \leq n + 1/2, \\ a_{n+1} + (2a_{n+1} - 4)(x - (n + 1)) & \text{if } n + 1/2 < x < n + 1. \end{cases}$$

Then $f(n) = a_n$, and $\int_1^\infty f(x)dx$ diverges.

26. For each positive integer n, let

$$b_n = \min(1/2,\, 1/n^2\, a_n).$$

For $x \in [n, n + 1)$, where n is a positive integer, define f as follows:

$$f(x) = \begin{cases} a_n - (a_n/b_n)(x - n) & \text{if } n \leqslant x \leqslant n + b_n, \\ 0 & \text{if } n + b_n < x < n + 1 - b_{n+1}, \\ a_{n+1} + (a_{n+1}/b_{n+1})\big(x - (n+1)\big) & \text{if } n + 1 - b_{n+1} \leqslant x < n + 1. \end{cases}$$

Then $\int_1^\infty f(x)dx$ converges.

27. The series converges for all values of $p > 1$ and diverges for all values of $p \leqslant 1$.

28. Let

$$a_n = \begin{cases} 1/2^k\, 3^k & \text{if } n = 2k,\ k \text{ a positive integer}, \\ 1/2^k\, 3^{k+1} & \text{if } n = 2k + 1,\ k \text{ a nonnegative integer}. \end{cases}$$

Then $a_{n+1}/a_n \leqslant 1/2$, but $\lim\limits_{n \to +\infty} \{a_{n+1}/a_n\}$ does not exist.

Problem Set 7.3.

1. Converges conditionally. **2.** Converges conditionally.
3. Converges conditionally. **4.** Converges conditionally.
5. Converges conditionally. **6.** Converges absolutely.
7. Converges absolutely. **8.** Converges absolutely.

9. The series $\sum\limits_{n=1}^\infty 1/n$ shows that the alternating signs are necessary, the series $\sum\limits_{n=1}^\infty a_n$, where

$$a_n = \begin{cases} 1/n & \text{if } n \text{ is even}, \\ -1/n^2 & \text{if } n \text{ is odd}, \end{cases}$$

shows that the decreasing condition is necessary, and the series $\sum\limits_{n=1}^\infty (-1)^n$ shows that the null sequence condition is necessary.

10. The convergent series $\sum\limits_{n=1}^\infty (-1)^{n+1}/n$ has the divergent subseries $\sum\limits_{n=1}^\infty 1/(2n - 1)$.

Problem Set 7.4.

1. Converges for $\alpha > 1$ and diverges for $0 < \alpha \leqslant 1$.
2. Converges. **3.** Diverges.
4. Diverges for $k = 1, 2$, and converges for $k = 3$.
6. The series converges for $\beta - \alpha > 1$ and diverges for $\beta - \alpha \leqslant 1$.

7. The series converges for $\gamma + \delta - \alpha - \beta > 1$ and diverges for $\gamma + \delta - \alpha - \beta \leqslant 1$.

8. The series have radius of convergence one for any value of α, and thus it converges $|x| < 1$ and diverges for $|x| > 1$. If $x = -1$, then the series converges for $\alpha \geqslant 0$ and diverges for $\alpha < 0$. If $x = 1$, then the series converges for $x > -1$ and diverges for $\alpha \leqslant -1$.

Problem Set 8.1.

7. For each positive integer n, let $f_n : [0, 1] \to R$ such that

$$f_n(x) = \begin{cases} 2n^2 x & \text{if } 0 \leqslant x \leqslant 1/2n, \\ -2n^2 (x - 1/n) & \text{if } 1/2n < x < 1/n, \\ 0 & \text{if } 1/n \leqslant x \leqslant 1. \end{cases}$$

Problem Set 8.2.

1. Converges uniformly on any set bounded away from 0.

2. Converges uniformly on any set bounded away from 0.

7. The sequence $\{f_n + g_n\}$ converges uniformly on S.

8. Converges uniformly on any set bounded away from 0.

19. Let $S = [0, 1]$ and for each positive integer n, let $f_n : S \to R$ such that

$$f_n(x) = \begin{cases} x^n & \text{if } n \text{ is even,} \\ x/n & \text{if } n \text{ is odd.} \end{cases}$$

Then $\{f_{2n+1}\}$ is a subsequence of $\{f_n\}$ which converges uniformly to the zero function on S, while $\{f_n\}$ does *not* converge uniformly on S.

21. Let $T = [0, 1/2]$ and $S = [0, 1]$. For each positive integer n, let $f_n : S \to R$ such that $f_n(x) = x^n$. Then the sequence $\{f_n\}$ converges uniformly on T, but does *not* converge uniformly on S.

23. For each positive integer n, let $f_n : R \to R$, $g_n : R \to R$ such that $f_n(x) = g_n(x) = x + 1/n$. Then the sequence $\{f_n\}$ and $\{g_n\}$ both converge uniformly on R, but the sequence $\{f_n g_n\}$ does not converge uniformly on R.

24. For each positive integer n, let $f_n : R \to R$ such that $f_n(x) = x + 1/n$ and let $g_n : R \to R$ such that $g_n(x) = 1/n$. Then the sequence $\{g_n\}$ is uniformly bounded on R, but the sequence $\{f_n g_n\}$ does *not* converge uniformly on R.

27. Let $g_1 : [0, 1] \to R$ such that $g_1(x) = 1$. For each positive integer $n > 1$, define $g_n : [0, 1] \to R$ such that

$$g_n(x) = \begin{cases} \max\{1/n, 1/q + 2n^2(x - p/q)\} & \text{if } 1 \leqslant q \leqslant n, 0 \leqslant p \leqslant q, p \\ & \text{and } q \text{ are relatively prime,} \\ & \text{and } x \in (p/2 - 1/2n^2, p/q], \\ \max\{1/n, 1/q - 2n^2(x - p/q)\} & \text{if } 1 \leqslant q \leqslant n, 0 \leqslant p \leqslant q, p \\ & \text{and } q \text{ are relatively prime,} \\ & \text{and } x \in (p/q, p/q + 1/2n^2), \\ 0 & \text{otherwise.} \end{cases}$$

For each positive integer n, let $f_n: R \to R$ such that $f_n(x) = g_n(y)$, where $y = x - [x]$. Then the functions f_n, $n = 1, 2, \ldots$, are continuous on R and the sequence $\{f_n\}$ converges pointwise to the function $f: R \to R$ such that

$$f(x) = \begin{cases} 1/q & \text{if } x = p/q, \text{ where } p \text{ and } q \text{ are relatively} \\ & \text{prime integers and } q > 0, \\ \\ 0 & \text{if } x \text{ is an irrational number.} \end{cases}$$

Since f is discontinuous on the set of rational numbers, the sequence $\{f_n\}$ can not converge uniformly on any interval.

Problem Set 8.3.

8. For each positive n, let $f_n: R \to R$ such that

$$f_n(x) = \begin{cases} 1/n & \text{if } x \text{ is a rational number,} \\ \\ 0 & \text{if } x \text{ is an irrational number.} \end{cases}$$

Then the sequence $\{f_n\}$ converges uniformly to the zero function on R.

Problem Set 8.4.

2. Yes.   **3.** No.   **4.** No.

6. For each positive integer n, let $g_n: [0, 1] \to R$ as defined in Exercise 27 of Problem Set 8.2. For each positive integer n, let $f_n: [-3, 1] \to R$ such that

$$f_n(x) = \begin{cases} -g_n(x + 3) & \text{if } x \in [-3, -2], \\ x + 1 & \text{if } x \in (-2, 0), \\ g_n(x) & \text{if } x \in [0, 1]. \end{cases}$$

Then each of the functions f_n, $n = 1, 2, \ldots$, is continuous on $[-3, 1]$, but the limit function is neither upper semicontinuous nor lower semicontinuous.

Problem Set 8.5.

6. $\sum\limits_{n=1}^{\infty} (\sin nx)/n^5$.   **7.** $\sum\limits_{n=1}^{\infty} (\cos (n\pi/2) - \cos (nx))/n^5$.

8. (a) $\lim\limits_{n} f_n(x) = \begin{cases} 0 & \text{if } x = 0, \\ \\ 1 & \text{if } x \in (0, 1]. \end{cases}$

The convergence is not uniform.

(b) $\lim\limits_{n} \int_0^1 f_n(x)dx = \int_0^1 \left[\lim\limits_{n} f_n(x) \right] dx = 1$.

9. (a) $\lim\limits_{n} f_n(x) = 0$.

(b) $\lim\limits_{n} \int_0^1 f_n(x)dx = 1/2, \int_0^1 \left[\lim\limits_{n} f_n(x) \right] dx = 0$.

11. For any $x \in (0, 2\pi)$. **12.** For any $x \in [0, 1)$.
13. For any $x \in (-\infty, 1/2)$. **14.** For any $x \in (1, +\infty)$.

Problem Set 9.1.

1. $(-\infty, +\infty)$. **2.** $[-1, 1)$. **3.** $(-\infty, +\infty)$.
4. $[-1, 1]$. **5.** $[-4, 4]$. **6.** $(-1/e^2, 1/e^2)$.
7. $(-4/3, 4/3)$. **8.** $(-27, 27)$. **9.** $(-k^k, k^k)$.
10. $(-\infty, +\infty)$. **11.** 0. **12.** $(-\infty, +\infty)$.
13. $(-1/4, 1/4)$. **14.** $(-\infty, +\infty)$. **15.** $[-1, 1]$.
40. $[1/2, +\infty)$. **41.** $(-\infty, -1] \cup (0, +\infty)$. **42.** $[0, +\infty)$.
43. The series converges for all x except $x = 3\pi/2 + 2k\pi$, where k is an integer. The series converges absolutely for all x except $x = \pi/2 + 2k\pi$ or $x = 3\pi/2 + 2k\pi$, where k is an integer. The series converges conditionally for $x = \pi/2 + 2k\pi$, where k is an integer.
44. Let ζ be the real number such that $\zeta = -e^\zeta$. Then the series converges on the interval $[\zeta, +\infty)$.

Problem Set 9.2.

1. $f'(x) = \displaystyle\sum_{n=0}^{\infty} x^n/(n + 1)^{n+1}$, with radius of convergence $+\infty$.

2. $f'(x) = \displaystyle\sum_{n=0}^{\infty} x^n/(n + 1)$, with interval of convergence $[-1, 1)$.

3. $f'(x) = \displaystyle\sum_{n=0}^{\infty} [(n + 1)!]^2 (n + 1) x^n/(2n + 2)!$, with interval of convergence $(-4, 4)$.

4. $f'(x) = \displaystyle\sum_{n=0}^{\infty} (n + 2) x^n$, with interval of convergence $(-1, 1)$.

8. $\displaystyle\sum_{n=1}^{\infty} x^{n+1}/(n + 1)^{n+1}$.

9. $\displaystyle\sum_{n=1}^{\infty} x^{n+1}/(n + 1)^3$.

10. $\displaystyle\sum_{n=0}^{\infty} (n!)^2 x^{n+1}/(2n)!(n + 1)$.

11. $\displaystyle\sum_{n=1}^{\infty} x^{n+1}/n$.

13. $1/(1 - x)^2$.

14. $1 - \ln(1 - x) + \dfrac{\ln(1 - x)}{x}$.

15. $x^2 \displaystyle\int_0^x \left[-\dfrac{\ln(1 - \zeta)}{\zeta^3} - \dfrac{1}{\zeta^2} - \dfrac{1}{2\zeta} \right] d\zeta$.

16. $1/(1 + x^2)^2$.

17. $x(1 + 2x)/(1 - x)^3$.

Problem Set 9.3.

1. $\sin x = \sum\limits_{k=0}^{n} a_k x^k/k! + R_n(x)$, where

$$a_k = \begin{cases} 0 & \text{if } k \text{ is even,} \\ (-1)^m & \text{if } k \text{ is odd and } k = 2m + 1, \end{cases}$$

and where

$$R_n(x) = \begin{cases} \dfrac{(-1)^{(n+1)/2} \sin(c_n)x^{n+1}}{(n+1)!} & \text{if } n \text{ is odd,} \\[3mm] \dfrac{(-1)^{n/2} \cos(c_n)x^{n+1}}{(n+1)!} & \text{if } n \text{ is even,} \end{cases}$$

with c_n between 0 and x for each positive integer n.
We have

$$R_5(x) = \frac{-\sin(c_5) x^6}{6!},$$

where c_5 is between 0 and x, and so

$$|R_5(x)| \leqslant |x|^6/(720).$$

3. $1/(1 + x) = \sum\limits_{k=0}^{n} (-x)^k + R_n(x)$, where $R_n(x) = (-x)^{n+1}/(1 + c_n)^{n+2}$, with

c_n between 0 and x for each positive integer n.
We have

$$R_5(x) = -x^6/(1 + c_5)^7,$$

where c_5 is between 0 and x, and so

$$|R_5(x)| \leqslant |x|^6/(1 - |x|)^7.$$

5. $e^{-x} = \sum\limits_{k=0}^{n} (-x)^n/n! + R_n(x)$, where

$$R_n(x) = e^{-c_n}(-x)^{n+1}/(n + 1)!$$

with c_n between 0 and x.
We have

$$R_5(x) = e-c_5x^6/6!,$$

where c_5 is between 0 and x, and so

$$|R_5(x)| \leqslant e^{|x|} |x|^6/(720).$$

7. $f(x) = \sum\limits_{k=0}^{n} a_k x^k + R_n(x)$, where $a_0 = 1$, $a_1 = 1$, $a_2 = 0$, $a_3 = 2$, $a_4 = 4$,

$a_5 = 1$, and $a_n = 0$ for $n \geqslant 6$, and where

$$R_0(x) = [1 + 6(c_0)^2 + 16(c_0)^3 + 5(c_0)^4]x,$$
$$R_1(x) = [12(c_1) + 48(c_1)^2 + 20(c_1)^3]x^2/2,$$
$$R_2(x) = [12 + 96(c_2) + 60(c_2)^2]x^3/6,$$
$$R_3(x) = [96 + 120(c_3)]x^4/24,$$
$$R_4(x) = 120x^5/120 = x^5,$$

and
$$R_n(x) = 0$$

for all $n \geqslant 5$, with c_0, c_1, c_2, and c_3 between 0 and x.

$$\text{Since } R_5(x) = 0, |R_5(x)| = 0.$$

8. $x - x^3/3! + x^5/5! - x^7/7! = x - x^3/6 + x^5/120 - x^7/5040.$

9. $1 - 4x^2/2! + 16x^4/4! - 64x^6/6! = 1 - 2x^2 + 2x^4/3 - 4x^6/45.$

10. $1 - x + x^2 - x^3.$

11. $x - x^2/2 + x^3/3 - x^4/4.$

12. $1 - x + x^2/2 - x^3/6.$

13. $1 + x^2 + x^4/2 + x^6/6.$

14. $1 + x + 2x^3 + 4x^4.$

28. $\cos(x + \pi/4) = \dfrac{\sqrt{2}}{2} \displaystyle\sum_{n=0}^{\infty} a_n x^n/n!$, where

$$a_n = \begin{cases} 1 & \text{if } n = 4k, \, k \text{ a nonnegative integer,} \\ -1 & \text{if } n = 4k + 1, \, k \text{ a nonnegative integer,} \\ -1 & \text{if } n = 4k + 2, \, k \text{ a nonnegative integer,} \\ 1 & \text{if } n = 4k + 3, \, k \text{ a nonnegative integer,} \end{cases}$$

with radius of convergence $+\infty$.

30. $e^x \log(1 + x) = \displaystyle\sum_{n=1}^{\infty} \left(\sum_{k=1}^{\infty} \dfrac{(-1)^{k+1}}{k(n-k)!} \right) x^n$, with radius of convergence one.

32. $\sec x = \displaystyle\sum_{n=0}^{\infty} a_n x^n$, where

$$a_n = \begin{cases} 0 & \text{if } n = 2k + 1, \, k \text{ a nonnegative integer,} \\ 1 & \text{if } n = 0, \\ \displaystyle\sum_{k=0}^{n/2-1} \dfrac{(-1)^k a_{n-2k-2}}{(2k+2)!} & \text{otherwise.} \end{cases}$$

35. $e^x = e^a \displaystyle\sum_{n=0}^{\infty} (x - a)^n/n!.$

36. $\cos x = \displaystyle\sum_{n=0}^{\infty} a_n(x - a)^n/n!$, where

$$a_n = \begin{cases} (-1)^k \cos a & \text{if } n = 2k, \, k \text{ a nonnegative integer,} \\ (-1)^k \sin a & \text{if } n = 2k + 1, \, k \text{ a nonnegative integer.} \end{cases}$$

Problem Set 10.1.

1. Mixed kind, converges to π. **2.** Second kind, diverges.
3. Mixed kind, diverges. **4.** Second kind, converges to 2.
5. Mixed kind, diverges. **6.** Mixed kind, diverges.
7. Mixed kind, diverges. **8.** First kind, diverges.
9. First kind, diverges. **10.** First kind, converges to 1.
11. First kind, converges for $p < 1$ and diverges for $p \geqslant 1$.
12. Mixed kind, diverges. **13.** Second kind, converges to $\pi/2$.
14. Mixed kind, diverges. **15.** Mixed kind, diverges.
16. Mixed kind, converges for $p < 1$ and diverges for $p \geqslant 1$.
17. First kind, diverges. **18.** Mixed kind, diverges.
19. Mixed kind, converges for $p < 1$ and diverges for $p \geqslant 1$.
20. First kind, converges to $2/e$.
27. Let $f: [2, +\infty) \to R$ such that for each integer $n \geqslant 2$, f is defined by

$$f(x) = \begin{cases} -n^2[x - (n + 1/n^2)] & \text{if } x \in [n, n + 1/n^2], \\ 0 & \text{if } x \in \left(n + 1/n^2, (n + 1) + 1/(n + 1)^2\right) \\ (n + 1)^2[x - \left((n + 1) - 1/(n + 1)^2\right)] & \\ & \text{if } x \in [(n + 1) + 1/(n + 1)^2, (n + 1)). \end{cases}$$

Then $\lim\limits_{x \to +\infty} f(x)$ does not exist, but the improper integral

$$\int_2^{+\infty} f(x)dx \text{ converges to } 1/8 + \sum_{n=3}^{\infty} 1/n^2.$$

31. Let $f: [1, +\infty) \to R$ such that f is zero on $[1, 2)$ and f is defined in Exercise 27 on $[2, +\infty)$.

Problem Set 10.2.

1. Converges.
2. Diverges for $\alpha = 1$ and converges for $\alpha > 1$.
3. Converges. **4.** Converges. **5.** Converges.
6. Diverges. **7.** Diverges for $\alpha \leqslant -1$ and converges for $\alpha > -1$.
8. Diverges. **9.** Diverges. **10.** Diverges.
11. $\alpha > 1$. **12.** No values of α. **18.** 0.
19. 0. **20.** 0. **21.** π.

Problem Set 11.1.

1. 0. **2.** 1. **3.** 0 if $k = 0$ and $|2^k - 1|$ if $k \neq 0$.
4. $e - 1$. **5.** 4. **6.** $\sin(1)$.
10. Let $f: [a, b] \to R$ such that $f(x) = x$, and let $g: [a, b] \to R$ such that $g(x) = -x$. Then

$$V(f + g; [a, b]) = 0,$$
$$V(f; [a, b]) = V(g; [a, b]) = b - a.$$

16. Let $f: [0,1] \to R$ such that

$$f(x) = \begin{cases} x & \text{if } 0 \leqslant x \leqslant 1/2, \\ 1 - x & \text{if } 1/2 < x \leqslant 1. \end{cases}$$

Then f is not differentiable at $x = 1/2$, but since $v_f(x) = x$ for each $x \in [0,1]$, v_f is differentiable at $x = 1/2$.

17. No. **18.** Yes. **19.** See Exercise 18. **20.** See Exercise 18.

Problem Set 11.2.

1. 2/3.   **4.** 5.   **6.** 2/e.

2. 3/5.   **5.** 3/2.   **7.** 1.

3. 3.

10. $V(g, [0, 2]) = 9$, DS $\int_0^2 f(x)dg(x) = -86/15$.

11. $V(g, [0, 2]) = 4$, DS $\int_0^2 f(x)dg(x) = -10/3$.

Problem Set 11.3.

1. $(\pi - 2)/2$.   **5.** 34.   **8.** 2.

2. $-\pi/2$.   **6.** $2e^4 + e^3 + e^2 + e - 1$.   **9.** $(\ln 101)/2$.

3. $(1 - e^{-\pi/2})/2$.   **7.** $2e^4 - e^3 - e^2$.   **10.** 3/2.

4. 1.

36. Let $f, g : [0, 1] \to R$ such that $f(x) = x$ and $g(x) = [x]$.

39. Let $f, g : [0, 1] \to R$ such that $f(x) = [x]$ and $g(x) = x$.

42. Let $f, h : [0, 1] \to R$ such that $f(x) = [x]$ and $h(x) = x$.

Problem Set 11.4.

1. 2.   **5.** 18.   **8.** $e^5 - 2e^2 + 112/3$.

2. $-\pi$.   **6.** $2e^4 + e^3 + e^2 + e + 1$.   **9.** 0.

3. $(e^{-\pi} - e^{\pi})/2$.   **7.** 9.   **10.** $e - 2 + e^{-1}$.

4. 0.

Index